ODOR EMISSIONS AND CONTROL FOR COLLECTION SYSTEMS AND WATER RESOURCE RECOVERY FACILITIES

WEF Manual of Practice No. 25

Second Edition

2020

Water Environment Federation
601 Wythe Street
Alexandria, VA 22314–1994 USA
http://www.wef.org

ISBN 978-1-57278-367-6

About WEF

The Water Environment Federation (WEF) is a not-for-profit technical and educational organization of 35,000 individual members and 75 affiliated Member Associations representing water quality professionals around the world. Since 1928, WEF and its members have protected public health and the environment. As a global water sector leader, our mission is to connect water professionals; enrich the expertise of water professionals; increase the awareness of the impact and value of water; and provide a platform for water sector innovation. To learn more, visit www.wef.org.

Prepared by **Odor Emissions and Control for Collection Systems and Water Resource Recovery Facilities** Task Force of the **Water Environment Federation**

Neil A. Webster, P.E., *Chair*

Dirk C. Apgar, P.E., PMP
Scott Cowden, P.E.
Alexander Demith
Christopher C. Easter
Shirley Edmondson, P.E.
Jordan C. Fahmie
Ken Galardi, P.E.
Sally Gordon, EIT
Dennis Froehlich
Samantha Henningsen
Lawrence Hentz
Brian Huang, P.E.
Christopher D. Hunniford, P.E.
Betty Jordan, P.E.
James R. Joyce, P.E.
Bart Kraakman, P.Eng.
David McEwen
Michael A. McGinley, P.E.
Ryan McKenna, P.E.
Lynne H. Moss, P.E., BCEE
Richard J. Pope, P.E., BCEE
Raymond C. Porter
Adrian Romero Flores, Ph.D.
Zachary Swartzbaugh
Robert B. Stallings, P.E.
Yuko Suda, P.Eng.
Lynsy Nagle Varner, P.E.
Samayyah Williams, P.E.
Todd Williams

Under the Direction of the **Air Quality and Odor Control Committee** of the **Technical Practice Committee**

Water Environment Federation
601 Wythe Street
Alexandria, VA 22314-1994 USA
http://www.wef.org

Manuals of Practice of the Water Environment Federation

The WEF Technical Practice Committee (formerly the Committee on Sewage and Industrial Wastes Practice of the Federation of Sewage and Industrial Wastes Associations) was created by the Federation Board of Control on October 11, 1941. The primary function of the Committee is to originate and produce, through appropriate subcommittees, special publications dealing with technical aspects of the broad interests of the Federation. These publications are intended to provide background information through a review of technical practices and detailed procedures that research and experience have shown to be functional and practical.

Water Environment Federation Technical Practice Committee Control Group

Contents

List of Figures xxi

List of Tables xxv

Preface xxvii

Chapter 1 Introduction 1
Neil A. Webster, P.E.

1.0 PURPOSE AND BACKGROUND 1

2.0 SCOPE AND ORGANIZATION 2

Chapter 2 Odors and Odorous Compound Measurement 5
Michael A. McGinley, P.E., and Samantha Henningsen

1.0 INTRODUCTION 6

2.0 ODORS AND ODORANTS 7

3.0 SAMPLING ODOR SOURCES 8

3.1 Sampling Exhaust Stacks and Vents 9

3.2 Sampling Surfaces 11

4.0 ODOR PARAMETERS 13

4.1 Background of Laboratory Olfactometry Standards 13

4.2 Odor Panels 14

4.3 Odor Concentration 15

4.4 Odor Intensity 18

4.5 Odor Persistency 18

4.6 Odor Characterization 21

4.7 Confidence Interval of Odor Results 22

4.8 Odor Reduction Efficiency 22

5.0 MEASURING AMBIENT ODORS 23

5.1 Ambient Odor Intensity 23

5.2 Grid Method 24

5.3 Ambient Odor Concentration—Field Olfactometers 24

5.4 Application of Field Olfactometry 25

6.0 SAMPLING AND CHEMICAL ANALYSIS OF SPECIFIC ODORANTS 27

6.1 Methods for Reduced Sulfur Compounds, Including Hydrogen Sulfide 27

6.2 Methods for Volatile Fatty Acids 28

6.3 Methods for Amine Compounds 28

6.4 Methods for Volatile Organic Compounds 29

6.5 Methods for Other Wastewater Odorants, Including Ammonia 29

7.0 MATRIX OF CHARACTER, THRESHOLD, AND ANALYSIS 30

8.0 SUMMARY 30

9.0 REFERENCES 34

Chapter 3 Odor Regulations and Policies 39

Lynne H. Moss, P.E., BCEE and Shirley Edmondson, P.E.

1.0 INTRODUCTION 39

2.0 ODOR STANDARDS 40

2.1 Concentration 40

2.2 Averaging Time 43

2.3 Frequency 43

2.4 Impact Location 43

3.0 ODORANT CONCENTRATION LIMITS 44

4.0 ODOR LIMITS 46

5.0 ODOR NUISANCE 51

6.0 ODOR NUISANCE RESPONSE PROTOCOLS 51

6.1 Intensity 52

6.2 Frequency, Intensity, Duration, Offensiveness, and Location (FIDOL) 53

6.3 Other Considerations 55

7.0 ODOR MANAGEMENT PLANS 55

7.1 Role 55

7.2 Examples 56

8.0 COMPARISON OF ODOR REGULATIONS AND APPROACHES 57

9.0 REFERENCES 58

Chapter 4 Source Sampling 61

Christopher D. Hunniford, P.E., Ryan D. McKenna, P.E., and Brian Huang, P.E.

1.0 INTRODUCTION 61

2.0 ODOR GENERATION AND RELEASE 62

2.1 Hydrogen Sulfide 62

2.2 Organic Reduced Sulfur Compounds 63

2.3 Ammonia and Nitrogen Compounds 63

2.4 Other Wastewater Odorants 63

3.0 SAMPLE METHODS 64

3.1 Liquid Phase 64

3.2 Vapor Phase 65

3.2.1 Continuous Hydrogen Sulfide Monitoring 65

3.2.2 Instantaneous Hydrogen Sulfide Sampling 66

3.2.3 Sensory Sampling 67

3.2.4 Sulfur Speciation Sampling 68

3.2.5 Ammonia and Amines Sampling 68

3.2.6 Pressure Monitoring 69

3.2.7 Airflow Measurement 69

3.2.8 Electronic Odor Monitoring 69

4.0 SAMPLE SOURCES 70

4.1 Collection Systems 70

4.1.1 Gravity Sewers and Force Mains 70

4.1.2 Lift Stations 71

4.1.3 Septage and Grease Wastes 71

4.2 Water Resource Recovery Facilities 72

4.2.1 Preliminary Treatment 72

4.2.2 Solids Settling and Nutrient Removal 72

4.2.3 Biosolids Handling 73

5.0 REFERENCES 76

6.0 SUGGESTED READING 77

Chapter 5 Collection Systems 79
Robert B. Stallings, Jordan Fahmie, Yuko Suda, and Dennis Froehlich

1.0 INTRODUCTION 80
- 1.1 Background 80
- 1.2 Purpose 80

2.0 CHARACTERIZATION OF ODORS 81
- 2.1 Liquid Sampling 81
- 2.2 Air Monitoring 83
- 2.3 Advanced Testing and Monitoring 83
- 2.4 Instrumentation 84

3.0 COLLECTION SYSTEM EVALUATION AND DESIGN 85
- 3.1 Flow Monitoring and Modeling 85
- 3.2 Hydraulic Effects on Odor Emissions in Gravity Sewer Systems 86
- 3.3 Force Main Design Considerations 89
- 3.4 Inverted Siphon Design Considerations 91
- 3.5 Odor Capture/Containment 91
- 3.6 Linings and Coatings Selection 92
- 3.7 Material Selection 94
- 3.8 Site Selection 94
- 3.9 Pumping Station Building and Ancillary Equipment Requirements 95
- 3.10 Maintenance and Access Requirements 97

4.0 COLLECTION SYSTEM AIR MOVEMENT AND VENTILATION 98
- 4.1 Physical Assessment of Air Movement 98
- 4.2 Airflow Measurement 99
- 4.3 Air Pressure Measurement 100
- 4.4 Airflow Modeling 100
- 4.5 Modeling of Hydrogen Sulfide Generation 101
- 4.6 Computational Fluid Dynamics Modeling 102

5.0 SOURCE CONTROL 103
- 5.1 Constituents Impacting Odor Emissions 103
- 5.2 Permits and Ordinances 107

5.3	Inspections, Monitoring, and Enforcement	108
5.4	Pretreatment of Industrial Waste	109
6.0	OPERATION AND MAINTENANCE PRACTICES	110
6.1	Safety Precautions	110
6.2	Visual and Closed Circuit Television Inspections	110
6.3	Radar, Ultrasonic, and Laser Scan Inspection Techniques	111
6.4	Smoke Testing	112
6.5	Cleaning and Housekeeping	112
6.6	Equipment Maintenance and Exercising	113
6.7	Protective Coatings and Corrosion Control	114
6.8	Recordkeeping	116
6.9	Budgeting and Financial Management	117
7.0	LIQUID-PHASE TREATMENT	117
7.1	Chemical Oxidants	119
7.2	Biological Treatment Methods	121
7.3	Iron Salts	124
7.4	Peroxide Regenerated Iron	127
7.5	Oxygenation	128
7.6	Adjustment of pH	129
8.0	VAPOR-PHASE TREATMENT	131
8.1	Biological	132
8.2	Chemical	133
8.3	Adsorption	134
8.4	Other	135
9.0	REFERENCES	136

Chapter 6 Water Resource Recovery Facilities 139
Richard J. Pope, P.E., BCEE

1.0	ODOR CHARACTERIZATION	140
1.1	Background	140
1.2	Odor Derivation from Collection Systems	141
1.3	Odor Derivation from Water Resource Recovery Facilities	141
1.4	Mass Transfer of Odorous Compounds	141

		1.4.1 Favorable Mass Transfer Conditions: Turbulence	142
		1.4.2 Favorable Mass Transfer Conditions: pH	143
		1.4.3 Favorable Mass Transfer Conditions: Temperature	143
		1.4.4 Favorable Mass Transfer Conditions: Seasonal and Diurnal Patterns of Wastewater Flows	144
		1.4.5 Favorable Mass Transfer Conditions: Air Pressure	145
2.0	WASTEWATER ODOROUS COMPOUNDS		145
3.0	ODOR EMISSIONS BY PROCESS UNIT		146
	3.1 Preliminary Treatment		146
	3.2 Primary Treatment		151
	3.3 Biological Treatment		152
	3.4 Intermediate and Final Settling		154
	3.5 Chlorine Contact Tanks		154
	3.6 Stabilization Ponds		154
4.0	PROCESS IMPROVEMENT AND ODOR MITIGATION		155
	4.1 Odor Prevention		155
	4.2 Physical System Layout		156
5.0	ODOR CAPTURE AND CONTAINMENT		158
	5.1 Introduction		158
	5.2 Odor Cover Alternatives		159
		5.2.1 Flat Covers	162
		5.2.2 Barrel Arch Covers	165
		5.2.3 Pitched Covers	168
		5.2.4 Building Extension or Enclosure	168
		5.2.4.1 Building or Penthouse	169
		5.2.4.2 Dome Enclosure	170
	5.3 Odor Capture Alternatives for Bar Screens		171
	5.4 Materials of Construction		175
		5.4.1 Concrete	176
		5.4.2 Aluminum	176
		5.4.3 Fiberglass Reinforced Plastic	177
		5.4.4 Stainless Steel	177

5.4.5 Canvas 180

5.4.6 Air-Supported Canvas 181

5.4.7 Wood 181

5.5 Access 181

5.6 Odor Containment 184

5.6.1 National Fire Protection Association 820 Ventilation Rates 184

5.6.2 Standard Ventilation Rates 186

5.6.2.1 Nonworker-Accessible Space 186

5.6.2.2 Worker-Accessible Space 189

5.6.2.3 Additional Considerations 190

5.6.3 Ventilation System Layout 191

5.6.4 Conveying Odors 193

5.6.4.1 Duct Materials of Construction 194

5.6.4.2 Shapes 195

5.6.4.3 Sizing Criteria 195

5.6.4.4 Additional Ductwork Notes 196

5.6.5 Fans 196

5.6.5.1 Fan Positioning 197

5.6.5.2 Fan Balancing 197

5.6.6 Dampers 198

6.0 CASE HISTORIES 199

6.1 Coney Island Water Resource Recovery Facility 199

6.2 North River Water Resource Recovery Facility 201

7.0 REFERENCES 206

8.0 SUGGESTED READING 206

Chapter 7 Solids Handling 207

Chris C. Easter, Todd Williams, Lawrence Hentz, and Zachary Swartzbaugh

1.0 BACKGROUND: WHY BIOSOLIDS ODOR AND VOLATILE ORGANIC COMPOUND EMISSIONS MATTER 207

1.1 Vector Attraction Reduction and Pathogen Kill 209

1.2 Federal and State Regulation Restrictions Versus Community Odor Impacts 209

2.0 ODOR SOURCE CHARACTERIZATION IN SOLIDS HANDLING 210

2.1 Odor Generation Mechanisms and Odorous Compound Summary 210

2.2 Thickening 211

2.3 Digestion 211

2.4 Thermal Hydrolysis 212

2.5 Dewatering 212

2.6 Lime Stabilization 213

2.7 Composting 213

2.8 Cake Storage 213

2.9 Odor Data Summary for Various Types of Biosolids 213

3.0 CONTROL METHODS AND TECHNOLOGY OPTIONS 215

3.1 Solids Process Configurations to Minimize Odors 217

3.2 Liquid-Phase Sludge Treatment Summary 218

3.3 Odor Capture, Containment, and Ventilation 219

3.4 Gas-Phase Treatment Summary 221

4.0 REFERENCES 223

Chapter 8 Biological Odor Control 227

Bart Kraakman, P.Eng, James R. Joyce, P.E., and Adrian Romero Flores, Ph.D

1.0 INTRODUCTION 228

2.0 BIOLOGICAL ODOR TREATMENT FUNDAMENTALS 229

2.1 The Treatment Process 229

2.2 Biochemical Processes 230

2.3 Critical Operating Parameters 232

2.3.1 Moisture 232

2.3.2 pH 233

2.3.3 Nutrients 234

2.3.4 Temperature 234

3.0 BIOFILTERS 235

3.1 Biofilter Reactor Configurations 235

3.1.1 In-Ground Biofilters 235

3.1.2 In-Vessel Biofilters 236

3.1.3 Modular Biofilters 237

3.2 Biofilter Media Types 237
3.2.1 Organic Biofilter Media 238
3.2.2 Inert Biofilter Media 239
3.2.3 Biofilter Media Additives and Coatings 240
3.3 General Biofilter Design Guidelines 240
3.3.1 Pressure Drop 240
3.3.2 Gas Contact Time 241
3.3.3 Air Distribution 241
3.3.3.1 Open Air Plenum 242
3.3.3.2 Media-Supported Air Plenum 242
3.3.3.3 Prevention of Preferential Airflows 242
3.3.4 Moisture Control 243
3.3.4.1 Pre-Humidification 243
3.3.4.2 Surface Irrigation 243
3.3.4.3 Subsurface Drip Irrigation 244
4.0 BIOSCRUBBERS AND BIOTRICKLING FILTERS 244
4.1 Defining Bioscrubbers and Biotrickling Filters 244
4.2 Bioscrubber and Biotrickling Filter Reactor Configurations 245
4.2.1 Recirculating and Once-Through Operating Mode 247
4.2.2 Single-Stage and Multi-Stage Treatment 248
4.2.3 Cocurrent or Countercurrent 248
4.3 Media Types 249
4.4 General Design Guidelines 250
4.4.1 Pressure Drop 250
4.4.2 Gas Contact Time 251
4.4.3 Gas Distribution 252
5.0 DESIGN AND OPERATIONAL CONSIDERATIONS 252
5.1 General Advantages and Limitations of Biological Odor Treatment Technologies 252
5.2 Best Practices and General Considerations 255
5.2.1 Startup 255
5.2.2 Water Supply Sources 255
5.2.3 Drain Water Considerations 256

5.2.4 *Instrumentation and Controls* 256

5.2.5 *Operations and Maintenance Considerations* 256

5.3 Performance Parameters 257

5.4 Reliability 257

5.4.1. *Defining Robustness and Resilience* 257

5.4.2. *Examples of Upset Conditions* 258

5.4.2.1 *Water Supply Failure* 258

5.4.2.2 *Fan Failure* 258

5.4.2.3 *Power Outage* 259

5.4.2.4 *Nutrient Dosing Failure* 259

5.4.2.5 *High Chlorine in Supply Water* 259

5.4.2.6 *Hypochlorite in Inlet Air* 259

5.4.2.7 *Toxics in Inlet Air* 260

5.4.2.8 *Low Air Inlet Concentrations* 260

5.4.2.9 *Sudden Air Inlet Peak Concentrations* 260

5.4.2.10 *High Inlet Concentration for Extended Period* 261

5.4.2.11 *Extreme Weather Conditions* 261

5.4.3 *Opportunities to Improve Reliability* 262

5.4.3.1 *Sustaining Biological Activity* 262

5.4.3.2 *Increasing the Buffering Capacity* 263

6.0 PROCESS INTEGRATED BIOTECHNOLOGY 264

6.1 Introduction 264

6.2 Activated Sludge Recycling 268

6.2.1 *Introduction* 268

6.2.2 *Critical Design and Operating Parameters* 270

6.3 Activated Sludge Diffusion 271

6.3.1 *Introduction* 271

6.3.2 *Critical Design and Operating Parameters* 272

6.4 Oxidized Ammonia Recycling 274

6.4.1 *Introduction* 274

6.4.2 *Critical Design and Operating Parameters* 274

7.0 REFERENCES 276

Chapter 9 Chemical or Oxidation Odor Control 279
Dirk Apgar, P.E., PMP, Samayyah Williams, P.E., and Sally Gordon, EIT

1.0 INTRODUCTION 279
2.0 WET CHEMICAL GAS ADSORPTION SYSTEMS 280
2.1 Mass Transfer Concepts and Chemistry of Wet Chemical Absorption Systems 280
2.2 Wet Chemical Scrubber Configurations 283
3.0 PACKED TOWER SCRUBBER DESIGN PARAMETERS AND SIZING 287
4.0 DETERMINING REQUIRED PACKING DEPTH 288
5.0 MAKEUP WATER REQUIREMENTS 292
6.0 ESTIMATING CHEMICAL USE 296
7.0 MIST ELIMINATION 301
8.0 MIST SCRUBBER SYSTEMS 303
9.0 IONIZATION SYSTEMS 304
10.0 CATALYTIC OXIDATION 305
11.0 REFERENCES 306

Chapter 10 Physical and Thermal Odor Control 307
Scott Cowden, P.E., Ken Galardi, P.E., and Alexander Demith

1.0 INTRODUCTION 308
2.0 DRY MEDIA ADSORPTION 308
2.1 Physical Adsorption 309
2.2 Chemical Adsorption 309
2.3 Reactive Adsorption 309
2.4 Media Types 310
2.4.1 Virgin Activated Carbon 310
2.4.1.1 Coal-Based Carbon 311
2.4.1.2 Coconut Shell Carbon 311
2.4.2 Impregnated Medias 311
2.4.3 High-Capacity Carbon 312
2.4.4 Water-Regenerable Carbon 312
2.4.5 Other Media 312
2.4.6 Advantages/Disadvantages 313

2.5 Adsorption Mechanisms and Performances 313
2.5.1 Volatile Sulfur Compounds 314
2.5.2 Volatile Nitrogen Compounds 314
2.5.3 Volatile Organic Compounds 314
2.5.4 Key Factors Affecting Media Performance 314
2.5.4.1 Media Characteristics 314
2.5.4.2 Adsorption Properties 315
2.5.4.3 Adsorption Competition in Multi-Compound Applications 315
2.5.4.4 Odorant Conversion and Transformation 315
2.5.4.5 Moisture/Humidity 315
2.6 Design Considerations 316
2.6.1 Configurations 316
2.6.1.1 Vertical 316
2.6.1.2 Horizontal 316
2.6.1.3 Radial 316
2.6.1.4 Single Bed 317
2.6.1.5 Multi-Bed 317
2.6.1.6 Other Configurations 317
2.6.2 Media Types 317
2.6.3 Bed Depth 317
2.6.4 Bed Velocity 318
2.6.5 Polishing Downstream of Primary Treatment 318
2.6.6 Bed Smoldering 319
2.6.7 Passive Versus Active Systems 320
2.6.8 Best Practices 320
2.7 Operation and Maintenance Considerations 321
2.7.1 Media Sampling 321
2.7.2 Monitoring 321
2.7.3 Media Change-Out Approaches 321
2.7.3.1 Manual 321
2.7.3.2 Outsource 322
2.7.3.3 Dense-Phase Conveyance and Vacuum 322

3.0 THERMAL OXIDATION 322
3.1 Applicability 323
3.2 Technologies Available 323
3.2.1 Direct-Fired Thermal Oxidizer 323
3.2.2 Recuperative Thermal Oxidizer 323
3.2.3 Regenerative Thermal Oxidizer 323
3.2.3.1 Catalytic Operation of Regenerative Thermal Oxidizers 324
3.2.4 Other Technologies 324
3.2.4.1 Flares 324
3.3.4.2 Incinerators 325
3.3.4.3 Catalytic Incinerators 325
3.3.4.4 Microturbines 325
3.3 Design Considerations 325
3.3.1 Minimum Destruction Efficiency 325
3.3.2 Combustion Chamber Temperature 326
3.3.3 Detention Time 326
3.3.4 Composition of Inlet Air 326
3.3.5 Pretreatment (Specific to Regenerative Thermal Oxidizers) 326
3.3.6 Media Type (Specific to Regenerative Thermal Oxidizers) 326
3.3.7 Materials of Construction 327
3.3.8 Air Permit Considerations 327
4.0 REFERENCES 327

Chapter 11 Odor Impact Assessments 329
David McEwen

1.0 INTRODUCTION 330
1.1 Odor Impact Assessment Purposes 330
1.2 Odor Control Goal Setting 330
2.0 ODOR SOURCE TYPES 331
2.1 Point Sources 331
2.2 Area Sources 332
2.3 Volume Sources 332

3.0 ODOR DISPERSION MODELING 332
3.1 Dispersion Modeling Theory 332
3.2 Dispersion Modeling Process 334
3.2.1 Odor Impact Assessment Applications 334
3.2.2 AERMOD 335
3.2.3 CALPUFF 335
3.2.4 Screening Models 336
3.3 Model Input Parameters 336
3.3.1 Fence Line Delineation and Model Area 336
3.3.2 Source Locations and Characteristics 337
3.3.3 Emission Rate Calculations 338
3.3.4 Meteorological Data 338
3.3.5 Terrain Data 338
3.3.6 Sensitive Receptors 339
3.3.7 Physical Structures and Buildings 339
3.4 Dispersion Model Output 340
3.4.1 Isotherm Plots 340
3.4.2 Individual Source Impacts 341
3.5 Dispersion Model Limitations 342
4.0 AIR CONTAINMENT AND EXTRACTION DESIGN 343
4.1 Odor Containment Methods 343
4.1.1 Covers 343
4.1.2 Enclosures 343
4.1.3 Hoods 344
4.2 Foul Air Capture Design Criteria 344
4.2.1 Room Ventilation for Safety and Comfort 344
4.2.2 Hood Airflow Rate Design 345
4.2.3 Covered Processes 346
5.0 FACILITY ODOR EVALUATIONS 347
5.1 Development of Candidate Technologies 347
5.2 Life Cycle Cost Analyses 348
5.3 Non-Economic Factors Analysis 349
5.3.1 Footprint 350
5.3.2 Operations and Maintenance Requirements 350

5.3.3 Safety 350

5.3.4 Familiarity 350

6.0 CASE STUDIES 351

6.1 Odor Impact Assessment Case Study 1 351

6.2 Odor Impact Assessment Case Study 2 354

7.0 REFERENCES 357

8.0 SUGGESTED READING 358

Chapter 12 Public Outreach and Involvement 361

Betty Jordan, P.E. and Lynsy Nagle Varner, P.E.

1.0 INTRODUCTION 361

2.0 STAKEHOLDER GROUPS 363

2.1 Personnel (Operations, Maintenance, Pretreatment/ Sampling Staff) 364

2.2 Elected Officials and Regulators 365

2.3 Local Residents/Homeowners Associations 365

3.0 ODOR COMPLAINT MANAGEMENT AND RESPONSE 366

3.1 Documentation and Response Procedures 366

3.2 Data Analysis and Use 366

3.3 Personnel Knowledge and Communication 368

3.4 Standard Operating Procedures 369

3.4.1 Improve Odor Complaint Data Level of Detail 370

3.4.2 Complete the Circle 370

3.4.3 Implement Standard Operating Procedures for Conducting and Documenting Site Investigations 370

3.4.4 Standardized Site Investigations 374

3.4.5 Collect Both Gas and Liquid Samples During Site Investigations 374

3.5 Coordination Activities 375

4.0 ENGAGING THE COMMUNITY 375

4.1 Public Outreach Meetings 375

4.2 Managing Expectations 376

4.3 Social Media 376

4.4 Coordination of Outdoor Activities 377

5.0 CASE STUDY—TRINITY RIVER AUTHORITY OF TEXAS, DENTON CREEK REGIONAL WASTEWATER SYSTEM 377

6.0 REFERENCES 380

Index 381

List of Figures

2.1 Vacuum chamber for odor sampling 9
2.2 Dynamic dilution sampling probe 10
2.3 Tall passive chimney sampler 11
2.4 Flux chamber surface sampler 12
2.5 Dilution of odor in the ambient air 15
2.6 Assessor (left) at olfactometer while test administrator (right) operates olfactometer and monitors the test. 16
2.7 Example odor intensity referencing scales (OIRS) shown using reference odorant n-butanol 19
2.8 Power law for a single odorant 20
2.9 Odor persistency or dose-response of an odor sample 20
2.10 Odor characterization of odor control inlet and outlet samples. 21
2.11 Dilution-to-threshold (D/T) with a field olfactometer. 24
3.1 Examples of state hydrogen sulfide limits in the United States 44
3.2 Citizen complaint pyramid. 52
3.3 Field quality intensity scale—New Jersey guidelines. 53
3.4 Texas Commission on Environmental Quality FIDO protocol excerpt. 54
4.1 Colorimetric sulfide test kit 65
4.2 Area and point source sensory sampling 67
4.3 Electronic odor monitoring system. 70
5.1 Typical manhole hydrogen sulfide monitor. 84
5.2 Idealized airflow velocity contours 99
5.3 Typical differential pressure data logger and housing. 101
5.4 Typical CFD model output. 103
5.5 Typical mobile odor treatment system 110
5.6 Typical temporary chemical dosing system. 118
5.7 Typical super oxygenation system 129
6.1 Diurnal hydrogen sulfide graphical representation 144
6.2 Progressive odor control process 160

6.3 Important factors in selecting a cover or enclosure system . . . 161
6.4 Flat headworks channel cover . . . 162
6.5 Flat influent channel cover . . . 163
6.6 Flat aeration tank cover . . . 163
6.7 Flat circular tank cover with external truss . . . 164
6.8 Flat rectangular tank cover with external support . . . 164
6.9 Barrel arch covers made of fiber reinforced plastic (FRP) . . . 166
6.10 Barrel arch covers made of aluminum and plexiglass . . . 167
6.11 Penthouse enclosure; occupied space . . . 168
6.12 Penthouse enclosure; unoccupied space . . . 169
6.13 Geodesic dome enclosure . . . 170
6.14 Geodesic dome enclosure with center walkway open to the atmosphere . . . 172
6.15 Geodesic dome enclosure with enclosed center walkway . . . 172
6.16 Aluminum geodesic dome cover . . . 177
6.17 Aluminum flat cover . . . 178
6.18 Fiberglass reinforced plastic (FRP) dome cover . . . 178
6.19 FRP flat covers . . . 179
6.20 FRP weir area covers . . . 179
6.21 Canvas flat covers . . . 180
6.22 Summary of the ventilation rate evaluation approaches for worker- and nonworker-accessible spaces . . . 190
6.23 Aerial view—Coney Island WRRF, New York . . . 199
6.24 Primary sedimentation basins barrel arch covers—Coney Island WRRF, New York . . . 203
6.25 Primary sedimentation basins barrel arch cover (end view with weighted damper [left] and access door [right])—Coney Island WRRF, New York . . . 203
6.26 Primary sedimentation basins barrel arch cover: Tank access point—Coney Island WRRF, New York . . . 204
6.27 Aerial view—North River WRRF, New York . . . 205
7.1 Biological ORP scale . . . 208
7.2 Odorant production pathways . . . 210
7.3 Belt filter press with integral enclosure system . . . 220
8.1 Typical custom-made in-ground biofilter system for 15 000 m^3/h . . . 236

8.2 Typical custom-made in-vessel biofilter system for 150 000 m^3/h . . . 236
8.3 Typical modular biofilter system for 10 000 m^3/h . . . 237
8.4 Schematic representation of traditionally classified biological vapor-phase treatment technologies . . . 246
8.5 Typical biotrickling filter system treating 12 500 m^3/h . . . 249
8.6 In-vessel biofilter system with dispersion stack treating 50 000 m^3/h . . . 265
8.7 Multi-stage biotrickling filter system treating 125 000 m^3/h . . . 265
8.8 Typical WRRF flow diagram including two different options for Activated Sludge Recycling (ASR) operation with (1) direct ASR from the aerobic activated sludge reactor (dotted line) and (2) ASR from the secondary clarifier (dashed line) . . . 269
8.9 Typical flow diagram of an ASD operation . . . 273
8.10 WRRF flow diagram including the sludge line and two different options for an OAR operation. 1) dotted line: OAR from centrate nitrification units. 2) dashed line: OAR from the nitrification stage in the biological reactor . . . 275
9.1 Sulfide equilibrium diagram . . . 282
9.2 Typical counter-current packed tower scrubber for wastewater odor control . . . 284
9.3 Wet chemical scrubber mass balance schematic . . . 285
9.4 Low-profile three-stage scrubber for hydrogen sulfide and odor removal . . . 286
9.5 Scrubber pollutant gas absorption diagram . . . 289
9.6 Simplified psychrometric chart for determining scrubbing liquid evaporation . . . 295
9.7 Mist eliminators commonly used in wet chemical scrubbers . . . 302
9.8 Typical mist scrubber system schematic diagram . . . 304
10.1 Macro, meso, and micro pore structure of carbon . . . 310
10.2 Dry media mass transfer breakthrough curve . . . 319
11.1 Gaussian plume horizontal and vertical dispersion . . . 333
11.2 Odor impact contour plot . . . 340
11.3 Threshold exceedance plot . . . 341

11.4 Dispersion model output: Case Study 1 353
12.1 Example odor complaint response process 363
12.2 Odor complaint data log 367
12.3 Typical odor complaint map 368
12.4 Example of prevailing winds wind rose 369
12.5 Odor complaint record. 371
12.6 Vapor-phase grab sample results 372
12.7 Odalog monitoring record 373
12.8 Liquid-phase grab sample results 374
12.9 Typical brochure for odor control 378

List of Tables

2.1 Matrix of WRRF odorants, their odor threshold values, and laboratory analysis methodology 31

3.1 Odorous compounds in wastewater 41

3.2 Select reduced sulfur and nitrogen compound odorant limits 45

3.3 Bay Area Air Quality Management District maximum allowable emission concentrations 46

3.4 Examples of ambient odor limits 48

3.5 IAQM benchmark odor criteria 50

3.6 Elements of an odor management plan 56

3.7 Odor regulation approach comparison 58

5.1 Typical field sampling program 82

6.1 Sulfur-based compounds at wastewater facilities 147

6.2 Amines, nitrogen compounds, and ammonia at wastewater facilities 149

6.3 Carboxylic acids at wastewater facilities 150

6.4 Comparison of process unit cover alternatives (walk-in vs. limited-to-no access) 173

6.5 Cover materials of construction comparison 182

6.6 Coney Island WRRF—primary sedimentation tank covering alternatives summary 200

6.7 Comparison of alternative cover and materials of construction costs 201

6.8 Coney Island WRRF—primary sedimentation tank covering costs 202

7.1 Odor source characterization 214

7.2 Odor detection threshold concentrations for anaerobically digested biosolids from a municipal WRRF in southeastern Virginia 215

7.3 Anaerobically digested WRRF biosolids reduced sulfur odor species data from a municipal WRRF in southeastern Virginia (concentrations in parts per billion by volume) 216

7.4 Amine-based compounds detected from anaerobically digested cake biosolids at a municipal WRRF in southeastern Virginia ... 217

7.5 Liquid-phase odor control methods ... 218

7.6 Summary of odorous air treatment alternatives ... 221

8.1 Advantages and disadvantages of vapor-phase biological odor treatment systems ... 254

8.2 Potential upset conditions and opportunities to improve reliability of biological odor treatment systems ... 266

9.1 Typical mass-transfer performance data for hydrogen sulfide removal ... 291

10.1 Advantages and disadvantages of different activated carbon media ... 313

10.2 Comparison of odor control dry adsorption medias ... 318

11.1 Example off-site impacts distribution ... 342

11.2 Odor control technology selection guide ... 348

11.3 Non-economic comparison of odor control technologies ... 351

11.4 Comparison of odor control improvement phases with baseline ... 356

Preface

This manual covers all aspects of odor control in wastewater conveyance and treatment including the basics of odor emissions, characteristics, regulations, and the use of technologies to treat these odors. Since the first edition of Manual of Practice (MOP) 25, *Control of Odors and Emissions from Wastewater Treatment Plants*, was published in 2004, there have been significant advancements in the field of understanding odor emissions and in applying technologies for effective treatment. This book is an embodiment of the current state of practice in the industry.

This second edition of this manual was produced under the direction of Neil Webster, *Chair*.

Authors' and reviewers' efforts were supported by the following organizations:

AECOM
ALS Environmental, Houston, Texas
Air Quality Engineering, Bellevue, Washington
Black & Veatch
Brown and Caldwell
Hazen and Sawyer
HDR Inc.
Jacobs, Dallas, Texas
Jacobs Engineering Group Inc., Dallas, Texas
King County—Wastewater Treatment Division, Seattle, Washington
Plummer Associates, Inc., Fort Worth, Texas
Porter Odor Science
St. Croix Sensory, Inc., Stillwater, Minnesota
V & A Consulting Engineers, Inc., Houston, TX
Webster Environmental Associates, Inc.

1

Introduction

1.0 PURPOSE AND BACKGROUND	1	2.0 SCOPE AND ORGANIZATION	2

1.0 PURPOSE AND BACKGROUND

Odor testing and control have undergone rapid advancements in the last 16 years, with an expanding body of knowledge and real-world experiences that this manual seeks to capture and bring to the reader. This manual replaces and updates Water Environment Federation's Manual of Practice 25, *Control of Odors and Emissions from Wastewater Treatment Plants*. It is intended to be the primary reference document for current standards of practice for professionals involved in odor control at wastewater treatment facilities and collections systems.

This manual covers all aspects of odor control in wastewater conveyance and treatment facilities, including the basics of odor testing and emissions, sensory analysis, local and state regulations, public outreach, and the technologies used to treat odors. The manual embodies the current state of practice in the industry. A new feature of this manual that was not included in the 2004 edition is the chapter entitled "Collection Systems", which addresses ventilation of sewers, air treatment, and chemical treatment.

The manual starts with examples of public policies to provide goals to prevent nuisance odors. The manual then proceeds through the origin of odors in sewer systems and treatment facilities, the emissions that can cause off-site odors, the many treatment methods and technologies to control odors, step-by-step methods to objectively quantify odor impacts in the community, and, finally, how to work with the community to resolve odor issues. Indeed, working with the community and establishing trust is a key element in implementing a successful odor control strategy. As such, a "how-to guide" is also introduced in the manual for the effective use of a complaint management system tool to track and record the cause of odor episodes while working with stakeholders and engaging the community through plant tours, meetings, educational materials, and social media.

Still, there is no substitute for good management practices to control odors, such as emissions testing, proper planning to meet the established goals, designs incorporating the best technologies for the application, and the ongoing task of operating and maintaining odor control systems. Monitoring the odor control systems, making adjustments to the equipment, replacing consumables in a timely manner, and assessing the potential impact in the community are all practices that are covered in this manual. Safety is a primary consideration when testing for hydrogen sulfide (H_2S) and when working in collection systems and water resource recovery facilities (WRRFs), and all personnel should carefully follow their organization's safety guidelines.

2.0 SCOPE AND ORGANIZATION

Each chapter in this manual serves a specific purpose so that each can be referenced as a stand-alone "document", whereas the entire manual provides a comprehensive look at all aspects of odor control in the industry today. Although specific vendors or brands are sometimes referenced in the manual, these references serve as examples only for the convenience of the reader; they do not represent an endorsement by the authors of a particular odor control product.

Chapter 2 addresses how odors are analyzed and measured, and Chapter 3 identifies the current regulations and policies. There is a need to know how to sample for odors and odorous compound emissions, and that is covered in Chapter 4. Chapters 5, 6, and 7 address odor control in collection systems, WRRFs, and solids handling facilities, respectively. This manual presents best available technologies for controlling odors, starting with the coverage found in Chapter 8 of bioscrubbers, biofilters, and other sustainable methods that are experiencing the highest growth in popularity in the industry because of significant advancements in design and effectiveness over the last 15 years. Chapter 9 describes chemical systems such as scrubbers, and Chapter 10 details physical odor control technologies such as carbon adsorbers.

The final two chapters of the book assess odor impacts in neighborhoods and communities served by WRRFs and wastewater collection systems. Chapter 11 provides information on assessment tools, such as atmospheric dispersion modeling, to predict impacts to the community and prioritize projects and the allocation of capital dollars to meet odor standards and goals. Chapter 12 covers public outreach and involvement as part of best management practices relating to being a "good neighbor".

In conclusion, this manual teaches the proper aspects of how to be a good neighbor and share clean air around a treatment facility, while still providing indispensable wastewater services to the larger community.

2

Odors and Odorous Compound Measurement

1.0 INTRODUCTION	6
2.0 ODORS AND ODORANTS	7
3.0 SAMPLING ODOR SOURCES	8
3.1 Sampling Exhaust Stacks and Vents	9
3.2 Sampling Surfaces	11
4.0 ODOR PARAMETERS	13
4.1 Background of Laboratory Olfactometry Standards	13
4.2 Odor Panels	14
4.3 Odor Concentration	15
4.4 Odor Intensity	18
4.5 Odor Persistency	18
4.6 Odor Characterization	21
4.7 Confidence Interval of Odor Results	22
4.8 Odor Reduction Efficiency	22
5.0 MEASURING AMBIENT ODORS	23
5.1 Ambient Odor Intensity	23
5.2 Grid Method	24
5.3 Ambient Odor Concentration—Field Olfactometers	24
5.4 Application of Field Olfactometry	25
6.0 SAMPLING AND CHEMICAL ANALYSIS OF SPECIFIC ODORANTS	27
6.1 Methods for Reduced Sulfur Compounds, Including Hydrogen Sulfide	27
6.2 Methods for Volatile Fatty Acids	28
6.3 Methods for Amine Compounds	28
6.4 Methods for Volatile Organic Compounds	29
6.5 Methods for Other Wastewater Odorants, Including Ammonia	29
7.0 MATRIX OF CHARACTER, THRESHOLD, AND ANALYSIS	30
8.0 SUMMARY	30
9.0 REFERENCES	34

1.0 INTRODUCTION

Odors remain one of the top air pollution complaints for regulators and government bodies in the United States and internationally. Likewise, nuisance odors also dominate the environmental concerns expressed to water resource recovery facilities (WRRFs). Ambient air holds a mixture of chemicals from everyday activities of industrial and commercial enterprises, including wastewater collection and conveyance systems and treatment facilities.

A person's olfactory system allows for the detection of some odorant compounds at extremely low concentrations (i.e., parts per trillion or even parts per quadrillion). When we experience odorous air emissions from facilities such as those for wastewater collection and treatment, at its most primitive level our sense of smell acts as an antenna for an early warning of compounds present outside our normal baseline. Most citizens are moved to complain when these odorants are unpleasant or are above their acceptable level of frequency and duration.

When wastewater collection and treatment odors affect air quality and cause citizen complaints, investigation requires that specific odorant chemicals be measured and all of the odorous air be measured using standardized methods of odor perception. Measurement and quantification of odors and odorant compounds from wastewater collection systems and treatment facilities is typically required for the following purposes:

- Monitoring for compliance as part of permit requirements or renewal;
- Determination of baseline status for facility expansion planning;
- Determination of specific odor sources during complaint investigation;
- Monitoring operations for management performance evaluation;
- Comparison of operating practices when evaluating operating alternatives;
- Monitoring specific events or episodes for defensible, credible evidence;
- Comparison of odor mitigation measures during testing and trials;
- Determination of odor control system performance; and
- Verification of estimated odor effects from dispersion modeling.

Odors and odorant compounds are measurable using quantitative, standardized scientific methods. Point, area, and volume emissions can be collected and the samples sent to a laboratory for testing of chemical concentration (e.g., gas chromatography–mass spectrometry [GC–MS]) or of odor parameters (e.g., odor concentration, intensity, characterization). Odorant concentrations and odor perception can also be measured and quantified directly in the surrounding air, at the property line, and in the community using standardized air sampling/analytical techniques as well as field odor assessment practices, respectively.

This chapter presents the basics of sampling and measuring odorous air. A brief explanation of nasal anatomy and how the olfactory systems work is also presented. Methods for sampling odorous air are explained and illustrated. Odor parameters are defined along with a presentation of the background history of odor testing standards. Field assessment methods of ambient odors are also discussed. Finally, the chapter presents chemical analyses of common odorants from WRRFs and comparisons are made between compounds, odor thresholds, and laboratory analyses associated with wastewater.

2.0 ODORS AND ODORANTS

Of the five senses, the sense of smell is the most complex and unique in structure and organization. As summarized by McGinley, McGinley, and McGinley (2000), the olfactory system plays a significant role as a defense mechanism by creating a natural aversion to malodorous and irritant compounds. Olfaction is a protective sense, protecting from tainted food and matter, such as rotting vegetables, putrefying meat, and fecal matter. This is accomplished by two main nerves: (a) the olfactory nerve (first cranial nerve) processes the perception of chemical odorants and (b) the trigeminal nerve (fifth cranial nerve) processes the irritation or pungency of chemicals, which may or may not include odorants.

During normal nose breathing, only a small percentage of inhaled air passes up and under the olfactory receptors in the top, back part of the nasal cavity. When a sniffing action is produced, either voluntarily or as an involuntary reflex, the inhaled air passed over the olfactory receptors more than doubles because of turbulent action in front of the nasal turbinates. In both nasal cavities, 10 to 25 million olfactory cells comprise the olfactory epithelium. Cilia on the surface of the epithelium have a receptor contact surface area of approximately 5 cm^2 (0.78 $in.^2$) because of the presence of many microvilli. Supporting cells surrounding these cilia secrete mucus, which acts as a trap for chemical odorants.

Chemical odorants pass by the olfactory epithelium and are dissolved (transferred) into the mucus at a rate dependent on their water solubility and other mass-transfer factors. The olfactory cells respond to the presence of these chemicals. If an action potential threshold of olfactory electrical signals is met, the pulse propagates along the olfactory nerve, through the ethmoidal bone between the nasal cavity and the brain compartment, where it synapses with the olfactory bulb.

All olfactory signals meet in the olfactory bulb where the information is distributed to two different parts of the brain. One major pathway of information is to the limbic system, which processes emotion and memory response of the body. This area also influences the signals of the hypothalamus and the pituitary gland, the two main hormone control centers of the human body. The second major information pathway is to the frontal cortex. This is where conscious sensations take place as information is processed with other information and compared to cumulative life experiences so that the individual can recognize the odor and make some decisions about the experience.

Frequently, the terms *odor* and *odorant* are used interchangeably. Odor refers to the perception experienced when one or more chemical substances in the air contact the various human sensory systems. Odor is the human response. Odorant refers to any chemical in the air that is capable of being part of the perception of odor by a human. The odorants are the chemical compounds. Odor perception may occur when a person experiences one or many chemical odorants.

3.0 SAMPLING ODOR SOURCES

Odorous air samples can be collected from point emission sources (i.e., stacks or vents) and from surface air emissions sources (i.e., liquid or solid surfaces). Whole-air samples for laboratory testing are typically collected in 10-L Tedlar® gas sample bags for transport to the sensory laboratory for odor analysis or in 1-L Tedlar gas sample bags for transport to the analytical laboratory for chemical analysis (St. Croix Sensory, Inc., 2003).

Odor and odorant sampling is often part of an odor study, odor control system performance test, or routine performance evaluation of the WRRF. The purpose of the sampling is often to compare odors and odorants from various processes at the facility or to determine if the control system is performing according to specifications.

Sampling protocols typically require the samples to be collected under normal operating conditions. The sampling planners determine when normal operating conditions exist or if sampling should take place during a specific timeframe, such as during worst-case conditions. The person conducting the sampling is responsible for documenting the conditions at the time of the sampling.

Before sampling, the sampler must gather the equipment that might be needed, including ladders, tools for opening sample ports, pitot tubes and related items, thermometers, vacuum case with pump, emission isolation flux chamber for quiescent surface sampling, and Teflon sample lines (St. Croix Sensory, Inc., 2003). Sample collection supplies must also be gathered, which may include Tedlar bags with labels, sorbent tubes, or inert-coated stainless steel canisters.

In addition to collecting samples for odor analysis, the protocol may require that a companion sample (i.e., duplicate) is collected in a Tedlar sample bag or canister for chemical analysis (e.g., reduced sulfur compound gas analysis or volatile organic compound [VOC] analysis). The protocol may also require testing for specific chemical odorants in the air with portable instruments (e.g., hydrogen sulfide analyzer).

3.1 Sampling Exhaust Stacks and Vents

Frequently, samples are collected from various point sources such as exhaust stacks or vents. Measurements should be taken for the velocity, pressure, and temperature of the air streams from the source. The sampler prepares the sample Teflon tubing, sample bag, vacuum case, and pump, as well as a labeled (i.e., field number and date) 10-L Tedlar sample bag. With the bag valve open, the bag is connected to the tubing inside the vacuum case. The vacuum case is then sealed. Acting as a sampling probe, the Teflon sample tubing is held in position inside the exhaust stack or ducting and connected to the bag inside the vacuum case. The pump is then connected to the case to gradually evacuate the air. This vacuum causes the sample bag to fill odorous air from the exhaust stack while only being in contact with the Teflon sample line. Figure 2.1 illustrates the sampling apparatus.

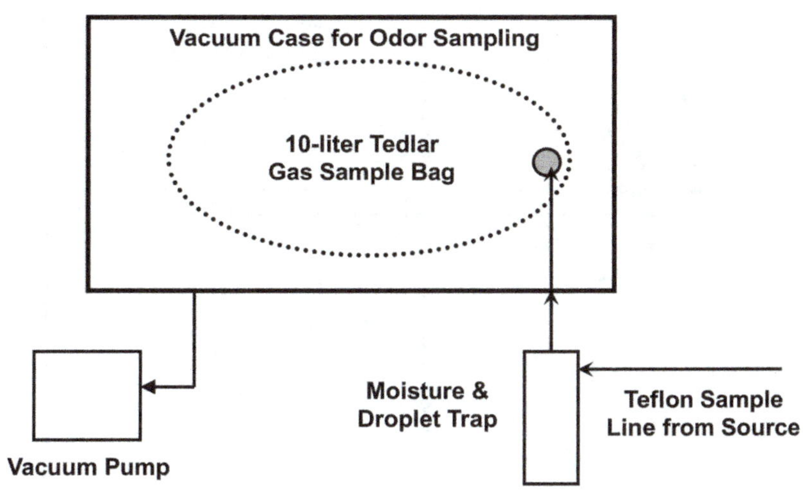

FIGURE 2.1 Vacuum chamber for odor sampling.

The Tedlar sample bag is first filled with the odorous air to about one-third capacity and held for one minute to condition the bag. The bag is then emptied using the pump to pressurize the vacuum case or by removing the bag from the vacuum case and manually squeezing the bag until it is empty.

The whole-air sample is then collected in the sample bag using the vacuum case. The sample bag should be no more than two-thirds full to allow sufficient room for approximately 20% expansion when air shipping is used. When the vacuum case is stopped, the sample bag is closed, the sample line is disconnected, and the sample bag is removed.

If exhaust air is saturated with moisture, or if the exhaust air dewpoint is above ambient temperatures, additional sampling procedures must be incorporated. Figure 2.1 illustrates the use of a water trap in the sampling line before the vacuum case for the purpose of preventing any water droplets from entering the sample bag. Further, the sample may require pre-dilution to prevent the warm, moist exhaust from condensing in the sample bag. This pre-dilution could be performed statically, by prefilling the bag with dry, zero air or high-purity nitrogen and then filling with a known volume of the odorous air. A preferred method would be the use of a dynamic dilution sampling probe (as shown in Figure 2.2) to simultaneously collect and mix the sample from the exhaust source with a diluting gas such as zero air or high-purity nitrogen. Additionally, the exhaust air that contains oxidizing chemicals, such as ozone or chlorine, may change with time. Extra sampling precautions or procedures may be necessary in these cases; thus, sampling specialists may be consulted for specialized equipment.

The collected sample bag must be protected from sunlight and from potential puncture with a durable shipping case or box. A chain-of-custody

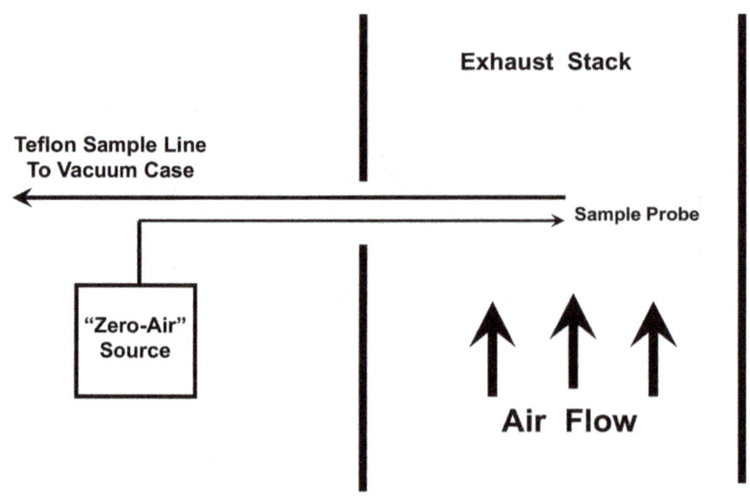

FIGURE 2.2 Dynamic dilution sampling probe.

record must be completed. The date, time, and description of the sample are recorded on the form, as well as the analysis requested from the laboratory. Each 10-L Tedlar sample bag must be protected by placing the bag inside the shipping box on its end. Sample bags must never be shipped on top of one another because the weight of one bag can damage another bag. Allow sufficient room in the shipping box for each bag to expand approximately 20% when aircraft shipping is involved.

The final step for the sampler, before dispatching the sample shipment, involves completing the shipping documents and calling the shipping company for pickup. Most odor sampling protocols require the odor evaluation to be conducted within 30 hours after sample collection, and analytical hold times range from 24 to 72 hours from a Tedlar bag, seven days in a canister, and 14 days from a sorbent tube. After a long and potentially expensive day of sampling, it is imperative to be sure the collected samples are picked up by the shipping courier before the end of the business day.

3.2 Sampling Surfaces

An odorous air sample can be collected from surfaces, sometimes called "area sources". Wind speed and direction, air temperature and relative humidity, and solar radiation all affect the odorous emission rate from the quiescent surface (e.g., influent channel of primary clarifier). Aerated surfaces are also affected by the aeration blower flowrate in a diffused air process or the airflow emission from the surface of a biofilter. Emission rates for aerated area sources are calculated by multiplying the odor concentration (OU/m^3) by the blower or exhaust fan flowrate (m^3/s).

A tall passive chimney or simulated stack is an apparatus used to collect aerated surface emissions samples. Figure 2.3 illustrates the sampling

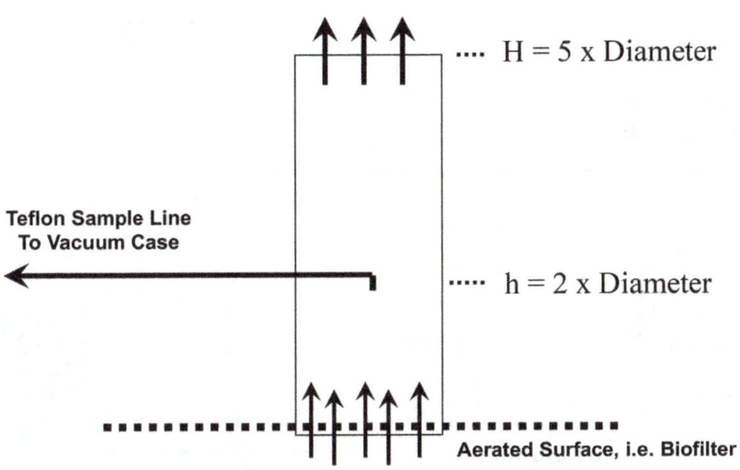

FIGURE 2.3 Tall passive chimney sampler.

method to isolate an aerated surface. The tall passive chimney sampler minimizes the effects of crossflow winds at the time of sample collection. A vacuum case is used to collect the whole-air sample of exhaust air from the biofilter surface using the same bag-filling procedure described for the point source sample collection.

Samples can also be collected using a flux chamber floating on liquid surfaces. The flux chamber, also called a surface emission isolation chamber, was originally developed in the 1970s to quantify emissions of inorganic gases from soils. In the 1980s, the U.S. Environmental Protection Agency (U.S. EPA) studied flux chambers for measuring the emissions of VOCs from contaminated soil and water surfaces at hazardous waste sites (Klenbusch, 1986). Figure 2.4 illustrates the method to collect whole-air samples from quiescent liquid or solid surfaces. The flux chamber uses a flotation collar to float the chamber on a liquid surface. A clean, odor-free carrier gas (e.g., dry zero air or high-purity nitrogen) is metered into the flux chamber at a known flowrate (e.g., 5LPM) as the sweep air. After an equilibration period of 3 to 4 residence times, a sample is withdrawn from the flux chamber at a flowrate less than the sweep airflow rate (e.g., 2LPM). Similar to sampling a point source, a vacuum chamber and Tedlar sample bag are used to collect the sample from the flux chamber (Schmidt, Card, & Goodwin, 2008).

The odorous emission rate (OU/s/m^2) for an area source is calculated by multiplying the odor concentration (OU/m^3) by a sweep air flowrate (m^3/s) and dividing by the area (m^2) of the flux chamber used to collect the surface emission odor samples.

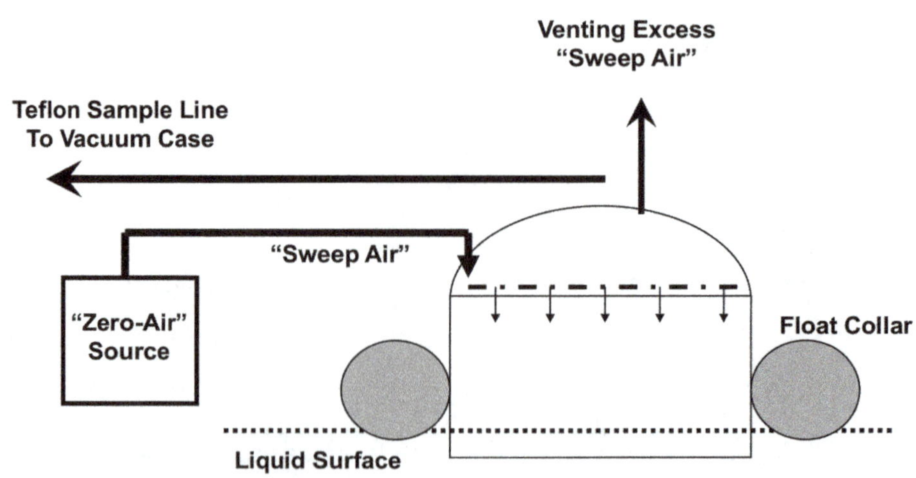

FIGURE 2.4 Flux chamber surface sampler.

4.0 ODOR PARAMETERS

The perception of odor can be quantified using objective, scientific methods. Odor testing has evolved over time with changes in terminology, methods, and instrumentation. The following are four measurable, objective parameters of perceived odor:

- Odor concentration—measured as a dilution ratio and reported as a detection threshold (DT) and recognition threshold (RT) or more generally as a dilution-to-threshold, (D/T) and commonly assigned the pseudo-dimensions of odor units per cubic meter (OU/m^3);
- Odor intensity—rating the perceived strength of odor using a word or number scale, typically linked to a referencing scale of discreet n-butanol concentrations in parts per million;
- Odor persistency—reported as the dose–response function, a relationship of odor concentration and odor intensity; and
- Odor character descriptors—what the odor smells like according to category scales.

These odor parameters are objective because they are measured by trained panels of assessors using validated techniques dealing with facts that are not distorted by personal feelings or prejudices. The following are additional measurable, but subjective parameters of perceived odor:

- Hedonic tone—pleasantness versus unpleasantness of the odor;
- Objectionable—causes avoidance or physiological effects in a person; and
- Annoyance—interference with comfortable enjoyment of life and property.

These odor parameters are subjective because they are reported by individuals relying on their interpretation of word scales and their personal feelings, beliefs, memories, experiences, and prejudices. Guidelines and legal definitions of these subjective odor parameter scales often assist individuals in reporting observed odor; however, the nature of these parameters remains subjective.

4.1 Background of Laboratory Olfactometry Standards

In the early years of odor testing in laboratories, the American Society for Testing and Materials (ASTM) D1391 syringe dilution technique measured odors in the laboratory from samples collected at the odor source (American Society for Testing and Materials, 1978). In 1979, ASTM D1391 was replaced with ASTM E679, Standard Practice for Determination of Odor and Taste Thresholds by a Forced-Choice Ascending Concentration Series Method of Limits. The current edition of ASTM E679 was published in 2019 as ASTM E679-19 (ASTM International, 2019).

In 1995, a working group was formed within the European Committee for Standardization (CEN) to develop a unified olfactometry standard. This working group recognized the need to help industry and regulators develop a consistent basis for monitoring and testing odors, and thus help determine the potential for odor nuisance. This was to be accomplished by developing a method that

- improved consistency within each laboratory (repeatability);
- achieved comparable results among laboratories (reproducibility); and
- connected the results to a traceable reference material, n-butanol.

To achieve these goals, the CEN committee focused on the following issues: sampling procedures; sample containers; olfactometer construction and operation; the olfactometer and assessor interface; the odor testing room; and methods of data processing and assessor selection, training, and performance (van Harreveld, Heeres, & Harssema, 1999).

The CEN olfactometry standard was released to the public at the end of 1999 as proposed CEN standard 13725, Air Quality—Determination of Odour Concentration by Dynamic Olfactometry (European Committee for Standardization, 2003). The final draft of the document was officially approved in 2003 (British Standards Institution, 2003). This standard, EN 13725, has become a de facto international standard. The CEN TG264 met in 2017 and 2018 to review and edit the EN 13725 standard to publish a revised document in 2019 or 2020.

4.2 Odor Panels

Odor evaluation of whole-air odor samples is conducted under controlled laboratory conditions following standard sensory evaluation practices using panels of trained assessors.

An odor laboratory is an odor-free, non-stimulating space. Each odor assessor, when working on odor evaluations, focuses on the task of observing the odor sample when presented. A comfortable and relaxing waiting area without distractions enhances a low-stress environment for assessors and nurtures their commitment and dedication to quality performance.

From a pool of on-call assessors, five to 10 assessors are selected for a scheduled odor panel. Odor panels consist of individuals that are recruited from the community and then selected and trained following Guidelines for the Selection and Training of Sensory Panel Members (American Society for Testing and Materials, 1981) and EN 13725 (European Committee for Standardization, 2003). Assessors are not allowed to smoke, use smokeless tobacco, or have significant allergies. Women who are pregnant or may be pregnant are also not allowed to be assessors because they are likely to have a heightened sensitivity to odors. Assessors must follow general odor panel rules including the following: being free of illness, no eating or drinking 30 minutes before or during a panel, being fragrance free, and so on.

Each odor assessor is tested to determine their individual olfactory sensitivity using standard odorants (e.g., n-butanol and hydrogen sulfide). The assessor receives training that consists of olfactory awareness, sniffing techniques, standardized odor descriptors, and olfactory responses.

4.3 Odor Concentration

The most common measure of odorous air samples is the odor threshold value (OTV), also referred to as the "odor concentration". The determination of the odor concentration is made using an instrument called an *olfactometer* and is expressed as a dilution-to-threshold (D/T) level that is typically expressed as the detection threshold (DT) and recognition threshold (RT).

The laboratory olfactometer simulates the dilution of odor in the ambient air. Figure 2.5 illustrates how the wind dilutes odorous air emissions. A large dilution ratio on the olfactometer (e.g., 20 000, log 20 000 = 4.3) represents a high dilution of the odor sample, which is similar to a person standing at a large distance from the odorous emission. A small dilution ratio (e.g., 20, log 20 = 1.3) represents a small dilution of the odor sample, which is similar to a person standing close to the odor source.

The detection threshold is an estimate of the number of dilutions needed to make the actual odor emission non-detectable. Recognition threshold is

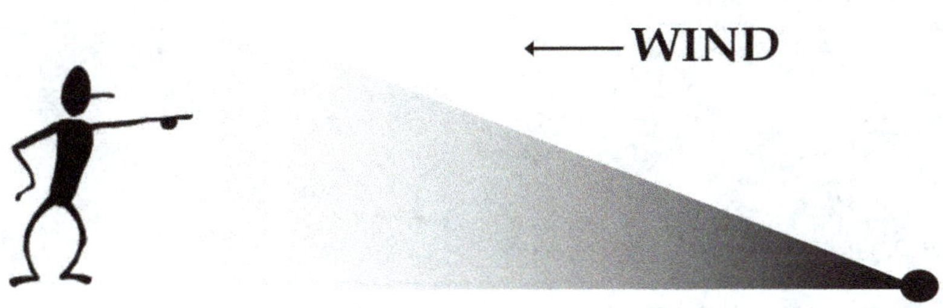

FIGURE 2.5 Dilution of odor in the ambient air.

also a dilution ratio and represents the number of dilutions needed to make the odor sample faintly recognizable. The recognition threshold value, in odor units per cubic meter, is always lower than the detection threshold value because it takes more dilution to bring an odor to its detection threshold (no odor present) compared to its recognition threshold (odor is not recognizable). Likewise, a large value for the odor concentration represents a strong odor, whereas a small odor concentration is a weaker odor.

During an odor test, the assessor sniffs a dilute sample of the odor as it is discharged from the olfactometer as one of three sample presentations (one presentation with the dilute odor and two with odor-free air). Figure 2.6 shows one assessor (left) sniffing from the olfactometer nasal mask with the test administrator (right) operating the olfactometer. The assessor sniffs the triad of the presentations and must select the one of the three that is different from the other two, even if the selection is a guess. This statistical approach is called "triangular forced-choice". The assessor declares to the test administrator if the selection is a guess, detection (i.e., the selection is different from the other two), or recognition (i.e., the selection smells like something).

The assessor is then presented with the next set of three presentation choices, one of which contains the diluted odor sample. However, this next set of three samples presents the odor at a higher concentration, lower amount of dilution (e.g., two times higher, one-half dilution ratio). The assessor continues to evaluate additional levels of higher concentration (i.e., lower dilution) presentations following the triangular forced-choice procedure

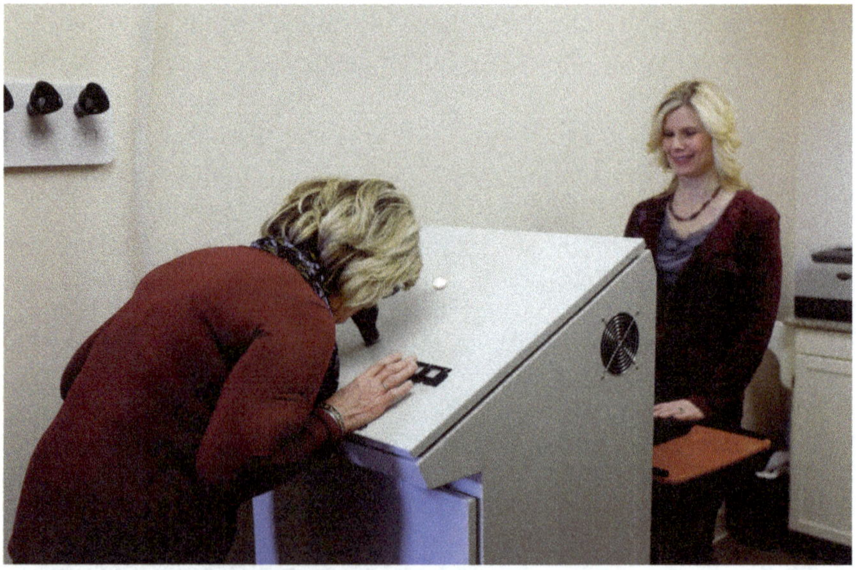

FIGURE 2.6 Assessor (left) at olfactometer while test administrator (right) operates olfactometer and monitors the test (permission granted by St. Croix Sensory).

and required designation of guess, detection, or recognition. This statistical approach of increasing levels of sample presentation is called "ascending concentration series". The ascending concentration series approach avoids sensitizing the assessors' olfactory sense to the sample odor.

Therefore, the odor concentration is a number derived from the laboratory dilution of sample odors. The dilution ratio (total presentation volume divided by the odor sample volume) of each sample presentation level is used to calculate the concentration of the evaluated sample. For example, consider an assessor that did not detect the odor presentation on the olfactometer level 5, which is a dilution ratio of 4000, but did detect the odor at the next odor concentration at level 6, which is a dilution ratio of 2000. The assessor's individual threshold estimate of the detection threshold is the geometric mean between these two dilution ratios, which is a result of 2820. This is called the "best-estimate threshold".

The geometric mean is used to estimate the individual threshold to balance the variance between dilution levels (Stevens, 1962). The ascending concentration series is a geometric progression, where each successive level is twice the previous odor level; therefore, the scale does not have equal spread between dilution levels. There is equal variance along the logarithmic scale (Dravnieks, Schmidtsdorf, & Mellgaard, 1986).

The individual estimate thresholds of the assessor responses are averaged on the log scale to determine the detection threshold for which 50% of the individuals observe the presence of the odor. This detection threshold value that is obtained from odor testing is actually derived from dilution ratios and is, therefore, dimensionless. However, the pseudo-dimensions of odor units (OU) or odor units per cubic meter (OU/m^3) are commonly applied.

Odor concentration is an estimate of the number of dilutions needed to make the actual odor emission non-detectable. The dilution of the actual odor emission is the physical process that occurs in the atmosphere downwind of the odor source. The receptor (citizen) sniffs the diluted odor. If they detect the odor, then the amount of odor in the atmosphere is higher than the receptor's detection threshold level.

The pseudo-dimension odor units per cubic meter is commonly used in odor dispersion modeling, taking the place of grams per cubic meter (g/m^3). The odor concentration can be multiplied by the airflow rate (m^3/s), resulting in odor units per second, analogous to grams per second. Odor concentrations from multiple odor sources cannot be added or averaged directly. Odor modeling must be conducted with caution by a professional with modeling experience. The modeling result showing an odor concentration of 1 represents the odor detection threshold level that was determined by the odor panel. A value less than 1 is no odor or subthreshold odor. A value greater than 1 represents odor at suprathreshold.

4.4 Odor Intensity

Odor intensity is the perceived strength of the odor above the recognition threshold (suprathreshold). Standard Practice for Referencing Suprathreshold Odor Intensity (ASTM E544-18) presents two methods for referencing the intensity of odors: the dynamic scale method (procedure A) and the static scale method (procedure B) (ASTM International, 2018). The dynamic scale method uses an olfactometer with continuous flows of a standard odorant (e.g., n-butanol) for presentation to the assessors. The assessor compares the observed intensity of an odor sample to a specific concentration level of the standard odorant. The static scale method uses a set of bottles with fixed dilutions of the standard odorant in water solution. Field investigators and odor laboratories commonly use the static scale method (Turk, Switala, & Thomas, 1980).

The odor intensity result is expressed as either the average scale level—for example, level 3 on the 10-point scale—or as the n-butanol concentration in parts per million of the matching level. A larger value of n-butanol concentration means a stronger odor. These n-butanol concentrations are referenced for purposes of documentation and communication in a reproducible format.

An important aspect of the n-butanol intensity reference scale is the variety of available scales. Figure 2.7 presents three common odor intensity referencing scales (OIRS). The specific olfactometer device determines the dilution levels of the dynamic scale method used by laboratories. The dilution levels of the static scale method used by laboratories and field inspectors are determined from interpretation of ASTM procedure B, which accepts numerous scale choices. The laboratory or field investigator selects the starting point of the scale and the geometric progression of the concentration series.

Common scales of n-butanol concentration in air and water have starting points between 10 and 25 ppmv. Many scales use a geometric progression of 2 (each level twice the concentration of the previous); however, some scales use geometric progressions of 1.5 to 3. All laboratories and investigators presenting the odor intensity data should reference an n-butanol concentration in air or water to allow comparison of results from different data sources.

4.5 Odor Persistency

Odor is a psychophysical phenomenon. Psychophysics involves the response of an organism to changes in the environment perceived by the five senses (Stevens, 1960). Some examples of psychophysical phenomenon include how the human body perceives sound loudness, light brightness, or odor intensity.

12 Point Scale	8 & 10 Point Scales	5 Point Scale
1 < 10 >	1 < 12 >	
2 < 20 >	2 < 24 >	1 < 25 >
3 < 40 >	3 < 48 >	
4 < 80 >	4 < 96 >	2 < 75 >
5 < 160 >	5 < 194 >	
6 < 320 >	6 < 388 >	3 < 225>
7 < 640 >	7 < 775 >	4 < 675 >
8 < 1280 >	8 < 1550 >	5 < 2025 >
9 < 2560 >	9 < 3100 >	
10 < 5120 >	10 < 6200 >	
11 < 10240 >		
12 < 20480 >		

KEY: < XXX > is Parts Per Million n-Butanol Equivalent Odor Intensity

FIGURE 2.7 Example odor intensity referencing scales (OIRS) shown using reference odorant n-butanol (permission granted by St. Croix Sensory).

These psychophysical phenomena lead to sensory responses, which follow a power law. Apparent odor intensity grows as a power function of the stimulus odorant. Stevens showed that this power law (Steven's Law) follows the equation

$$I = \mathrm{k}C^{\mathrm{n}}$$

where I is intensity, C is the mass concentration of the odorant, and k and n are constants that are different for every specific odorant or mixture of odorants (Stevens, 1962).

As shown in Figure 2.8, this equation is a straight line when plotted on a log–log scale. The x-axis is the mass concentration of the single odorant. The upward slope of the graph illustrates that the perceived intensity of the odorant increases as the mass concentration increases.

Odor persistency is a term used to describe the rate at which an odor's perceived intensity decreases as the odor is diluted (in the atmosphere downwind). Figure 2.9 illustrates how odor intensity decreases as the odor is diluted in this dose-response function. This is determined from odor intensity measurements at various dilutions and at full strength (i.e., no dilution).

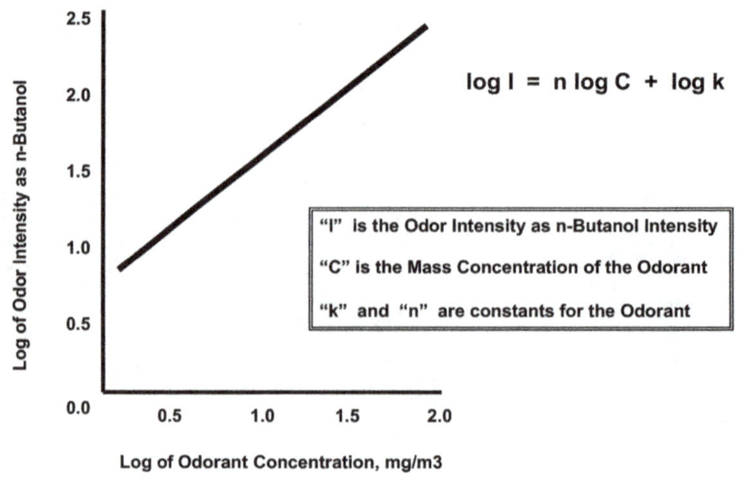

FIGURE 2.8 Power law for a single odorant.

The odor concentration (dose), expressed as the log of the presented dilution ratio, and the odor intensity (response) produce the linear relationship specific to each odor sample (Dravnieks & Jarke, 1980). The negative slope of the line represents the relative persistency of the odor. The logarithm of the constant k is related to the intensity of the odor sample at full strength (i.e., the y-axis intercept) (Dravnieks et al., 1986).

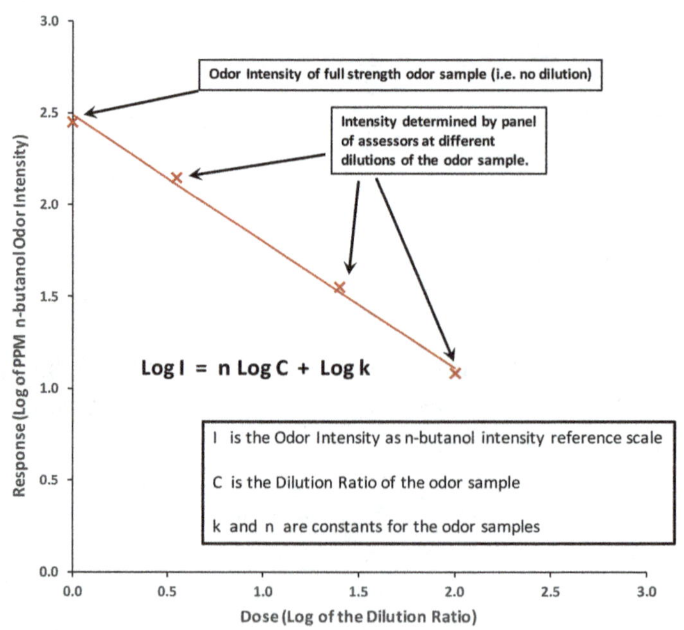

FIGURE 2.9 Odor persistency or dose-response of an odor sample.

4.6 Odor Characterization

Odor characterization is a nominal (category) scale of measurement. Odors can be characterized using reference vocabulary for taste, sensation, and odor descriptors. Numerous standard odor descriptor lists are available. In 1986, the International Association of Water Pollution Research and Control proposed eight major odor descriptor categories and illustrated them on an odor wheel as "vegetable", "fruity", "floral", "medicinal", "chemical", "fishy", "offensive", and "earthy" (Mallevialle & Suffet, 1987). Each of the major categories can have specific descriptors, which can be presented in training using exemplars. In Suffet, Burlingame, Rosenfeld, and Bruchet (2004), a wastewater odor wheel was presented for categorizing wastewater odors and related odorant compounds (Suffet et al., 2004).

In addition to identifying the presence of specific odor descriptors, assessors can also provide an intensity rating for the individual descriptors. It is then possible to compare the odor characters from multiple sources by plotting odor descriptor intensity data on a spider plot. Figure 2.10 shows descriptors for an odor control inlet (red circle) and outlet (blue diamonds). The inlet has predominant sulfur, decay, animal, earth, and sulfidic vegetable characteristics, while the outlet shows reduced levels of these malodorous characteristics and predominantly report herbal, medicinal, decay, and earth descriptors.

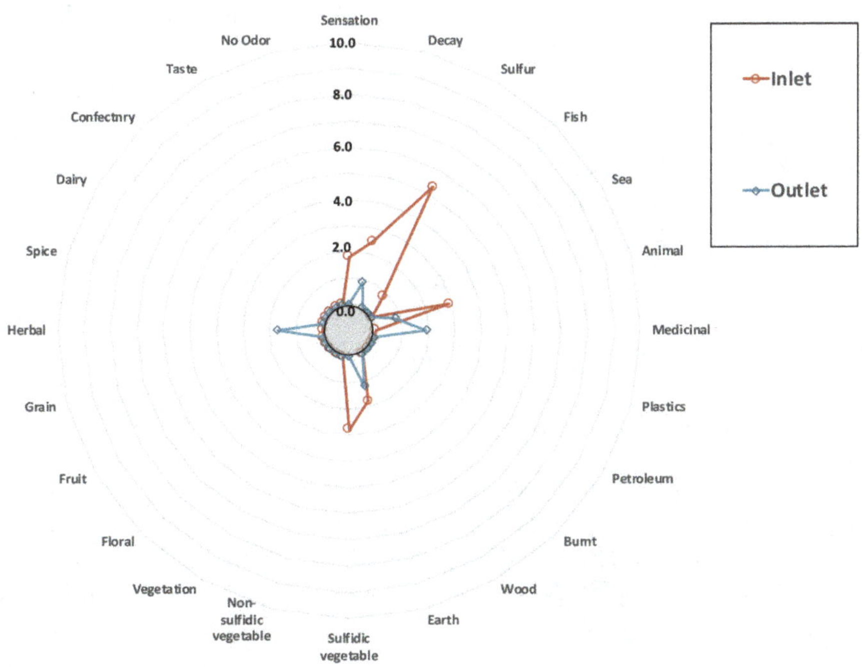

FIGURE 2.10 Odor characterization of odor control inlet and outlet samples.

A more advanced look at odor characterization has been described by Yubin et al. by looking at the odor profile (characterization) of environmental odor samples at multiple dilution ratios and comparing those results to an odor profile analysis of pure compounds (Yubin et al., 2018).

4.7 Confidence Interval of Odor Results

It is important to highlight the necessity of logarithm base-10 transformations used in odor testing calculations. These transformations are used to make the nonlinear dilution ratio scale a linear scale in log base-10. More specifically, the transformations are performed to stabilize (i.e., make uniform) the variance. With the uniform variance, the linear transformed data show symmetry around the group average. However, these data will be asymmetrical around the reported dilution factor values of detection and recognition threshold (MacBerthouex & Brown, 1994).

When odor testing is conducted on a number of odor samples, with replicates, the data produce odor results with a standard deviation. The standard deviation from replicate sampling represents the odor testing reproducibility. From the reported standard deviation, confidence limits can be calculated for odor testing data (European Committee for Standardization, 2003).

An olfactometry laboratory may develop a record of replicate repeatability with a log standard deviation of 0.05. This is the standard deviation on the transformed scale of logarithms. A confidence interval can be calculated for a typical odor concentration value using this known standard deviation (McGinley & McGinley, 2010).

As an example, a detection threshold value of 2100 has a transformed log value of 3.32. The 95% confidence interval, where t is 2.0 and standard deviation is 0.05, is calculated to be ±0.10. This yields a symmetric confidence interval of 3.32 ±0.10, or a range of 3.22 to 3.42. Taking antilogarithms to transform back to the original scale, the final asymmetrical confidence interval is 1660 to 2630 around the 2100 detection threshold. Note that narrower confidence intervals can be achieved with replicate sampling for each odor source (McGinley & McGinley, 2010).

4.8 Odor Reduction Efficiency

Testing of an odor control system may be used to determine the odor reduction efficiency of the system. If the inlet to the odor control system has a detection threshold of 1560 and the treated outlet from the system has a detection threshold of 100, then the odor reduction efficiency can be determined by

$$E = ([1560 - 100] \div 1560) \times 100\% = 94\%$$

The efficiency calculations can be conducted using the odor result values and not require logarithmic transformations (European Committee for Standardization, 2003).

5.0 MEASURING AMBIENT ODORS

Odors can also be measured and quantified directly in the ambient air by trained inspectors using standard practices. The first method uses a standard OIRS of n-butanol. A second method, a grid method standardized in Europe, uses the randomized short-time continuous observations of many odor judges. A third methodology uses a field olfactometer, which dynamically dilutes the ambient air with carbon-filtered air in distinct D/Ts. These three field olfactometry methods can be used for many different applications.

5.1 Ambient Odor Intensity

Odor intensity of the ambient air can be measured objectively using the OIRS (ASTM E544-18) discussed previously. The air pollution inspector, plant operator, or community odor monitor observes the odor in the ambient air and compares it to the selected OIRS. The person making observations should use a carbon-filtering mask to refresh their olfactory sense between observations. Without the use of a carbon mask, the observer's olfactory sense may become adapted to the surrounding ambient air or become fatigued from any odor in the surrounding air. The adaptation of an observer's olfactory sense is a common phenomenon when attempting to evaluate ambient odors. Practicing the procedures of ASTM E544 is nearly identical to the standard method of quantifying the opacity of smoke plumes (Missen & Stein, 1975). As with opacity measurement training, an odor training course provides the field inspector with an understanding of odors and the ability to calibrate their intensity judgements to the reference scale.

Using the OIRS, the intensity of the observed ambient air is expressed as a scale level or the equivalent concentration of n-butanol in water (ppmv). Common n-butanol static scales, shown in Figure 2.7, were discussed previously. The reference scale serves as a standard practice to quantify the odor intensity objectively. Field air pollution inspectors, using the standard OIRS, can provide measured, dependable, and repeatable observations of ambient odor intensity. To allow for a comparison of results from different data sources and maintain a reproducible method, the equivalent n-butanol concentration or scale points are reported along with the scale range and the starting point.

5.2 Grid Method

The grid method, originally developed in Germany, is now a European method (EN 16841) (British Standards Institution, 2016). A grid is prepared over the assessment area in the community with grid connections that are 100 m (328 ft) to 1000 m (3281 ft) apart. Inspectors visit the grid connection points at least 26 times during the assessment time, which is typically 6 to 12 months. This provides 104 total assessments for each grid box. At each assessment, the inspector sniffs the air every 10 seconds over a 10-minute period and documents whether an odor is observed and what the odor smells like. The percentage of observations when odor is detected can then be documented at each point and within each grid box to determine odor frequency.

5.3 Ambient Odor Concentration—Field Olfactometers

In 1958, 1959, and 1960, the United States Public Health Service sponsored the development of an instrument and procedure for field olfactometry through project grants. The first field olfactometer, called the "Scentometer", was manufactured by the Barnebey-Cheney Company (Barnebey and Sutcliffe Corporation, 1974).

A field olfactometer creates a series of dilutions by mixing the odorous ambient air with odor-free, carbon-filtered air. The U.S. Public Health Service method defined the dilution factor as the D/T, the number of dilutions needed to make the odorous ambient air non-detectable.

The method of producing D/Ts with a field olfactometer consists of mixing two volumes of carbon-filtered air with specific volumes of odorous ambient air. Figure 2.11 illustrates the mixing of carbon-filtered air with the odorous air. The D/T for a field olfactometer is calculated as follows:

Dilution Factor = Volume of carbon filtered air/volume of odorous air = D/T

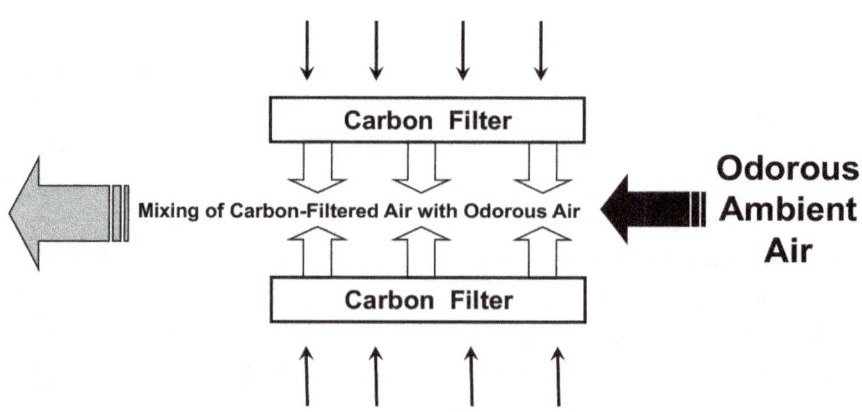

FIGURE 2.11 Dilution-to-threshold (D/T) with a field olfactometer.

Note that the calculation method for field olfactometry is slightly different from laboratory olfactometry. As mentioned in Section 4.3 "Odor Concentration", the laboratory olfactometry dilution factor is calculated as follows:

Dilution Factor = Volume of total presentation air/volume of odorous air = Z

where Volume of total presentation air = the addition of the volume of carbon filtered air to the volume of odorous air.

The calculation difference is illustrated in the following example:

- A field olfactometer uses the ratio of 7 volumes of carbon-filtered air to 1 volume of odorous air: 7/1 = 7 D/T
- A laboratory olfactometer uses the ratio of total presentation air volume to 1 volume of odorous air: (7 + 1)/1 = 8 (Z value)

The field olfactometer instrument, the D/T terminology, and the method of calculating D/T are referenced in a number of existing state and local agencies' odor regulations and permits (State of Colorado, 2013; State of Missouri, 2013). Therefore, a field olfactometer instrument, in the hands of trained air pollution investigators or monitors, is a realistic and proven method for quantifying ambient odors.

5.4 Application of Field Olfactometry

Field olfactometry using an OIRS, grid monitoring or a calibrated field olfactometer is a cost-effective means to quantify odors. Facility operators, community inspectors, and neighborhood citizens can confidently monitor odor strength at specific locations around a facility's property line and within the community.

The following types of odor monitoring are presented in brief exemplary form as an application guide for field olfactometry:

- On-site monitoring—operators have the unique ability to monitor odors through the day with field olfactometry. Operators can observe odors of arriving materials, outdoor process activities, and fugitive air emissions. Monitoring odors on-site may involve following a logical pathway around a facility to determine where odors exist or making odor observations at predetermined locations.
- Random off-site monitoring—the random inspection approach is a frequently used method for ambient odor monitoring. Random monitoring leads to a compilation of data that can be correlated with meteorological information and on-site activities.

- Scheduled monitoring—well-planned, scheduled monitoring can be limited to a daily walk or drive around an area or be structured with several visits to predetermined monitoring locations. Data from field olfactometry can be used to correlate the many parameters that influence odor episodes, including meteorological conditions, and on-site operating activities.
- Intensive odor survey—an in-depth evaluation of on-site odor generation and off-site odor effects may be needed for permit renewal or facility expansion. Extensive data collection using field olfactometry will identify which sources or operations do or do not cause odor off-site. All potential odor sources and operations could be ranked and their relative contributions determined. Short-term trials or tests of odor mitigation measures (e.g., odor counteractants) would also require an intensive period of data collection.
- Citizen monitoring—the implementation of citizen odor monitoring with field olfactometry can be part of an interactive community outreach program. The primary function of citizen odor monitoring is to collect information through accurate recordkeeping, which represents real conditions in the community. Citizens recruited and trained to measure odors using OIRSs or field olfactometers would also report odor descriptors. Citizen odor monitoring assists in determining prevalent times and weather conditions of odor episodes. Citizen odor monitoring with field olfactometry also helps in understanding the strength at which an odor first becomes a nuisance.
- Complaint response—the use of odor complaint hotlines is a common method used by facilities and communities to respond to odor episodes. A complaint response plan, with designated on-call responders, creates opportunities for verifying odor episodes, tracking odor sources, and quantifying odor strength with field olfactometry.
- Plume profiling—standard and specialized air dispersion modeling predicts the transport and dilution of odors by the wind. A protocol, known as "plume profiling", supplements and calibrates air dispersion modeling. Several inspectors using OIRS or field olfactometers, spaced crosswind and downwind from an odor source, measure and record the odor strength as n-butanol intensity or D/T values. The odor plume profile is then documented and overlaid on the local terrain map. Therefore, the air-dispersion modeling and the local topography are integrated with actual odor measurements from field olfactometry.

Note that these methods are not mutually exclusive and are often combined into a comprehensive odor management program.

6.0 SAMPLING AND CHEMICAL ANALYSIS OF SPECIFIC ODORANTS

Once onsite olfactometry has been assessed, addressing odor concerns from a chemical perspective may present many challenges. Because these odors are typically a complex mixture of chemicals, the constituents that are significant contributors to the perceived odor often have very low odor thresholds (parts per billion). Although sensory testing is necessary when conducting nuisance investigations, examination beyond sensory analysis may be necessary to isolate and identify the root cause of the odor. By complementing the sensory analysis with a quantitative analytical chemical approach, source and mechanism identification can be accomplished and emissions control facilitation can be implemented.

Typical odorous constituents in biogenic processes are either nitrogenous or sulfur-containing; however, some VOCs such as volatile fatty acids (VFAs), higher molecular weight alcohols, and aldehydes may also significantly contribute to the perceived odor. Because of the varied chemical and physical properties of these odorous compounds, multiple methods must be administered to adequately address all compounds of concern. To quantify a majority of these odorous compounds, at least four independent methods need to be used for comprehensive source identification.

6.1 Methods for Reduced Sulfur Compounds, Including Hydrogen Sulfide

Historically, the primary focus of odor regulation and control fell upon a classification of chemicals known as reduced sulfur compounds, an odorous chemical family of compounds generated as byproducts of anaerobic decomposition that are known to emit a wide variety of unpleasant odors such as those characterized by rotten eggs, cabbage, onion/garlic, or burnt matches.

Typical odorous target compound lists include 21 compounds for sulfur chemical quantification. The list was designed to accommodate a variety of applications beyond that of a wastewater treatment, including petroleum refining, pulp and paper processes, municipal landfills, composting operations, and general agricultural activities. Target compounds include hydrogen sulfide, mercaptans, and other thiol compounds. Samples are collected and analyzed according to ASTM D5504, Standard Test Method for Determination of Sulfur Compounds in Natural Gas and Gaseous Fuels by Gas Chromatography and Chemiluminescence (ASTM International, 2012), which uses a Tedlar bag or inert-coated canister, with analysis using gas chromatography with sulfur chemiluminescence detection. Analysis yields results on the same order of magnitude as the odor threshold of most compounds (low parts per billion volume range); however, there is no analytical method with the sensitivity to reach the odor threshold of hydrogen sulfide (0.5 ppbv). With the proliferation of practical sulfur removal systems, attention has now been shifted toward other constituents that were once masked by the potent and concentrated sulfur odors.

6.2 Methods for Volatile Fatty Acids

Volatile fatty acids (also known as "carboxylic acids" or "organic acids") are generated during anaerobic decomposition. Typical target analytes from this family include acetic, propanoic, butyric, valeric, and isovaleric acids.

Volatile fatty acids are among some of the most unpleasant odors encountered. Odor descriptors from this classification include sour, fecal, and "dirty sweat sock"; while these compounds are unpleasant to experience, their permissible exposure limits are approximately 10 times higher than their odor threshold (National Institute for Occupational Safety and Health, n.d.). Because promulgated methods are traditionally developed to address public health concerns, current state and federally circulated methods do not reach the sensitivity necessary to reach the odor threshold of most VFAs. There are, however, environmental laboratories that specialize in odor testing that use proprietary analytical capabilities for quantification of VFAs at the odor threshold. Samples are typically collected onto a sorbent tube for chemical analysis and analyzed by GC–MS (Henningsen & McGinley, 2016), or by using headspace solid phase micro-extraction (Feng, Huang, & Wang, 2008).

6.3 Methods for Amine Compounds

Amines are nitrogen-containing compounds most commonly produced as a byproduct of high-temperature processes or through the microbial decomposition of amino acids. Amines often accompany ammonia, and have a pervasive fishy odor at even very low analyte concentrations. The most common odorous amines are trimethylamine, dimethylamine, and butylamine. Samples are typically collected on a sorbent tube for chemical analysis, and analyzed by gas chromatography with a nitrogen-phosphorus detector.

Like VFAs, because the odor threshold of amines are appreciably lower than their corresponding permissible exposure limits, typical published methods do not reach the required analytical sensitivity. There are, however, environmental laboratories that specialize in odor testing that use proprietary analytical capabilities for quantification of amines at the odor threshold (Henningsen & McGinley, 2016).

6.4 Methods for Volatile Organic Compounds

The final common quantitative method offered for the odorous suite is a targeted mass spectral library search, which is performed to identify odorous VOCs. This analysis is the most comprehensive in terms of variety and quantity of target analytes available, and the search may focus on compounds with low odor thresholds such as terpenes, phenols, and higher molecular weight aldehydes and alcohols. These odorous compounds have typical odor descriptors that include "earthy" and "sweet", or have descriptors such as being "gasoline-like" or "having a solvent odor". Samples can be collected on either thermal desorption tubes packed with application-specific sorbents or via canister collection. At the laboratory, the samples are thermally desorbed, if necessary, and analyzed by GC–MS. Methods cited may include, but are not limited to, the latest editions of U.S. EPA TO-15, U.S. EPA TO-17 (1999), or ASTM D5466 (2015).

6.5 Methods for Other Wastewater Odorants, Including Ammonia

Other odorous compounds include ammonia, diamines, indoles, and aldehydes, including formaldehyde.

Ammonia is a pungent and common compound for wastewater applications. Its analytical technique may be varied because the promulgated methods are robust enough to reach the compound odor threshold. One common analytical technique is OSHA ID-188, where samples are analyzed with ion chromatography.

Indoles, which include both indole and skatole, are heterocyclic organic compounds with a chemical structure similar to that of benzene, with one methine group replaced by a nitrogen atom. These compounds, which are naturally found in fecal matter, also emit a strong fecal odor. Because there are no established exposure limits associated with indoles, there are currently no promulgated methods for these compounds. However, laboratories specializing in odor analysis may have established in-house capabilities or may cite similar methodologies for other VOCs.

Formaldehyde is a naturally occurring organic compound that is most commonly found in the production of industrial resins. It is not a commonly requested analyte for wastewater applications, but is odorous at relatively low thresholds and has a faintly sweet odor descriptor. Samples are collected using dinitrophenylhydrazine-coated silica gel tubes, and analyzed by U.S. EPA Determination of Formaldehyde in Ambient Air Using Adsorbent Cartridge Followed by High Performance Liquid Chromatography or a comparable ASTM or National Institute for Occupational Safety and Health method, depending on application and regulatory agency.

Diamines, the most commonly requested being putricine and cadaverine, are byproducts of the bacterial breakdown of amino acids. These compounds are of interest because they are produced during the putrefaction of animal tissue or the spoilage of food, and so the smell is not only unpleasant, but also alarming. There are currently no analytical methods published for quantification, but specialty laboratories may have proprietary quantification methods using sample derivatization and analysis by gas chromatography or high-performance liquid chromatography.

7.0 MATRIX OF CHARACTER, THRESHOLD, AND ANALYSIS

When designing a site-specific odor study, an initial olfactory investigation may be useful to adequately identify the proper analytical approach. The matrix in Table 2.1 cross-references common compounds found on the WRRF odor wheel with odor descriptors, thresholds, and general laboratory applicability for reference. Primary contributors to the general odor quality have also been annotated (Suffet et al., 2004).

8.0 SUMMARY

Odors and their related odorants can be measured using objective, quantitative, and standardized scientific methods in both odor testing and analytical laboratories as well as in the ambient air by trained inspectors. Point, area, and volume emission sources can be sampled and tested for chemical constituents as well as odor parameters such as odor concentration, intensity, persistence, and characterization. Odor can also be measured and quantified directly in the ambient air by trained inspectors, as described in Section 5.0, "Measuring Ambient Odors". The stakeholders for standardized odor sampling and measurement are regulators, researchers, industry, manufacturers, consultants, and citizens.

Armed with the knowledge of fundamental odor testing, an objective approach can be taken to addressing community nuisance odors and problematic odorous emissions. The subjectivity of nuisance odors can be replaced with objective, scientific methods of odor measurement using both laboratory and field olfactometry.

TABLE 2.1 Matrix of WRRF odorants, their odor threshold values, and laboratory analysis methodology.

Compound Name	Odorant Characteristics	Odor Threshold (ppmV)	Citation	Primary Odor Quality Parameters*	Promulgated Method to Reach Odor Threshold?	Comments & Typical Standard Reporting Limits (ppmV)
Sulfur Compounds - ASTM D5504						
Hydrogen sulfide	Rotten eggs	0.0005	a,e	*	Y	These are compounds listed on the standard ASTM D5504 compound list. Typical ranges are >0.005ppmV
Methyl mercaptan	Rotten cabbage	0.00002	a		Y	
		0.0005	e			
Ethyl mercaptan	Rotten cabbage	0.00001	a	*	Y	
	Decayed cabbage	0.0003	e			
Dimethyl sulfide	Rotten cabbage	0.001	a	*	Y	
	Decayed vegetables, decayed cabbage	0.0006–0.04, 0.001	c, d, e			
Carbon disulfide	Disagreeable, sweet	0.0077–00.96	c, d		Y	
	Vegetable sulfide, Aromatic				Y	
Dimethyl disulfide	Rotten cabbage,	0.000026	a		Y	
	Putrefaction	0.0001–0.0036	c,d			
Methyltrisulfide	Rotten cabbage	0.0012	a	*	Y	
n-Propyl Mercaptan	Unpleasant	0.0001	a	*	Y	
		0.0005	e			
Sulfur dioxide	Irritating	0.449	a	*	Y	
	Pungent, irritating	2.7				
Allyl mercaptan	Garlic, coffee	0.0001	e	*	N	Not on a standard ASTM D5504 list
	Disagreeable, garlic	0.0001	a			
Amyl mercaptan	Putrid	0.00002	a	*	N	Not on a standard ASTM D5504 list
	Unpleasant, putrid	0.0003	e			

(continues)

TABLE 2.1 Matrix of WRRF odorants, their odor threshold values, and laboratory analysis methodology (*Continued*).

Compound Name	Odorant Characteristics	Odor Threshold (ppmV)	Citation	Primary Odor Quality Parameters*	Promulgated Method to Reach Odor Threshold?	Comments & Typical Standard Reporting Limits (ppmV)
Sulfur Compounds - ASTM D5504						
Benzyl mercaptan	Unpleasant	0.0003	a	*	N	Not on a standard ASTM D5504 list
	Unpleasant, strong	0.0002	e			
Diphenyl sulfide	Unpleasant	0.0001	e	*	N	Not on a standard ASTM D5504 list
Phenyl mercaptan	Putrid, garlic	0.0003	e	*	N	Not on a standard ASTM D5504 list
Thiocresol	Skunky, irritating	0.0001	e	*	N	Not on a standard ASTM D5504 list
Volatile Fatty Acids - Laboratory In-House						
Formic acid	Biting	0.024	a		N	No current methodologies to address OT of compounds. Specialized laboratories can reach typical RLs range 0.0004 - 0.008ppmV depending on compound
Acetic acid	Vinegar	1.019	a		N	
Propionic acid	Rancid, pungent	0.028	a		N	
Isobutyric and butyric	Rancid	0.0003	a		N	
Isovaleric acid	Unpleasant	0.0006	a		N	
Valeric acid	Unpleasant	0.0006	a		N	
Amine Compounds - Laboratory In-House Methodologies						
Methyl amine	Fishy	3.2	b	*	N	No current methodologies to address OT of compounds. Specialized laboratories can reach typical RLs range <0.006ppmV
	Putrid fishy	4.7	e			
Ethyl amine	Ammonia like	0.27	e	*	N	
Triethyl amine	Fishy	0.48	b		N	
Dimethyl amine	Putrid, fishy	0.34	e	*	N	
Trimethyl amine	Fishy	0.00044 - 0.0004	b, e	*	N	
	Pungent fishy					
n-Butyl amine	Sour, ammonia	0.08	e	*	N	
Dibutyl amine	Fishy	0.016	e	*	N	
Diisopropyl amine	Fishy	0.13	e	*	N	

Compound Name	Odorant Characteristics	Odor Threshold (ppmV)	Citation	Primary Odor Quality Parameters*	Promulgated Method to Reach Odor Threshold?	Comments & Typical Standard Reporting Limits (ppmV)
Common VOCs - EPA or ASTM Methodologies						
Acetone	Pungent, fruity	0.067	e		Y	Typical VOC ranges 0.005 - 0.01ppmV
Acrolein	Burnt, sweet	0.0228	a		Y	
Other Wastewater Odorants - Various Methodologies						
Ammonia	Pungent	0.038	a	*	Y	Typical range in low ppmV
	Pungent, irritating	17	e			
Indole	Fecal, nauseating	0.0001	e	*	N	Specialized laboratories may analyze, but typical RLs are unknown.
Skatole	Fecal, nauseating	0.001	e	*	N	
Pyridine	Pungent, irritating	0.66	e	*	N	
Formaldahyde	Unpleasant	1.199	a		N	Typical ranges are <0.005ppmV
Acetaldehyde	Green sweet	0.0001	a	*	Y	
Butanaldehyde	Sweet	9.5	a		Y	
Crotonaldehyde	Pungent, suffocating	0.037	a		Y	
Methyl ethyl ketone	Sweet, minty	0.25	a		Y	
Propionaldehyde	Sweet, ester	0.011	a		Y	
Valeraldehyde	Pungent	0.028	a		Y	

a, Ruth (1986) (lowest OTC); b, Amoore and Hautala (1983); c, De Zwart and Kuenen (1992); d, De Vos et al. (1990); e, Sullivan (1969); * Identified in waste
*Suffet, et al. (2004)
Odor thresholds mentioned in this manual may vary and are not intended to be definitive because of the lack of agreement among experts. The odor thresholds are included to give the reader a sense of what odors are potentially an issue relative to the particular chapter of this manual in which they are included, such as Table 3.1 in Chapter 3.

9.0 REFERENCES

American Society for Testing and Materials. (1978). ASTM D1391, Standard test method for measurement of odor in atmospheres (Dilution Method). West Conshohocken, PA: ASTM International.

American Society for Testing and Materials. (1981). STP758, Guidelines for the selection and training of sensory panel members—Special technical publication 758. West Conshohocken, PA: ASTM International.

Amoore, J. E., & Hautala, E. (1983). Odor as an aid to chemical safety: Odor thresholds compared with threshold limit values and volatilities for 214 industrial chemicals in air and water dilution. *Journal of Applied Toxicology, 3*(6), 272–290.

ASTM International. (2012). ASTM D5504-12, Standard test method for determination for sulfur compounds in natural gas and gaseous fuels by gas chromatography and chemiluminescence. West Conshohocken, PA: ASTM International.

ASTM International. (2015). ASTM D5466-15, Standard test method for determination of volatile organic compounds in atmospheres (Canister Sampling Methodology). West Conshohocken, PA: ASTM International.

ASTM International. (2018). ASTM E544-18, Standard practices for supra-threshold odor intensity measurement. West Conshohocken, PA: ASTM International.

ASTM International. (2019). ASTM E679-19, Standard practice for determination of odor and taste threshold by a forced-choice ascending concentration series method of limits. West Conshohocken, PA: ASTM International.

Barnebey and Sutcliffe Corporation. (1974). Scentometer: An instrument for field odor measurement. Columbus, OH: Barnebey and Sutcliffe Corporation.

British Standards Institution. (2003). BS EN 13725:2003, Air quality—Determination of odour concentration by dynamic olfactometry. London, UK: British Standards Institution.

British Standards Institution. (2016). BS EN 16841-1, Ambient air—Determination of odour in ambient air by using field inspection—Part 1: Grid Method. London, UK: British Standards Institution.

Devos, M., Patte, F., Rouault, J., Laffort, P., & Gemert van, L. J. (1990). *Standardized human olfactory thresholds*. Oxford, UK: Oxford University Press.

De Zwart, J. M. M., & Kuenen, J. G. (1992). C_1-cycle of sulfur compounds. *Biodegredation, 3*, 37–59.

Dravnieks, A., & Jarke, F. (1980). Odor thresholds measurement by dynamic olfactometry: Significant operational variables. *Journal of the Air Pollution Control Association, 30*, 1284–1289.

Dravnieks, A., Schmidtsdorf, W., & Mellgaard, M. (1986). Odor thresholds by forced-choice dynamic triangle olfactometry: Reproducibility and methods of calculation. *Journal of the Air Pollution Control Association, 36*, 900.

European Committee for Standardization. (2003). EN 13725, Air quality—Determination of odour concentration by dynamic olfactometry. Brussels, Belgium: European Committee for Standardization.

Feng, L., Huang, Y., & Wang, H. (2008). Solid-phase microextraction in combination with GC-FID for quantification of the volatile free fatty acids in wastewater from constructed wetlands. *Journal of Chromatographic Science, 46*(7), 577–584.

Henningsen, S., & McGinley, C. (2016). Look beyond odor in to odor chemistry. *Proceedings of the Annual Water Environment Federation Technical Exhibition and Conference*, New Orleans, Louisiana, United States (pp. 149–158). Alexandria, VA: Water Environment Federation.

Klenbusch, M. R. (1986). Measurement of gaseous emission rates from land surfaces using an emission isolation flux chamber: User's guide (EPA/600/8-86/008. PB86-223161). Washington, D.C.: U.S. Environmental Protection Agency.

MacBerthouex, P., & Brown, L. C. (1994). *Statistics for environmental engineers*. Boca Raton, FL: Lewis Publishers (CRC Press).

Mallevialle, J., & Suffet, I. H. (1987). *Identification and treatment of taste and odors in drinking water* (pp. 102–120). Denver, CO: American Water Works Association Research Foundation.

McGinley, C. M., McGinley, M. A., & McGinley, D. L. (2000). *Odor basics, understanding and using odor testing*. Presented at the 22nd Annual Hawaii Water Environment Association Conference, Honolulu, Hawaii, United States, June 6–7. https://pdfs.semanticscholar.org/e851/d01699caf4629437f1badef51732d4e2ad8d.pdf

McGinley, M. A., & McGinley, C. M. (2010, March 21–24). Odor assessor performance to reference and non-reference odorants. *Proceedings of the Water Environment Federation Conference on Odors and Air Pollutants*, Charlotte, North Carolina, United States (pp. 240–251). 2010. doi: 10.2175/193864710802768064

Missen, R., & Stein, A. (1975). Guidelines for the evaluation of visible emissions (EPA-340/1-75-007). Part of Stationary Source Enforcement Series Training Course—Visible Emissions Field Certification and Classroom Lecture Program. Washington, D.C.: U.S. Environmental Protection Agency.

National Institute for Occupational Safety and Health. (n.d.). *Acetic acid.* Centers for Disease Control and Prevention. Retrieved April 11, 2016, from www.cdc.gov/niosh/npg/npgd0002.html

Ruth, J. H. (1986) Odor thresholds and irritation levels of several chemical substances: A review. *American Industrial Hygiene Association Journal, 47,* 142–151.

Schmidt, C. E., Card, T. R., & Goodwin, J. (2008, February). *Validation of the modified USEPA flux chamber technology for assessing high advective flow sources such as biofilters at compost sites.* 16th Annual US Composting Council Conference, Oakland, California, United States.

St. Croix Sensory, Inc. (2003). A detailed assessment of the science and technology of odor measurement. https://www.pca.state.mn.us/sites/default/files/p-gen2-01.pdf

State of Colorado. (2013). Regulation no. 2 odor emission (5 CCR 1001-4) official publication of the state administrative rules (24-4-103(11) C.R.S.). Department of Public Health and Environment. https://www.colorado.gov/pacific/sites/default/files/5-CCR-1001-4.pdf

State of Missouri. (2013). Restriction of emission of odors (10 CSR 10-6.165). Rules of Department of Natural Resources (Division 10, Chapter 6). Department of Natural Resources. https://www.sos.mo.gov/adrules/csr/current/10csr/10csr

Stevens, S. S. (1960). The psychophysics of sensory function. *American Scientist, 48*(2), 226–253.

Stevens, S. S. (1962). The surprising simplicity of sensory metrics. *American Scientist, 17*(1), 29–39.

Suffet, M., Burlingame, G., Rosenfeld, P., & Bruchet, A. (2004). The value of an odor-quality-wheel classification scheme for wastewater plants. *Water Science & Technology, 50*(4), 25–32.

Sullivan, R. J. (1969). Preliminary air pollution survey of odorous compounds. Raleigh, NC: National Air Pollution Control Administration, United States Department of Health, Education, and Welfare.

Turk, A., Switala, E. D., & Thomas, S. H. (1980). Suprathreshold odor measurement by dynamic olfactometry: Principles and practice. *Journal of the Air Pollution Control Association, 30*(12), 1289–1294. doi: 10.1080/00022470.1980.10465183

U.S. Environmental Protection Agency. (1999). Air Method, Toxic Organics-11A (TO-11A): Determination of formaldehyde in ambient air using adsorbent cartridge followed by high performance liquid chromatography (HPLC) (EPA/625/R-96/010b). Washington, D.C.: U.S. Environmental Protection Agency.

U.S. Environmental Protection Agency. (1999). Air Method, Toxic Organics-15 (TO-15): Compendium of methods for the determination of toxic organic compounds in ambient air, second edition: Determination of volatile organic compounds (VOCs) in air collected in specially-prepared canisters and analyzed by gas chromatography/mass spectrometry (GC/MS) (EPA 625/R-96/010b). Washington, D.C.: U.S. Environmental Protection Agency.

U.S. Environmental Protection Agency. (1999). Air Method, Toxic Organics-17 (TO-17): Determination of volatile organic compounds in ambient air using active sampling onto sorbent tubes (EPA 625/R-96/010b). Washington, D.C.: U.S. Environmental Protection Agency.

van Harreveld, A. Ph., Heeres, P., & Harssema, H. (1999). A review of 20 years of standardization of odor concentration measurement by dynamic olfactometry in Europe. *Journal of the Air & Waste Management Association*, *49*(6), 705–715. doi: 10.1080/10473289

Yubin, Z., Tadeo, V., Cowden, S., & Suffet, I. H. (2018). Identification of musty odor nuisance at wastewater treatment plants. *Proceedings of Water Environment Federation Conference on Odors and Air Pollutants*, Portland, Oregon, United States (pp. 270–287). doi: 10.2175/193864718822846760

3

Odor Regulations and Policies

1.0	INTRODUCTION	39
2.0	ODOR STANDARDS	40
	2.1 Concentration	40
	2.2 Averaging Time	43
	2.3 Frequency	43
	2.4 Impact Location	43
3.0	ODORANT CONCENTRATION LIMITS	44
4.0	ODOR LIMITS	46
5.0	ODOR NUISANCE	51
6.0	ODOR NUISANCE RESPONSE PROTOCOLS	51
	6.1 Intensity	52
	6.2 Frequency, Intensity, Duration, Offensiveness, and Location (FIDOL)	53
	6.3 Other Considerations	55
7.0	ODOR MANAGEMENT PLANS	55
	7.1 Role	55
	7.2 Examples	56
8.0	COMPARISON OF ODOR REGULATIONS AND APPROACHES	57
9.0	REFERENCES	58

1.0 INTRODUCTION

Both regulation and policy-based approaches have been implemented at locales around the globe to minimize off-site odor impacts from wastewater systems. The specific approaches can vary significantly in content, but generally fall into the following categories:

- Odor standards—Numeric odor standards have been established by agencies across the globe for both individual odorous compounds (odorants) and for odors. The latter is typically set via olfactometry and defined in terms of dilution-to-threshold (D/T) or odor units (OU, OU/m^3). Olfactometry is discussed at length in Chapter 2. Odor standards can apply at the odor source, but in the United States they are more often applied at property lines or off-site sensitive receptors, with dispersion modeling used to define odor treatment levels needed for compliance.

- Nuisance levels—Field evaluation methods, such as quantitative or qualitative field intensity ratings (based on a point-scale system or minimum length of time) can be used to determine whether odor impacts constitute a nuisance. These methods have been used by some regulatory agencies to both define odor impacts and provide a framework to enforce nuisance limits.
- Best available control technology (BACT)—BACT and similar source-based approaches prescribe that proven control technologies be used to control odors at certain types of new or upgraded facilities. This approach is used to control the potential generation of off-site nuisance conditions by mandating controls at the source.

Odor standards and nuisance-based approaches are discussed in Section 2.0 of this chapter (BACT is addressed in Chapter 6), along with key considerations for the adoption of each approach.

2.0 ODOR STANDARDS

Numeric odor standards include not only odorant or odor concentrations, but also specify the averaging time over which the concentration is measured, frequency of compliance, and where the standard is to be applied. Each element is described in the following subsections, along with a more detailed discussion of how these are used in odor regulations.

2.1 Concentration

Efforts to define a quantitative odor standard typically start with discussions of which odor levels are detectable to the affected population.

Odorant/odor detectability (or threshold) refers to the minimum concentration that produces an olfactory response or sensation. Detection thresholds have been defined for a selection of water resource recovery facility (WRRF)-related odorants and are presented in Table 3.1. It is critical to note, however, that data in the literature on odorant threshold concentrations may differ significantly—in many cases by tenfold or more. The values in Table 3.1 should be approached accordingly.

TABLE 3.1 Odorous compounds in wastewater[a,b] (AIHA, 1989; AIHA 2019; Moore & Hautala, 1983; Sullivan, 1969).

Compound Name	Formula	Molecular Weight	Volatility at 25 °C, ppm (V/V)	Detection Threshold[a], ppm (V/V)	Recognition Threshold[a], ppm (V/V)	Odor Description
Acetaldehyde	CH_3CHO	44	Gas	0.067	0.21	Pungent, fruity
Allyl Mercaptan	$CH_2{:}CHCH_2SH$	74		0.0001	0.0015	Disagreeable, garlic
Ammonia	NH_3	17	Gas	17	37	Pungent, irritating
Amyl Mercaptan	$CH_3(CH_2)_4SH$	104		0.0003	—	Unpleasant, putrid
Benzyl Mercaptan	$C_6H_5CH_2SH$	124		0.0002	0.0026	Unpleasant, strong
N-Butyl Amine	$CH_3(CH_2)NH$	73	93 000	0.080	1.8	Sour, ammonia
Chlorine	Cl_2	71	Gas	0.080	0.31	Pungent, suffocating
Dibutyl Amine	$(C_4H_9)_2NH$	129	8000	0.016	—	Fishy
Diisopropyl Amine	$(C_3H_7)_2NH$	101		0.13	0.38	Fishy
Dimethyl Amine	$(CH_3)_2NH$	45	Gas	0.34	—	Putrid, fishy
Dimethyl Sulfide	$(CH_3)_2S$	62	830 000	0.001	0.001	Decayed cabbage
Diphenyl Sulfide	$(C_6H_5)_2S$	186	100	0.0001	0.0021	Unpleasant
Ethyl Amine	$C_2H_5NH_2$	45	Gas	0.27	1.7	Ammonia-like
Ethyl Mercaptan	C_2H_5SH	62	710 000	0.0003	0.001	Decayed cabbage
Hydrogen Sulfide	H_2S	34	Gas	0.0005	0.0047	Rotten eggs
Indole	$C_6H_4(CH)_2NH$	117	360	0.0001	—	Fecal, nauseating
Methyl Amine	CH_3NH_2	31	Gas	4.7	—	Putrid, fishy

(*continues*)

TABLE 3.1 Odorous compounds in wastewater[a,b] (AIHA, 1989; AIHA 2019; Moore & Hautala, 1983; Sullivan, 1969) (*Continued*).

Compound Name	Formula	Molecular Weight	Volatility at 25 °C, ppm (V/V)	Detection Threshold[a], ppm (V/V)	Recognition Threshold[a], ppm (V/V)	Odor Description
Methyl Mercaptan	CH_3SH	48	Gas	0.0005	0.0010	Rotten cabbage
Ozone	O_3	48	Gas	0.5	0.0015	Pungent, irritating
Phenyl Mercaptan	C_6H_5SH	110	2000	0.0003	0.0015	Putrid, garlic
Propyl Mercaptan	C_3H_7SH	76	220 000	0.0005	0.020	Unpleasant
Pyridine	C_5H_5N	79	27 000	0.66	0.74	Pungent, irritating
Skatole	C_9H_9N	131	200	0.001	0.050	Fecal nauseating
Sulfur Dioxide	SO_2	64	Gas	2.7	4.4	Pungent, irritating
Thiocresol	$CH_3C_6H_4SH$	124		0.0001	—	Skunky, irritating
Trimethylamine	$(CH_3)_3N$	59	Gas	0.0004	—	Pungent, fishy

[a]Different sources report a range of values for odor detection or recognition thresholds, particularly for compounds such as hydrogen sulfide or ammonia. The range of values used for hydrogen sulfide generally varies from 0.47 to 9 ppb. Odor thresholds mentioned in this manual may vary and are not intended to be definitive because of the lack of agreement among experts. The odor thresholds are included to give the reader a sense of which odors are potentially an issue relative to the particular chapter of this manual in which they are included. (See additional referenced information in Table 2.12 of Chapter 2.)

[b]Most of the odorous substances are gaseous under normal atmospheric conditions or at least have a significant volatility. The volatility is shown in the table as parts per million (ppm) and is equal to the vapor pressure. The molecular weights of these substances typically range from 30 to 150. Usually, the lower the molecular weight of a compound, the higher its vapor pressure and potential for emission to the atmosphere. Substances of high molecular weight are generally less volatile and, thus, typically have less effect for causing odor complaints.

For odors, the detection threshold is typically determined by an odor panel consisting of a specified number of people, and with the threshold value occurring when 50% of the panel correctly detects the odor (Chapter 2 provides additional information on this methodology). Similarly, a recognition threshold is the value when 50% of the panel is able to detect the odor and also identify some odor descriptor or character.

Defining concentration-based standards is a challenging task because of the difficulty in reaching agreement on what constitutes an unacceptable odor. The olfactory sensory system is exceedingly complex and people's responses to odors are variable. This variability makes it difficult (but not impossible) to designate an odor standard that is applicable to a wide variety of situations.

2.2 Averaging Time

Averaging time is also a critical consideration in defining an odor standard or limit. Peak odor levels can be significantly higher than the mean levels if the averaging time used is lengthy. The concentration of odors can exceed the odor threshold several times during a period when the hourly mean is well below the odor threshold. This issue is particularly important when analyzing odor dispersion modeling output results.

2.3 Frequency

Some odor standards include a minimum frequency for compliance, often expressed as a percentage, and generally between 98% and 100%. The lower compliance frequencies can reflect, to some extent, the reality that process upsets do occasionally occur with wastewater and solids processing.

2.4 Impact Location

Odor standards can be applied at the odor source, property boundary, or off-site receptor. For off-site odor events, some agencies differentiate between types of receptors, typically applying a more stringent standard to more sensitive receptors (e.g., residences as opposed to industrial areas).

Standards applied at the odor source can typically be measured, but demonstrating compliance with odorant limits via measurements at WRRF boundaries or receptors beyond those boundaries may be difficult. Consequently, dispersion modeling is often used to "back-calculate" source concentrations required to meet a specified fence-line or receptor limit.

In addition to the aforementioned elements, the following questions should be considered when establishing a new odor- or odorant-based numerical standard:

- Which segment of the population should the standard protect from nuisance conditions—the average resident, the most sensitive, or the 90th percentile?
- Should existing facilities that develop problems have less stringent standards than new facilities, and should new facilities be required to incorporate an extra level of conservatism to their approach?
- Should new facilities be designed only to prevent odor complaints (distinct odors), or should the goal be, particularly for new facilities, to prevent even weak odors?

3.0 ODORANT CONCENTRATION LIMITS

Odorant limits established by regulatory or other agencies frequently reflect the dominant odorant in wastewater systems: hydrogen sulfide (H_2S). For hydrogen sulfide, limits typically range from 1 to 100 ppbv, with the variance reflecting considerations such as location, available buffer zone, and/or receptor types. Figure 3.1 presents examples of this type of criteria. The figure also shows another key area of variance: averaging periods. Averaging times (or the duration over which odors are measured for compliance purposes) vary significantly from as short as 15 minutes to as long as 1 year. It should be noted that while some of the limits shown were derived to address agricultural odors, they are applied to water resource recovery systems as well.

Although hydrogen sulfide is considered the most prevalent odorous compound associated with wastewater, other odorous compounds can

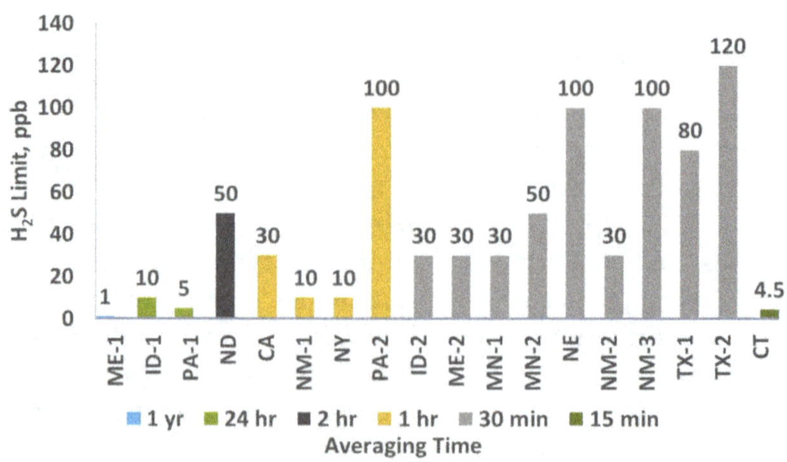

FIGURE 3.1 Examples of state hydrogen sulfide limits in the United States (derived from RWDI Air Inc., 2005).

contribute to odors as well and should also be considered when defining odorant-based limits. Compounds such as mercaptans, organic sulfides, and amines may be significant contributors. Depending on the types of chemicals present in industrial discharges, other odorous compounds may include organic acids, aldehydes, and ketones. Like hydrogen sulfide, some of these other compounds can have very low odor detection and recognition thresholds (as shown in Table 3.1).

To date, only a few agencies have established odorant limits for compounds other than hydrogen sulfide. Table 3.2 highlights a few of these agencies, a selection of odorants they regulate, and the limits established for those odorants. Note that the list in the table is by no means exhaustive, but is meant to provide an idea of the relative limits for some reduced sulfur and nitrogen compounds.

TABLE 3.2 Select reduced sulfur and nitrogen compound odorant limits (RWDI, 2005). Permission granted by Province of B.C., Ministry of Environment and Climate Change Strategy.

Agency	Odorant	Averaging Time	Limit
New South Wales (Australia)	Methyl mercaptan	3 min	0.42 ppb
	Triethylamine	3 min	90 ppb
Japan	Dimethyl disulfide	N/A[a]	9–100 ppbv
	Methyl mercaptan	N/A	2–10 ppbv
	Butyric acid	N/A	1–6 ppbv
	Triethylamine	N/A	5–70 ppb
Korea	Methyl mercaptan	N/A	4 ppb (industrial), 2 ppb (residential)
	Dimethyl sulfide	N/A	50 ppb (industrial), 10 ppb (residential)
	Dimethyl disulfide	N/A	30 ppb (industrial), 9 ppb (residential)
	Trimethylamine	N/A	20 ppb (industrial), 5 ppb (residential)
Connecticut (USA)	Methyl mercaptan	N/A	2.2 mg/m^3
Nebraska (USA)	Total reduced sulfur	30 min	100 ppb

[a]N/A = Not available

The Bay Area Air Quality Management District (BAAQMD) based in California, United States, goes a step further than some agencies, providing not only emission limits for odorants other than hydrogen sulfide, but also defining different limits based on source type (see Table 3.3).

The use of odorant-based limits, while relatively common for hydrogen sulfide, is not without challenges. First, the great number of odorants associated with WRRF emissions creates problems when using individual compounds as the basis for assessing odors because it is unlikely that all odorants could be adequately quantified. In addition, detection and odor annoyance thresholds cited in literature and regulations vary widely for many compounds, as noted earlier.

However, there is at least one advantage to an odorant-based regulatory approach, particularly when focused on hydrogen sulfide: measurement is both quick and relatively easy (measurement methods are described in Chapter 2). Some research has been performed to preliminarily correlate hydrogen sulfide to odor concentrations on a process-by-process basis for WRRFs (McGinley & McGinley, 2008). More research would be helpful to further refine preliminary correlations because they vary significantly.

4.0 ODOR LIMITS

Odors are often complex mixtures of numerous odorants. Analytical monitoring of individual odorants present in such emissions is typically not practical and, as such, a number of agencies across the globe have moved to a sensory method to quantify odors based upon olfactometry. The concentrations for odor are expressed in different terms; typical expressions include

TABLE 3.3 Bay Area Air Quality Management District maximum allowable emission concentrations (BAAQMD, 1982).

Compound or Family of Compounds	Type A (Point Source) Emission Point (ppm)	Type B (Area Source) Emission Point (ppm)
Dimethyl sulfide $(CH_3)_2S$	0.1	0.05
Ammonia NH_3	5000	2500
Mercaptans calculated as methyl mercaptan CH_3SH	0.2	0.1
Phenolic compounds calculated as phenol C_6H_5OH	5.0	2.5
Trimethylamine $(CH_3)_3N$	0.02	0.02

odor unit (OU), odor unit derived using European odor testing standard (OU_E), odor units per cubic meter (OU/m^3), and dilution-to-threshold (D/T). In many cases, however, the basic methods to determine the odor concentrations are similar, and for the purposes of this overview, the terms are considered to be synonymous.

Odor limits are typically applied at plant boundaries or receptors and vary considerably, in some cases reflecting factors such as receptor sensitivity, buffer zones, and other considerations. For instance, in South Australia, the ambient odor concentration limits range from 2 OU for communities greater than 2000 people to 10 OU for communities with fewer than 12 people (de Melo Lisboa, Sivret, & Stuetz, 2014). In the United Kingdom, many planning authorities use a benchmark of 5 OU_E/m^3 as a 98th percentile of hourly averages for setting site boundary limits. The 98th percentile is the hourly odor concentration that is achieved for 98% of the year. Table 3.4 presents examples of odor limits applied at plant boundaries or sensitive receptors.

As indicated by the variability in odor standards shown in Table 3.4, there is no consensus on which D/T (OU/m^3) level or averaging time represents the most effective and fair odor standard to mitigate off-site impacts, although the issue has been investigated. A study conducted in the United States for the California Air Resources Board reviewed six published studies that related to recognizability, unpleasantness, and annoyance associated with a variety of unpleasant odors. The analyses concluded that for unpleasant odors, the threshold of annoyance is approximately 5 times the threshold of detection. California's South Coast Air Quality Management District states that at 5 D/T (OU/m^3) people become consciously aware of the presence of an odor, and that at 5–10 D/T odors are strong enough to evoke registered complaints (California South Coast Air Quality Management District, 1993). However, this conclusion was based on test methods (ASTM E679) that have been largely replaced with the newer EN 13725 method (European Committee for Standardization, 2003). In many cases, the latter method recommends a higher presentation rate (20 L/min) of the odor sample to panelists than the rate often used for ASTM E679 analyses (0.5 to 3 L/min). This in turn can result in a higher D/T value for samples analyzed using EN 13725 than for samples analyzed using the older methods. The ramification of this method shift on annoyance thresholds has not been documented.

It should also be noted that may of the U.S. odor standards presented in Table 3.4 were developed before the advent of EN 13725, and thus may be overly conservative. That being said, European and other agencies have adopted lower thresholds reflecting EN 13725 analyses, and the ASTM-based values are in the public domain. Thus, there has been no move to

TABLE 3.4 Examples of ambient odor limits.

Location		Odor Limit	Averaging Times & Frequency of Perception
United States	Colorado[a]	7 D/T (Scentometer)	N/A
	Connecticut	7 D/T	N/A
	Missouri[a]	7 D/T	Two trials, 15 minutes apart, within 1-hour period
	New Jersey[a]	5 D/T at sensitive receptor with highest impact	5 minutes or less
	North Dakota[a]	2 D/T (Scentometer)	N/A
	Oregon[a]	1 to 2 D/T	15 minutes
	California		
	– Oakland[a]	50 D/T	3 minutes
	– San Diego[a]	5 D/T	—
	– Bay Area Air Quality District (San Francisco)[a]	5 D/T	5 minutes (99.5% compliance)
	– Orange County Sanitation District[b]	20 D/T	Applied after at least 10 complaints in a 90 day period <100 hrs/yr noncompliance
	Washington		
	– City of Seattle[b]	5 D/T	5 minutes
	– King County[b]	0–3 D/T	<50 hrs/yr exceedance
		0–5 D/T	<100 hrs/yr exceedance
	Iowa[b]	15 D/T	2 hours
	Kentucky[b]	7 D/T (Scentometer)	N/A
	Massachusetts[b]	5 D/T	1 hour
	North Carolina[b]	4 D/T	30 seconds
	Wyoming[b]	7 D/T	N/A
	Florida		
	– Palm Beach County Solid Waste Authority[b]	7 D/T	N/A

TABLE 3.4 Examples of ambient odor limits (*Continued*).

Location		Odor Limit	Averaging Times & Frequency of Perception
Australia[c]	New South Wales	2 OU/m^3 – Urban point source	1 second;
		10 OU/m^3 – Rural point source	1 second;
		15 OU/m^3 – Area source	1 hour
	South Australia	2–10 OU/m^3	3 minutes
	Western Australia	2 OU 4 OU	3 minutes, 99.5th percentile 3 minutes, 99.9th percentile
United Kingdom[d]	Newbiggin-by-the-Sea WRRF	5 OU/m^3	98% compliance
	Debby WRRFs	5 OU/m^3	98% compliance
	Mogden Sewage Treatment Works	5 OU/m^3	98% compliance
	Swindon Sewage Treatment Works	5 OU/m^3	98% compliance
Canada[b]	Ontario	1 OU/m^3	10 minutes
	Quebec	1 OU at most sensitive receptor	10 minutes
Singapore[d]	Changi Water Reclamation Plant	5 OU/m^3	—
Hong Kong[b]	Siu Ho Wan WRRF	5 OU/m^3	5 seconds

[a]Mahin (2001)
[b]RWDI Air Inc. (2005)
[c]de Melo Lisboa et al. (2014)
[d]Koh et al. (2016)

revisit the values in the table. Furthermore, the values noted are representative and can vary by locale; for applicable regulations in a particular area, the local state/jurisdiction should be consulted.

Standards are continuing to evolve beyond D/T concentrations alone, however, albeit slowly and on a limited basis. Specifically, some agencies are considering the odor offensiveness in setting standards, whereas others are looking to recognition thresholds (rather than detection thresholds). For example, the Institute of Air Quality Management (IAQM) recommends lowering odor limits by 75% for more offensive odors (IAQM, 2014). Table 3.5 presents IAQM recommendations. Germany has incorporated offensiveness into its odor standards as well, applying a factor of 0.5 to pleasant odors before assessing compliance with odor impacts (RWDI Air Inc., 2005). The term "recognizable odors" in German regulations provides an example of the use of recognition, rather than detection thresholds. The recognition threshold differs by compound, but is believed to be approximately 2 to 5 times the detection thresholds for most odorants.

In addition to the ambient standards shown in Table 3.4, some countries have adopted source-based emission standards. For example, Austria has defined an emission limit of 500 OU/m^3 for biological treatment sources (as well as a buffer of 500 m from residences), and China has developed source-based odor emission standards that vary according to emission height for both odor and eight odorants. Maximum emission criteria range from 2000 OU for a 15-m-tall stack to 60 000 OU for a stack exceeding 60-m tall (Brancher et al., 2017).

TABLE 3.5 IAQM benchmark odor criteria (Institute of Air Quality Management, 2014).

Criterion (C_{98} OU_E/m^3)[a]	Offensiveness	Odor Emission Sources
1.5	Most offensive	– Processes involving decaying animal or fish remains – Processes involving septic effluent or sludge – Biological landfill odors
3.0	Moderately offensive	– Intensive livestock rearing – Fat frying (food processing) – Sugar beet processing – Well-aerated green waste composting
6.0	Less offensive	– Brewery – Confectionary – Coffee

[a]C98 OU_E/m^3 refers to the 98th percentile value of hourly average odor concentration, expressed in OU_E/m^3 for the nearest receptors.

5.0 ODOR NUISANCE

Odor nuisance regulations are relatively common, and can be applied by state, county, and municipal authorities. Pope (2000) reported that as many as 40 of the 50 states apply a nuisance law. These regulations are often qualitative, such as Alabama's requirement that "No person shall cause air pollution . . . presence in the outdoor atmosphere of one or more contaminants (including odors) in such quantities and duration as are . . . injurious to human health or welfare, animal or plant life, or property, or would interfere with the enjoyment of life or property" (ADEM Admin. Code r. 335-3-1-.02(e)).

However, determining the level of odors that rise to the nuisance level can be problematic because nuisance odors—and the complaints that they generate—can be impacted by a variety of factors, including odor frequency, intensity, quality, and duration. Odors typically considered to be pleasant at moderate concentrations (i.e., a coffee-roasting aroma) may become unpleasant when present at a high intensity and for a long duration of time. Conversely, an odorous compound such as hydrogen sulfide may be judged as unpleasant when present even at relatively low concentrations. Duration, frequency, and time of day an odor is present are also important factors in the evaluation of odor nuisance. An unpleasant odor that occurs only once a week, and then only briefly, may not disturb the average person, whereas a more frequent and persistent odor is more likely to cause an adverse reaction. However, a persistent odor causes adaptation and a lessening of the olfactory response, whereas an odor that occurs briefly at certain intervals within a day triggers an olfactory response close to its maximum sensitivity.

A conceptual model for what leads to an odor nuisance is the citizen complaint pyramid (Figure 3.2), which starts with an odor episode and builds with odor character, odor intensity, duration, and frequency. The cumulative effect of these four building blocks creates the nuisance experience that may yield a citizen complaint. Because of their relationship to complaints, these elements form the basis of one approach used by regulators to address odor complaints called the Frequency, Intensity, Duration, Offensiveness, and Location (FIDOL) protocol. ("Offensiveness" is a descriptor for odor character.)

6.0 ODOR NUISANCE RESPONSE PROTOCOLS

Enforcement of an odor nuisance regulation typically starts with the origin of a complaint. An agency should have established procedures for receiving and investigating odor complaints to confirm the existence of an objectionable odor. Again, the basic problem in applying an odor nuisance ordinance

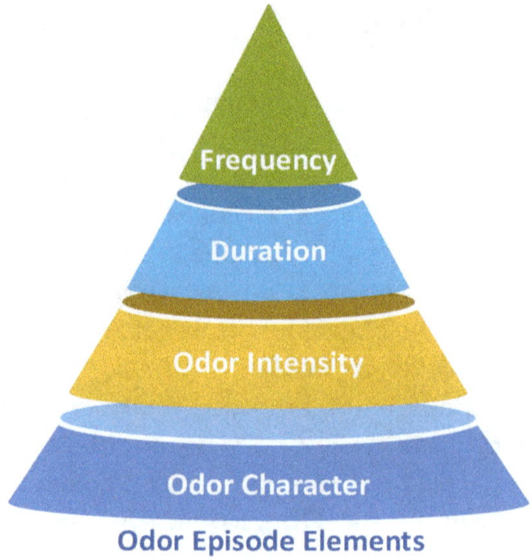

FIGURE 3.2 Citizen complaint pyramid (adapted from McGinley, Pope, & Mahin, 2000).

is defining objective criteria for establishing the odor's presence. These criteria should be related to the odor's annoyance threshold. The criteria may be based on field evaluation methods, such as the intensity and FIDOL protocols described in the following subsections.

6.1 Intensity

Odor intensity is a measure of the strength of the odor sensation and is related to the odorant concentration, which is a different category of measurement. The intensity of an odor is perceived directly without any knowledge of the odorant concentration or degree of air dilution of the odorous sample needed to eliminate the odor.

The n-butanol odor intensity reference scale E544-18 from ASTM International (2018) can be used to measure odor intensity in ambient air. Panelists match the perceived intensity of an ambient odor by comparing it with the butanol reference scale. Panelists' judgments should be reinforced by referring to butanol standards between odor intensity measurements of the ambient air.

An n-butanol scale consists of a series of numbers that correlate to n-butanol concentrations and corresponding verbal descriptors. For example, a scale value of n = 1 (15 ppm) could be described as "barely perceptible", while a scale value of n = 8 (2000 ppm) could be described as "overpowering" and "intolerable" (ASTM E544-18). However, some agencies have developed intensity scales that are not necessarily tied to n-butanol. The state of New Jersey, for example, uses a five-point odor intensity scale to respond to odor complaints, as shown in Figure 3.3. Other states have adopted similar approaches.

Using this method, the investigator (air pollution inspector, plant operator, or odor monitor) observes the odor in the ambient air and compares it to the odor intensity scale. The investigator must use a carbon-filtering mask to refresh the olfactory sense between observations (sniffing). Without the use of a carbon-filtering mask, the observer's olfactory sense becomes adapted to the surrounding ambient air or becomes fatigued from any odor in the surrounding air when attempting to evaluate ambient odors (i.e., a WRRF plant operator monitoring treatment plant odors off-site).

Intensity scales are simple to use, but they have certain issues that must be considered. First, the numbers are not proportional to the perceived intensities of the odors; instead, they are proportional to the logarithms of the perceived intensities. Second, people interpret a given category scale differently; this necessitates specialized training of the panelists. Despite these issues, several municipalities use intensity scales to monitor downwind odor emissions from WRRFs.

6.2 Frequency, Intensity, Duration, Offensiveness, and Location (FIDOL)

The FIDOL protocol addresses four discrete elements of nuisance odor:

- Frequency—refers to how often an odor episode occurs.
- Intensity—refers to the strength of an odor. Measured using an odor intensity referencing protocol such as the butanol intensity scale described above.
- Duration—refers to the elapsed time of each separate odor episode. The duration of an odor episode needs to be based on the recordkeeping

1	Very Light	Odorant present in the air, which activates the sense of smell but the characteristics may not be distinguishable
2	Light	Odorant present in the air, which activates the sense of smell and is distinguishable and definite, but not necessarily objectionable in short durations but may be objectionable in longer durations
3	Moderate	Odorant present in the air, which easily activates the sense of smell, is very distinct and clearly distinguishable and definite and may be objectionable and/or irritating
4	Strong	Odorant present in the air, which would be objectionable and cause a person to attempt to avoid it completely, could indicate a tendency to possibly produce physiological effects during prolonged exposure
5	Very Strong	Odorant present which is so strong it is overpowering and intolerable for any length of time and could tend to easily produce some physiological effects

FIGURE 3.3 Field quality intensity scale—New Jersey guidelines.

of the observer(s), such as citizen complaint calls, citizen notes or log-books, or air pollution inspector observations and data collection.

- Offensiveness—refers to the character of an odor at a given odor concentration and intensity.
- Location—refers to the location of the odor. This element, which is not included in all protocols, reflects the character of the location where the odor is noted, with greater weight accorded to complaints in sensitive areas (e.g., residential areas) than industrial or sparsely populated areas.

Traditionally, FIDOL has been used by regulators to respond to odor complaints. The FIDOL factors have been widely used in countries such as Germany, New Zealand, and the United States to assess odors (RWDI Air Inc., 2005). For instance, the Texas Commission on Environmental Quality (TCEQ) has developed detailed odor complaint investigation procedures using a FIDO chart (location is not included in their protocol). The TCEQ protocol categorizes odors as "Highly Offensive", "Offensive", "Unpleasant", and "Not Unpleasant", and incorporates five qualitative rankings ranging from "Very Strong" to "Very Light". Figure 3.4 illustrates how the agency uses FIDO to determine if a violation has occurred (white boxes illustrate conditions that are not deemed to violate nuisance criteria).

ODORS CHARACTERIZED AS HIGHLY OFFENSIVE

DURATION	FREQUENCY				
	Single Occurrence	Quarterly	Monthly	Weekly	Daily
1 minute	NA	NA	VS	S	M
10 minutes	NA	VS	S	M	L
1 hour	VS	S	M	L	VL
4 hours	S	M	L	VL	VL
12 hours+	M	L	VL	VL	VL

ODORS CHARACTERIZED AS NOT UNPLEASANT

DURATION	FREQUENCY				
	Single Occurrence	Quarterly	Monthly	Weekly	Daily
1 minute	NA	NA	NA	NA	NA
10 minutes	NA	NA	NA	NA	NA
1 hour	NA	NA	NA	NA	VS
4 hours	NA	NA	NA	VS	S
12 hours+	NA	NA	VS	S	M

Intensity Legend
VS
Very Strong
S
Strong
M
Moderate
L
Light
VL
Very Light

FIGURE 3.4 Texas Commission on Environmental Quality FIDO protocol excerpt (TCEQ, 2016). TCEQ has given permission to use this table. The use of TCEQ's FIDO table does not reflect TCEQ's views regarding any findings, opinions, conclusions, statements or recommendations in this publication or with the author's use of the table. This table may be updated by TCEQ and therefore TCEQ should be contacted to verify if this table has been revised. WEF's copyrights in this publication do not assert copyrights in this table.

6.3 Other Considerations

When adopting or implementing nuisance response protocols to respond to odor complaints, a number of factors should be considered. These protocols are advantageous because they are simple and can be measured in the field. However, odor nuisance conditions occur much more frequently in the evening and early morning when regulatory staff are typically not working; therefore, a disadvantage is the potential that odors might be "missed" by inspectors. Additionally, the sensitivity of investigators may vary. The following factors affect the variability among individuals evaluating nuisance conditions:

- Differences in the ability to detect odor (perception may vary with different classes of odorous compounds);
- Subjective acceptance or rejection of an odor based on experience;
- Circumstances under which the odor is detected; and
- Age, health, and attitude of the human receptor.

Generally, any nuisance-based regulations should also consider whether the objective should include eliminating isolated complaints that originate from relatively few, sensitive individuals. Some odor ordinances assume that requiring that no odor be detected beyond the property line may be neither technically nor economically feasible. One alternative approach is to eliminate a community odor nuisance when a specified number of valid complaints are received from the community during a designated time period. For a given situation, the degree of odor control required should be adequate to reduce the odor emissions to a level that avoids a continuing community odor nuisance episode.

7.0 ODOR MANAGEMENT PLANS

7.1 Role

Odor management plans are developed and implemented by facility owners whose processes generate odors, and are intended to provide a holistic, system-wide strategy for the control of odorous air emissions. The plan serves as a tool to communicate the approach for managing on-site odors to ensure regulatory compliance and limiting off-site impacts of nuisance odors. When the plan is executed and enforced as intended, it establishes a protocol for receipt of any odor complaints and subsequent action to be taken to address the concern, thus reinforcing responsiveness and commitment to being a good neighbor.

Implementation of an odor management plan empowers the owner to be proactive instead of reactive regarding fugitive emissions and foul air exhaust. An effective plan will address both short-term and long-term strategies for odor control. Table 3.6 summarizes key elements that comprise a typical odor management plan.

Once an odor management plan has been established, it is imperative that the plan is revisited at regular intervals to ensure applicability with current processes, regulations, and target goals for the facility.

7.2 Examples

As advancements in technology have allowed for better assessment, characterization, and treatment of odor, more agencies are considering implementation of an odor management plan to document their current and planned

TABLE 3.6 Elements of an odor management plan.

Element	Description
Facility/Site Description	• Describes the facility and systems or processes that generate odor
Odor Characterization	• Defines odor sources • Characterizes odorants • Identifies sensitive receptors
Odor Control: Mitigation and Management	• States permit limits and BACT requirements • Establishes feasible technologies for each process area • Identifies performance criteria • Summarizes best management practices for odor control
Monitoring and Reporting	• Identifies routine maintenance and monitoring activities • Establishes procedure for recordkeeping
Contingency Plans	• Identifies procedure to be followed in an emergency or in a non-routine (planned or unplanned) change in operations
Procedure for Public Complaints	• Addresses public outreach • Establishes protocol for receipt and response to complaints • Includes survey and complaint forms to be retained for records

approach for odor control. A key distinction is that a plan can be specific to a particular district or agency, or more general to encompass all operations within a country or region. Examples of such wide-scale odor management plans around the world include the following:

- New Zealand—The Ministry for the Environment has developed a guide to provide a practical and reasonable approach for managing odorous discharges to the atmosphere (Ministry for the Environment, 2016). The guide is one in a series of good practice guides for managing air quality and is targeted toward those responsible for assessing odor impacts along with other stakeholders who have an interest in odor management.
- United Kingdom—The Environment Agency has established a management plan to facilitate compliance with regulatory permits (Environment Agency, 2011). Although less comprehensive than a facility-specific plan, it provides guidelines for the assessment, control, and monitoring of odors.

8.0 COMPARISON OF ODOR REGULATIONS AND APPROACHES

Each of the three regulatory approaches discussed throughout this chapter have been used globally to reduce off-site odor impacts from a WRRF, but their relative advantages and disadvantages differ, as summarized in Table 3.7.

Generally, when developing an odor regulatory approach, the selected approach should be as simple as possible while also being effective. Overly complicated approaches that are poorly understood typically reduce compliance rates and can reduce public confidence in the ability of the regulatory agency to adequately prevent nuisance conditions in the community.

TABLE 3.7 Odor regulation approach comparison.

Approach	Advantage	Disadvantage
Odorant Concentration Limits	Relatively easy to measure, based on standard compounds (e.g., H_2S). Provides a quick method of detection in the field using portable meters (for limited parameters).	Does not account for complexity/subjectivity of odors.
Odor Limits	Best representation of odor impacts. Potentially the most rigorous and scientific approach. Best suited for sensitive sites.	Difficult to quantify at plant boundary. Air dispersion modeling typically used to determine compliance with targets.
Nuisance Quantification Protocols	Relatively easy to measure. Reflects all elements that trigger odor complaints. Does not require extensive modeling.	Most subjective approach, potentially impacted by sensitivity of field evaluators.

9.0 REFERENCES

ADEM Admin. Code r. 335-3.-1-.02. Air pollution. http://www.adem.state.al.us/alEnviroRegLaws/files/Division3.pdf

American Industrial Hygiene Association. (1989). *Odor thresholds for chemicals with established occupational health standards*. Akron, OH: American Industrial Hygiene Association.

American Industrial Hygiene Association. (2019). *Odor thresholds for chemical with established occupational health standards*, 3rd ed. Falls Church, VA: American Industrial Hygiene Association.

American Society for Testing and Materials. (2004). E679-04, Standard practice for determination of odor and taste thresholds by a forced choice ascending concentration series method of limits. Conshohocken, PA: ASTM International.

ASTM International. (2018). ASTM E544-18, Standard practice for referencing suprathreshold odor intensity. Conshohocken, PA: ASTM International.

Bay Area Air Quality Management District. (1982). Regulation 7, Odorous Substances. San Francisco, CA: Bay Area Air Quality Management District. http://www.baaqmd.gov/~/media/dotgov/files/rules/reg-7-odorous-substances/documents/rg0700.pdf?la=en

Brancher, M., Griffiths, K. D., Franco, D., & de Melo Lisboa, H. M. (2017). A review of odour impact criteria in selected countries around the world. *Chemosphere, 168*, 1531–1570.

California South Coast Air Quality Management District. (1993). California Environmental Quality Act (CEQA) Air Quality Handbook. Diamond Bar, CA: California South Coast Air Quality Management District.

de Melo Lisboa, H., Sivret, E., & Stuetz, R. M. (2014). Odour regulations—Experiences from Australia. *Chemical Engineering Transactions, 40*, 181–186.

Environment Agency. (2011). *H4 odour management*. Bristol, England: Environment Agency.

European Committee for Standardization. (2003). EN 13725, Air quality—Determination of odour concentration by dynamic olfactometry. Brussels, Belgium: European Committee for Standardization.

Institute of Air Quality Management. (2014). Guidance on the assessment of odour for planning. Institute of air quality management. London, UK. http://www.iaqm.co.uk/text/guidance/odour-guidance-2014.pdf

Koh, S.-H., Moss, L., Shaw, A., Coleman, P., Currie, J., & Koodie, T. (2016, March 21–24). Review of global odor control strategies and identifying a suitable approach for a WWTP. *Proceedings of the Odors and Air Pollutants Conference*, Milwaukee, Wisconsin, United States. doi: 10.2175/193864716821122784

Mahin, T. D. (2001). Comparison of different approaches used to regulate odors around the world. *Water Science & Technology, 44*(9), 87–102.

McGinley, M., & McGinley, C. (2008, April 6–9). Odor threshold emission factors for common WWTP processes. *Proceedings of the Water Environment Federation/Air & Waste Management Association Odors and Air Emissions Conference*, Phoenix, Arizona, United States (pp. 530–546). Alexandria, VA: Water Environment Federation.

McGinley, C., Pope, R., & Mahin, T. (2000, April 16–18). Elements of successful odor/odor laws. *Proceedings of the Water Environment Federation Odors and VOC Emissions 2000 Conference*, Cincinnati, Ohio, United States (pp. 937–963). Alexandria, VA: Water Environment Federation.

Ministry for the Environment (Manatū Mō Te Taiao). (2016). Good practice guide for assessing and managing odour (ME 1278). Wellington, New Zealand: Ministry for the Environment.

Moore, J. E., & Hautala, E. (1983). Odor as an aid to chemical safety: Odor thresholds compared with threshold limit values and volatilities for 214 industrial chemicals in air and water dilution. *Journal of Applied Toxicology, 3*(6), 272–290.

Pope, R. (2000, April 16–18). Establishing limits and defining odor control needs. *Proceedings of the Water Environmental Federation Odors and VOC Emissions 2000 Conference*, Cincinnati, Ohio, United States (pp. 973–990). Alexandria, VA: Water Environment Federation.

RWDI Air Inc. (2005). Final Report Odour Management in British Columbia: Review and Recommendations for the Ministry of Water, Land and Air Protection, British Columbia, Canada. https://olores.mma.gob.cl/wp-content/uploads/2019/03/Final-Report-Odour-management-in-British-Columbia.pdf

Sullivan, R. J. (1969). Preliminary Air Pollution Survey of Odorous Compounds. Raleigh, NC: National Air Pollution Control Administration, U.S. Department of Health, Education, and Welfare.

4

Source Sampling

1.0 INTRODUCTION 61
2.0 ODOR GENERATION AND RELEASE 62
2.1 Hydrogen Sulfide 62
2.2 Organic Reduced Sulfur Compounds 63
2.3 Ammonia and Nitrogen Compounds 63
2.4 Other Wastewater Odorants 63
3.0 SAMPLE METHODS 64
3.1 Liquid Phase 64
3.2 Vapor Phase 65
3.2.1 Continuous Hydrogen Sulfide Monitoring 65
3.2.2 Instantaneous Hydrogen Sulfide Sampling 66
3.2.3 Sensory Sampling 67
3.2.4 Sulfur Speciation Sampling 68
3.2.5 Ammonia and Amines Sampling 68
3.2.6 Pressure Monitoring 69
3.2.7 Airflow Measurement 69
3.2.8 Electronic Odor Monitoring 69
4.0 SAMPLE SOURCES 70
4.1 Collection Systems 70
4.1.1 Gravity Sewers and Force Mains 70
4.1.2 Lift Stations 71
4.1.3 Septage and Grease Wastes 71
4.2 Water Resource Recovery Facilities 72
4.2.1 Preliminary Treatment 72
4.2.2 Solids Settling and Nutrient Removal 72
4.2.3 Biosolids Handling 73
5.0 REFERENCES 76
6.0 SUGGESTED READING 77

1.0 INTRODUCTION

Odors associated with wastewater are considered objectionable, and odor mitigation is important for all phases of wastewater collection, treatment, and the transport and disposal of biosolids and other residuals. Because odors are inescapably linked to the conveyance and processing of wastewater, odor issues must be addressed at each step in the collection and treatment process. This chapter discusses odor generation and release in wastewater systems, the types of odors and odorous compounds that can be formed, and methods for sampling, including liquid-phase as well as vapor-phase techniques.

2.0 ODOR GENERATION AND RELEASE

Fresh, aerobic wastewater contains many odorous compounds such as indole, skatole, organic acids, alcohols, and aldehydes. Odors from wastewater and its residuals become significantly more intense and develop much higher concentrations of odorous compounds when oxygen in the waste is consumed and anaerobic conditions develop.

2.1 Hydrogen Sulfide

Hydrogen sulfide is the most common odorous compound in wastewater collection and treatment systems. Its characteristics are well known, and concentrations of hydrogen sulfide in air are easily measured using portable instruments. Hydrogen sulfide is generated from the biological reduction of sulfate (SO_4^{2-}) or thiosulfate under anaerobic conditions. Desulfovibrio bacteria, which are strict anaerobes, are responsible for the majority of the reduction of sulfate to sulfide.

In collection systems, most of the sulfide generation occurs in the slime layer on the pipe wall or in sludge deposits on the pipe invert. The contribution of sulfide from the bulk liquid is relatively small. Key parameters that affect sulfide generation include

- dissolved oxygen concentration,
- concentration of organic material and nutrients,
- sulfate concentration,
- temperature, and
- detention time.

Hydrogen sulfide is a colorless, potentially toxic gas with a characteristic rotten egg odor. Death can result from exposure to hydrogen sulfide concentrations of 500 ppmv in air (Occupational Safety and Health Administration, 2018). Hydrogen sulfide gas is moderately soluble in water and dissociates to other ionic (non-odorous) forms of sulfide depending on pH.

2.2 Organic Reduced Sulfur Compounds

Organic sulfur-based compounds include odorous compounds such as mercaptans, dimethyl sulfide, and dimethyl disulfide. Biological matter in wastewater subjected to anaerobic conditions often results in the generation of reduced sulfur compounds, which can contribute to the characteristic odor. Although removing hydrogen sulfide can often result in significant odor reduction, the organic-reduced sulfur compounds become much more prominent. Removing just hydrogen sulfide from these emissions may result in only marginal odor reduction if the reduced sulfur organic compounds are present at elevated concentrations. Some of these compounds are less soluble than hydrogen sulfide and can be difficult to treat.

As with hydrogen sulfide, turbulence promotes the release of these odorous compounds from the liquid to the air. Consequently, areas of high emission potential are often turbulent zones such as at junction chambers, weirs, mixed holding tanks, and many dewatering and mixing processes.

2.3 Ammonia and Nitrogen Compounds

Wastewater and its residuals contain various forms of nitrogen. Much of it is present as ammonia or organic nitrogen. The small quantity of ammonia in wastewater that off-gasses in neutral pH conditions contributes little to odor emissions, and these odors are typically less detectable than those of the sulfur-based compounds. However, ammonia and nitrogen-based compounds such as trimethylamine typically appear in the dewatering processes and in the solids created from dewatering. When wastewater solids are treated in a high-pH process, such as lime stabilization, or a high-temperature process, such as composting, the release of ammonia and other nitrogen-based odorants can become the significant factor in odor emissions.

Anaerobic digestion and autothermal thermophilic aerobic digestion can also cause the generation and release of ammonia and compounds such as amines. Amines are a class of nitrogen-based odorants that are often characterized as having fishy odors.

2.4 Other Wastewater Odorants

Although most odor characterization and control efforts focus on sulfur-based compounds and, to a lesser degree, nitrogen-based compounds, there are many other compounds present in wastewater that contribute to its odor. Other classes of odorous compounds include, but are not limited to, carboxylic acids, volatile organic compounds (VOCs), and aldehydes and ketones. Although these compounds may somewhat contribute to the odor of wastewater and sludge, they are not typically the dominant odor or the target of odor control efforts.

3.0 SAMPLE METHODS

Sampling typically begins with a walk-through of the area of interest to survey potential odor sources. Based on the results of the odor survey, as well as collection and analysis of relevant background data and information, a sampling plan is developed to provide the proposed framework for field sampling efforts related to both liquid-phase and vapor-phase data collection, including

- sampling locations,
- type of samples and data to be collected,
- sampling equipment required, and
- sampling frequency and duration.

Data gathered as part of sampling activities are used to assess the liquid- and vapor-phase conditions leading to the potential generation and release of odors from wastewater collection, treatment, and disposal. The data are used to develop recommendations in an effort to mitigate the potential for off-site odor effects to the surrounding community. Quality control of the data collection procedures, including equipment calibration and adherence to manufacturer instructions, is critical to ensure these recommendations are based on representative data.

3.1 Liquid Phase

Wastewater characteristics provide information on the potential for the generation of sulfides from the anaerobic transformation of sulfate. Liquid-phase samples also provide information on the concentration of sulfides present in the wastewater, the speciation, and the potential for release into the vapor phase.

Liquid-phase grab samples are often collected to measure dissolved/total sulfide concentrations, dissolved oxygen, pH, temperature, and oxidation–reduction potential (ORP). ORP, for example, is a measurement of the wastewater conditions present and the potential for various biological reactions. Sulfide formation in wastewater occurs when the ORP is between −50 to −250 mV.

Total and dissolved sulfides are typically measured with a colorimetric sulfide test kit (Figure 4.1), which uses the methylene blue method, whereas the other liquid-phase parameters previously mentioned are typically measured with handheld water quality probes.

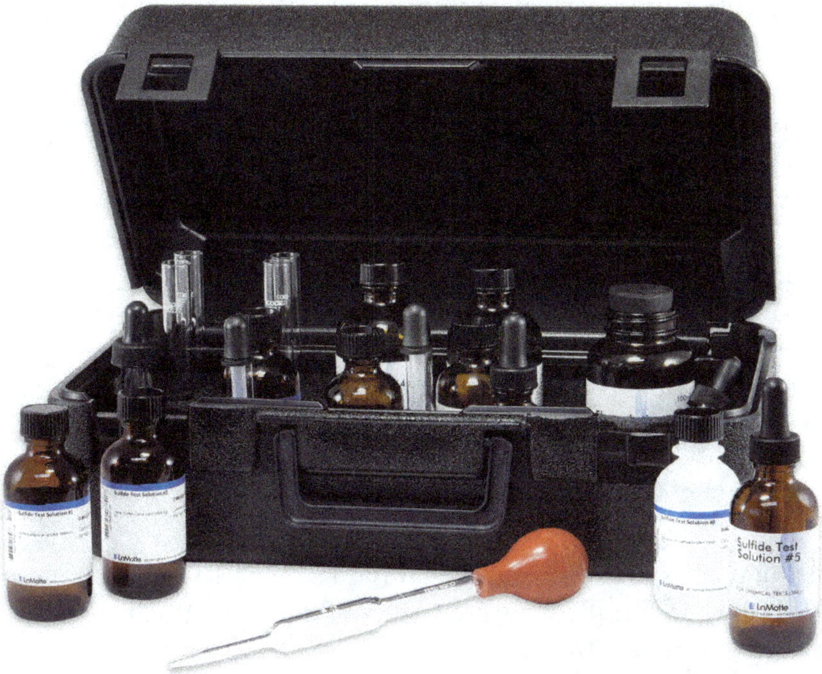

FIGURE 4.1 Colorimetric sulfide test kit (Photo courtesy of LaMotte Company).

3.2 Vapor Phase

For vapor-phase sampling, data for the following parameters are often collected:

- Continuous hydrogen sulfide monitoring
- Instantaneous hydrogen sulfide sampling
- Sensory sampling
- Sulfur speciation sampling
- Ammonia and amines sampling
- Pressure monitoring
- Airflow measurement

3.2.1 Continuous Hydrogen Sulfide Monitoring

OdaLogs® and AcruLogs™ are intrinsically safe, portable, battery-powered devices that are used to measure and record hydrogen sulfide levels over a continuous period of time. The range of hydrogen sulfide concentrations able to be detected depends on the model of the unit used. The units are typically hung in or near certain hydrogen sulfide area sources for accurate monitoring of process areas.

For point source locations where negative pressures or high-humidity conditions are present, such as the inlet duct of an existing odor control system (upstream of the fan), the unit is often housed in a sampling system enclosure, whereby foul air is transferred via tubing from a duct tap through a moisture trap on the outside of the enclosure to the OdaLog/AcruLog unit by way of an air pump located within the enclosure. There is also a wide range of permanent/fixed point source hydrogen sulfide monitoring products available from numerous manufacturers that are primarily designed for health and safety purposes or ambient monitoring.

3.2.2 Instantaneous Hydrogen Sulfide Sampling

For instantaneous hydrogen sulfide sampling, devices such as the Jerome® 631-X Hydrogen Sulfide Analyzer or the PortaSens II can be used. The Jerome 631-X is a handheld ambient air analyzer with a hydrogen sulfide range of 0.003 ppmv to 50 ppmv. The device operates by use of a thin gold film, which, in the presence of hydrogen sulfide, undergoes an increase in electrical resistance proportional to the mass of hydrogen sulfide in the sample. The standard units display instantaneous measurements that can be recorded by the user. Data loggers can also be purchased as an optional accessory to the unit. The data loggers record and store hydrogen sulfide values that can be downloaded onto a computer for storage and analysis.

The PortaSens II is a handheld device used to detect hydrogen sulfide, toxic gases, or oxygen levels by attaching the desired gas sensor module. The device uses a battery-powered intake pump with a flexible wand for sampling. Readings and settings are displayed on a backlighted, graphics LCD (liquid crystal display), which may be recorded and/or transferred to a computer.

Portable personal gas monitors used for monitoring atmospheric hazards often contain a hydrogen sulfide sensor. These monitors are typically operated to measure ambient instantaneous gas concentrations, but they can also be equipped with a sample pump and data logger. The available range, resolution, and accuracy of personal gas monitor sensors are typically less than equipment designed specifically for testing purposes. It is important to consider how the collected data will be used, anticipated concentrations, and regulatory enforcement standards when selecting a hydrogen sulfide sampling device.

3.2.3 Sensory Sampling

Odor sensory samples, discussed in detail in Chapter 2, are collected in individual 10-L Tedlar® bags and sent via overnight air to an odor laboratory, where the odor samples are analyzed using the "triangular forced-choice test" by a trained odor panel (ASTM E679-19 and EN 13725 methods) within the required 24-hour holding period. Sensory samples can be collected for both point and area sources to determine the odor detection threshold (D/T) and recognition threshold (R/T), odor intensity, odor characterization, and odor persistency (Figure 4.2).

For the area sources, an isolation flux chamber is used. This is a U.S. Environmental Protection Agency sponsored device that is designed to capture otherwise fugitive emissions from soil/solids and water surfaces to determine the emission rate of odors and/or VOCs. It is approximately 45 cm (18 in.) in diameter and 28 cm (11 in.) high, and it is used to isolate a portion of the emission surface, enabling air samples to be collected without interference from the effects of wind, other odors in the area, and other environmental factors.

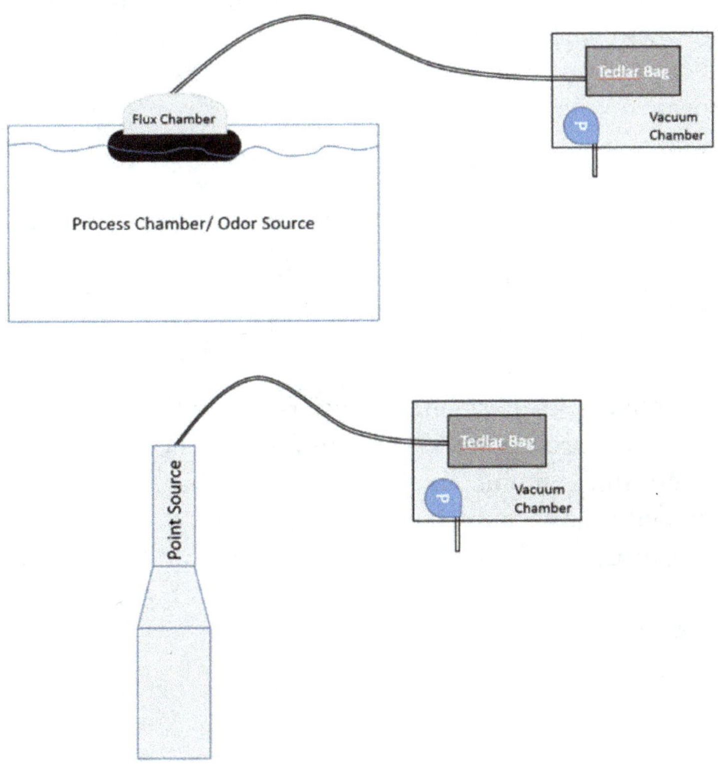

FIGURE 4.2 Area and point source sensory sampling.

For point sources, such as a stack or vent, samples are collected using a vacuum chamber and inert Tedlar bags. A small battery-operated vacuum pump removes air from the airtight vacuum chamber to create negative pressure, causing the Tedlar bag, connected by tubing to the desired point source, to inflate with odorous air.

The sensory data collected from the selected point and area sources can be used to determine the odor emission rate (OER) for each source. The OER is the mass emission rate and is a product of the D/T multiplied by the exhaust rate. This important parameter can then be used as an input for air-dispersion modeling. The data would be modeled to determine the theoretical transport distances for each of the individual odor sources, or multiple sources to determine their combined impact from the site to the surrounding communities.

3.2.4 Sulfur Speciation Sampling

Tedlar bags or glass-lined SilcoCan™ canisters are used to collect samples for analysis at an independent certified laboratory able to test for ASTM D5504-12. This laboratory procedure is used to analyze for 20 sulfur/reduced sulfur compounds, including many prevalent in wastewater collection and treatment, such as the following:

- hydrogen sulfide,
- mercaptans,
- dimethyl sulfide,
- dimethyl disulfide,
- carbonyl sulfide, and
- carbon disulfide.

A benefit of SilcoCan glass-lined canisters is that they have a 7-day hold time for laboratory processing versus the 24-hour hold time required for regular stainless steel Summa canisters or Tedlar bags. SilcoCan canisters are also provided in a vacuum state and, therefore, do not require the use of sampling pumps, evacuation chambers, or other additional equipment.

3.2.5 Ammonia and Amines Sampling

Sorbent detection tubes are used to sample for nitrogen-based odor compounds such as ammonia and amines. Ammonia can be sampled using Dräger-Tubes® and Accuro® hand pumps. Amine compounds can be sampled with the use of laboratory-provided sorbent tubes and battery-powered air pumps.

3.2.6 Pressure Monitoring

If a process area is not adequately ventilated to maintain a negative air pressure within the enclosed space, positive pressure from the space can be the source of fugitive odor emissions, leading to nuisance odor complaints, corrosion, and safety issues. Smoke candles can be used to provide visual confirmation of positive or negative pressure for selected process areas. AcruLog units are intrinsically safe sensors that can be hung in a covered process area to determine the differential pressure between the atmosphere and an enclosed process space. These units are deployed to continuously record pressure to determine if the process areas remain under negative pressure during all operating conditions.

3.2.7 Airflow Measurement

A hot wire anemometer is used to measure the air velocity in a duct or wall ventilation fan associated with existing odor control and HVAC systems. The anemometer's probe is inserted in a sample tap hole and the flowrate of air through is calculated using the velocity measurement and the circumference of the airflow (to determine the cross-sectional area). The velocity measurement is taken at multiple points across the diameter of the duct or fan opening to account for variations in the velocity caused by friction along the sidewalls. The velocity is then determined by averaging the measurements of the multiple data points collected. The airflow measurement results can also be used to determine if an existing system requires rebalancing to achieve design airflow rates.

3.2.8 Electronic Odor Monitoring

Advancements in sensor technology have led to the development of electronic odor monitoring devices (Figure 4.3) commonly referred to as "electronic noses" or "E-Nose". These devices typically consist of one or more sensors targeted to specific odor compounds. The devices are generally calibrated to odor-causing compounds that are known to be associated with odor emissions for a specific application. The calibration is based on the relationship between the observed odor compound concentration measured by the sensor for a given sample compared to the measured odor threshold value for that same sample. Once calibrated, the concentration of a particular odor-causing compound can be correlated to an odor threshold value. Changes in the specific odor-causing compounds that contribute to odor emissions may require recalibration. These devices are most commonly used to provide real-time odor exceedance alerts, often in combination with dispersion modeling.

FIGURE 4.3 Electronic odor monitoring system.

4.0 SAMPLE SOURCES

4.1 Collection Systems

Odors from wastewater collection systems are a concern for many municipalities. Although small gravity collection systems generate and release little odor, larger, more complex systems often emit nuisance odors. This condition results from a combination of factors, including the presence of force mains; large gravity interceptors constructed at minimum slope; long detention times resulting from oversizing designed for future growth; turbulent drops; inverted siphons; and industrial discharges. Further, the headspace air can become pressurized, forcing odorous air from manholes, vents, and other structures. Solutions to such problems can be costly and may include the addition of chemicals to the wastewater, injection of oxygen, and collection and treatment of the headspace air to prevent odor release into sensitive neighborhoods.

4.1.1 Gravity Sewers and Force Mains

Many gravity sewers that convey septic sewage containing hydrogen sulfide have no nuisance odor problems until the wastewater reaches the treatment facility. Other gravity sewers that convey relatively fresh wastewater may be the subject of odor complaints. The difference is the escape of odorous gas from the sewer atmosphere to the ambient air, where it can be detected by the public. The intensity of the odor, which is affected by wastewater characteristics and conditions, can affect the degree to which the odor is considered a nuisance.

Air pressure within a collection system is highly variable. For instance, some manholes are reported to "inhale" during certain times of day and "exhale" during others. Where odor emissions from a gravity sewer system are a problem, it may be valuable to monitor air pressure using differential pressure sensors to better understand why and where odors are likely to escape. In some cases, the most cost-effective approach may be to mechanically ventilate the sewer headspace at these locations and treat the odorous air.

Force mains are notorious for their generation of hydrogen sulfide and other odorous compounds, with most sulfide generation occurring in the slime layer on the pipe wall. Because a force main is under pressure and completely filled with wastewater, there is no opportunity for entrainment of oxygen from the headspace and, therefore, anaerobic conditions can quickly develop. Wastewater treatment facilities receiving a significant portion of their flow via force mains almost always have odor problems. Even force mains with relatively short detention times (e.g., 1 hour) can cause localized odor and corrosion problems. Where a force main discharges to a manhole, junction chamber, or wet well, the resulting turbulence and drop in pressure promote the rapid release of hydrogen sulfide and other odorants, such as reduced sulfur compounds, from the wastewater into the sewer atmosphere.

4.1.2 Lift Stations

Lift stations and pumping stations can be a source of odor complaints because they represent a point of potential odor release. For small package stations, the odor release point may be in the form of a simple vent that discharges odorous air when the wet well is filling. In larger stations, the wet well may be continuously ventilated to the atmosphere. If incoming wastewater contains dissolved sulfide, elevated hydrogen sulfide concentrations in the wet well atmosphere may increase the odor concentration or intensity, thus increasing the potential for odor complaints. This is true where force mains discharge to, or upstream of, another pumping station. Although wet wells are sometimes considered a source of sulfide generation, most sulfide generation occurs in the downstream force main.

4.1.3 Septage and Grease Wastes

Many facilities accept trucked wastes in the form of septage, grease trap pumpings, landfill leachate, and commercial and industrial wastewater and sludge. Accepting these wastes can be a significant source of income for the facility, and the discharge of these materials at a water resource recovery facility (WRRF) that is able to handle them makes sense environmentally. Some municipalities use remote dump stations in the collection system.

The effect of trucked waste discharges on odor emissions from a collection system or WRRF is dependent on the volume of the discharge relative to total flow (i.e., how much dilution occurs) and the characteristics of the material (i.e., how odorous it is). Clearly, the more odorous the material and the greater its percentage of total wastewater flow, the more potential for increased odor emissions from the sewer or plant headworks. In most cases, these wastes should be stored and metered into the wastewater at a controlled rate. Unless there is a direct connection to a receiving pipe, the physical process of discharging odorous wastes can release substantial odor.

4.2 Water Resource Recovery Facilities

4.2.1 Preliminary Treatment

For purposes of this discussion, preliminary treatment includes influent pumping stations, flow equalization basins, influent channels, screens, grit chambers, preaeration basins, and associated distribution boxes. The characteristics of the air emissions from these processes are highly dependent on the characteristics of the influent wastewater, detention time in the collection system, level of turbulence, and characteristics and volume of process sidestreams returned upstream of the preliminary treatment processes.

For plants that experience high odor emissions from preliminary treatment processes, source monitoring typically shows that while reduced sulfur compounds are often present, hydrogen sulfide dominates the observed odor. Other odorous reduced sulfur compounds may be present, but typically at low concentrations relative to hydrogen sulfide. Nevertheless, these compounds can contribute to odors that are released. Turbulent processes are often the first processes for which significant release of odorous gases can be expected.

4.2.2 Solids Settling and Nutrient Removal

Primary clarifiers can be a significant source of odor emissions, particularly if the influent wastewater contains significant levels of dissolved sulfide. Primary clarifiers represent a large odor-emitting surface area, and the character of the odor is considered objectionable. The turbulent effluent launders and, to a lesser extent, the turbulent influent feed wells often represent a significant portion of the total clarifier emissions. Again, this is attributable to the high turbulence levels that promote the transfer of dissolved odorous gas from the liquid phase to the vapor phase. The greater the freefall over the effluent weirs, the greater the release of odorants. Odor emissions from the effluent launder can represent 50 to 90% of the total clarifier odor emissions.

Biological nutrient removal processes typically generate odors that are considered less objectionable than those from preliminary and primary treatment systems or sludge handling processes. Although reduced sulfur compounds may still be components of the offgas, the odor character is often described as "earthy" or "musty" as opposed to "sewer" or "rotten egg". Atmospheric hydrogen sulfide levels at the surfaces of biological reactors are typically less than 1 ppmv. However, the strength of the odor (as measured in dilutions to threshold) and the odor emission rate can vary depending on the type of process, loading rate, and how the process is operated (e.g., sludge blanket depth).

The activated sludge process is a common biological wastewater treatment process. Emissions from activated sludge basins are directly affected by the type of aeration basins used. Mechanical aerators create very high turbulence and typically produce the highest emission rate, followed by coarse-bubble diffusers. Because of the low level of turbulence, fine-bubble diffusers likely have the lowest emission rate assuming all other factors (wastewater characteristics, mixed liquor suspended solids, and surface area) are equal.

Anoxic basins are expected to have relatively low odor emission rates because an oxygen source in the form of nitrate is present. However, emission rates are likely affected by the quality of the primary effluent and the dissolved sulfide concentration. Anaerobic suspended growth processes have a higher potential for odor production. Unfortunately, little data are available to support any distinction in odor emission rates.

Final clarifiers or settling tanks have low odor emission rates, and the odor is of a faintly musty or fresh character. Chlorine contact tanks have limited odor potential. Occasionally, a faint bleach odor can be detected, but this odor is not typically detected off-site.

4.2.3 Biosolids Handling

Odors from biosolids handling processes are typically complex mixtures of reduced sulfur compounds including hydrogen sulfide, mercaptans, dimethyl sulfide, dimethyl disulfide, and others. The incidence and concentrations of the reduced sulfur compounds are dependent not only on the type of sludge treatment process (particularly the stabilization process), but also on how the unit processes are operated. For example, excessive detention times in sludge holding or blend tanks can cause high levels of non-hydrogen-sulfide sulfur compounds to be formed. Some of these compounds, such as dimethyl disulfide, have extremely low odor thresholds and solubilities in water, and can be difficult to treat in conventional odor control systems. Odor control systems designed and operated to treat hydrogen sulfide often poorly treat dimethyl disulfide because it is a liquid at room temperature and is not very soluble in a scrubbing solution. Ammonia amine-based compounds may also be present from biosolids handling processes.

Gravity thickeners treating primary sludge can be significant sources of objectionable odors, not only from the settled solids but also from the grease and scum that accumulates on the surface. Rotary drum thickeners, when used for primary or primary–secondary sludge blends, can also be significant odor sources.

Processes such as dissolved air flotation, gravity belt, and rotary drum thickeners used on secondary sludge are typically less odorous because of the characteristics of the material being treated. Secondary sludge, specifically waste activated sludge, has an earthy or musty odor that is lower in intensity than the odor of primary sludge. However, if the secondary sludge is stored in the clarifiers or in holding tanks and allowed to become septic, odor emissions from these processes can be significant.

Blending and holding processes are not only significant sources of odor, their operation can dramatically affect the character and intensity of odor emissions from downstream dewatering, stabilization, and disposal processes. Holding times in excess of 24 hours can cause a significant increase in reduced sulfur compounds. The performance of an odor control system used for these processes can be adversely affected by the presence of compounds like dimethyl sulfide, dimethyl disulfide, and longer chain mercaptans. Other odorous compounds such as volatile fatty acids may also increase, yielding a sour odor.

Most dewatering processes currently in use are mechanical systems such as belt filter presses or centrifuges. Some facilities, typically smaller operations, use drying beds. The characteristics of the odor released from these processes are a function of the characteristics of the feed sludge. For example, a belt press dewatering aerobically digested solids from an oxidation ditch plant (no primary clarifiers) has significantly less odor than one handling a blend of thickened primary and waste activated sludge that has been stored for several days.

Belt filter presses are often significant sources of odor. This is because of the turbulent discharge of filtrate below the press that promotes the release of odors and the exposed surface area of feed sludge. The majority of odor emissions occur at the filtrate discharge. Some belt filter presses can be purchased completely enclosed to facilitate capture and treatment of the odorous air.

Centrifuges can release high levels of odor in the centrate vent, depending on the characteristics of the feed sludge. Although the odor is intense and objectionable, the volume of air emitted from the vent is low. Another odor release point is a cake discharge chute. One advantage of centrifuges is that, with the exception of the centrate vent and cake discharge chute, the device is sealed so that capturing odors can be accomplished with a relatively low air volume, resulting in less odorous air to be treated.

Plate-and-frame presses typically release little sulfur-based odor because the level of turbulence is low. Further, sludge dewatered in plate-and-frame presses is often conditioned with lime and ferric chloride, suppressing hydrogen sulfide release. The elevated pH resulting from lime addition may cause the release of ammonia, however.

Sludge drying beds can be significant sources of odor because of the large exposed surface area of odorous material. Odor releases are a particular concern during the application of sludge to beds and during removal of the dewatered material. Drying beds are suitable for rural areas in which the plant is remote from major receptors. Controlling odor emissions from sludge drying beds can be quite difficult. Many sludge drying beds have been abandoned in favor of mechanical dewatering to reduce labor requirements, increase sludge throughput, and mitigate odor emissions.

Thermal drying processes are advantageous because of the high-volume reduction they achieve. However, typical drying temperatures are often only sufficient to volatilize many of the odorous compounds present in the feed material, but inadequate to oxidize these compounds to less odorous byproducts. Therefore, control of both odor and particulate emissions is likely necessary for most drying processes. Thermal oxidation processes are often used to treat these offgases because of the relatively high temperature and the complex constituents in the airstream.

Odor emissions from cake conveyors, storage hoppers, and trucks are a function of the characteristics of the material being handled. Stabilized solids, for example, typically have less odor then raw sludge. The degree of agitation and the sludge temperature are other important variables. Movement of stored solids by front-end loaders, for example, exacerbates odor emissions. Some facilities have had significant numbers of odor complaints directed at trucks hauling dewatered biosolids through residential areas. Even with tarps over the trailer bed, odor released from a moving vehicle can be difficult to control. To eliminate odors from dewatering and cake handling and transport operations, some facilities have opted to remove biosolids from the plant as a liquid in sealed tanker trucks.

Odor emissions from the land application of biosolids are a function of the characteristics of the material and the method of application. The greater the odor of biosolids, the higher the odor emissions rate and potential for downwind complaints. Highly stable, aerobically digested biosolids from an extended aeration plant have less odor potential than even lime-stabilized biosolids from a conventional activated sludge plant with primary clarification. The method of biosolids application is also a significant factor affecting the odor emission rate. Spray application of liquid biosolids is likely to have the highest emission rate of the application processes. Application of dewatered biosolids using a manure spreader causes moderate emissions. The lowest odor emissions are associated with application methods such as subsurface injection or plow–furrow–cover techniques in which the material is immediately incorporated to the soil. Even with surface application of biosolids, the highest emissions rates occur during application. After that, the odor tends to dissipate to background levels within 24 to 48 hours.

Grit, screenings, and scum are miscellaneous residuals from WRRFs that have objectionable odors. Typical unprocessed grit and screenings often contain putrescible organics that emit foul odors, particularly during storage. Devices that provide washing and compaction of these materials reduce the content of organic matter and water, allowing less decomposition and odor production during storage. Although grit and screenings do cause localized odor conditions, these materials seldom make a major contribution to total plant odor emissions. Scum and grease from clarifiers and thickeners can be a source of odors, particularly if allowed to remain on tank surfaces or held for prolonged periods. Many facilities experience some difficulty handling these materials because they can clog weirs and collection troughs and moving them often requires assistance from operators equipped with high-pressure hoses. As with grit and screenings, scum pits or processing systems typically cause localized odors that are unlikely to be a cause of off-site odor complaints.

5.0 REFERENCES

ASTM International. (2012). D5504-12, Standard test method for determination for sulfur compounds in natural gas and gaseous fuels by gas chromatography and chemiluminescence. Conshohocken, PA: ASTM International.

ASTM International. (2019). ASTM E679-19, Standard practice for determination of odor and taste thresholds by a forced-choice ascending concentration series method of limits. Conshohocken, PA: ASTM International.

European Committee for Standardization. (2003). EN 13725, Air quality—Determination of odour concentration by dynamic olfactometry. Brussels, Belgium: European Committee for Standardization.

Occupational Safety and Health Administration. (2018). *Hydrogen sulfide hazards*. Washington, D.C., United States: Occupational Safety and Health Administration, U.S. Department of Labor.

6.0 SUGGESTED READING

Klenbusch, M. R. (1986). Measurement of gaseous emission rates from land surfaces using an emission isolation flux chamber (EPA/600/8-86/008). Las Vegas, NV: U.S. Environmental Protection Agency.

Smith, C., & Daniels, S. (2018). Proactively monitoring and managing odors using real-time odor plume prediction. *Proceedings of the Annual Water Environment Federation Technical Exhibition and Conference,* New Orleans, Louisiana, United States (pp. 580–590). Alexandria, VA: Water Environment Federation.

5

Collection Systems

1.0 INTRODUCTION 80
- 1.1 Background 80
- 1.2 Purpose 80

2.0 CHARACTERIZATION OF ODORS 81
- 2.1 Liquid Sampling 81
- 2.2 Air Monitoring 83
- 2.3 Advanced Testing and Monitoring 83
- 2.4 Instrumentation 84

3.0 COLLECTION SYSTEM EVALUATION AND DESIGN 85
- 3.1 Flow Monitoring and Modeling 85
- 3.2 Hydraulic Effects on Odor Emissions in Gravity Sewer Systems 86
- 3.3 Force Main Design Considerations 89
- 3.4 Inverted Siphon Design Considerations 91
- 3.5 Odor Capture/Containment 91
- 3.6 Linings and Coatings Selection 92
- 3.7 Material Selection 94
- 3.8 Site Selection 94
- 3.9 Pumping Station Building and Ancillary Equipment Requirements 95
- 3.10 Maintenance and Access Requirements 97

4.0 COLLECTION SYSTEM AIR MOVEMENT AND VENTILATION 98
- 4.1 Physical Assessment of Air Movement 98
- 4.2 Airflow Measurement 99
- 4.3 Air Pressure Measurement 100
- 4.4 Airflow Modeling 100
- 4.5 Modeling of Hydrogen Sulfide Generation 101
- 4.6 Computational Fluid Dynamics Modeling 102

5.0 SOURCE CONTROL 103
- 5.1 Constituents Impacting Odor Emissions 103
- 5.2 Permits and Ordinances 107
- 5.3 Inspections, Monitoring, and Enforcement 108
- 5.4 Pretreatment of Industrial Waste 109

6.0 OPERATION AND MAINTENANCE PRACTICES 110
- 6.1 Safety Precautions 110
- 6.2 Visual and Closed Circuit Television Inspections 110
- 6.3 Radar, Ultrasonic, and Laser Scan Inspection Techniques 111
- 6.4 Smoke Testing 112

6.5 Cleaning and Housekeeping 112
6.6 Equipment Maintenance and Exercising 113
6.7 Protective Coatings and Corrosion Control 114
6.8 Recordkeeping 116
6.9 Budgeting and Financial Management 117
7.0 LIQUID-PHASE TREATMENT 117
7.1 Chemical Oxidants 119
7.2 Biological Treatment Methods 121
7.3 Iron Salts 124
7.4 Peroxide Regenerated Iron 127
7.5 Oxygenation 128
7.6 Adjustment of pH 129
8.0 VAPOR-PHASE TREATMENT 131
8.1 Biological 132
8.2 Chemical 133
8.3 Adsorption 134
8.4 Other 135
9.0 REFERENCES 136

1.0 INTRODUCTION

The control and reduction of odor and emissions from wastewater collection systems is important because these facilities are typically located out in the community surrounded by residential neighborhoods, commercial businesses, public gathering locations, and transportation corridors.

1.1 Background

Most modern conveyance system components are enclosed, but fugitive odors can still escape via manholes, plumbing vents, wet wells, and other openings that are not readily apparent. Although individual emission points are typically relatively low volume, their proximity to receptors increases the likelihood of detection. Large municipal collection systems cover wide areas, and their potential impacts on the public may be greater than at treatment facilities.

1.2 Purpose

This chapter provides information to help identify collection system odors and mitigate complaints. There are numerous odor assessment and control strategies available to minimize collection system emissions. Some odors can be reduced through source control/industrial pretreatment, whereas others require either liquid-phase or vapor-phase treatment. Corrosion of steel or concrete features such as pipelines, structures, access stairs/ladders, and pipe supports are also a significant concern within collection systems. Proper design of collection systems and good operation and maintenance practices can reduce problems, but additional liquid or vapor phase treatment may be needed to rectify some problems.

This chapter discusses characterization of odors, collection system evaluation and design, collection system air movement and ventilation, source controls, operation and maintenance practices, liquid-phase treatment considerations, and vapor-phase treatment for collection systems.

2.0 CHARACTERIZATION OF ODORS

2.1 Liquid Sampling

Given that physical and chemical conditions in the sewer are dynamic—driven by diurnal flow, temperature, and atmospheric conditions—"average" measurements are often of the same order of magnitude as their 24-hour variation. It follows then that collecting a wastewater sample and measuring parameters that represent the physical–chemical properties and wastewater characteristics at the sampling manhole require several preliminary and fundamental considerations. Generally, these include the following:

- specific purpose and requirements for subsequent analyses;
- type of sample grab/discrete or average/composite;
- flow measurement needed for flow-proportional sampling;
- analysis-specific sampling methods and testing for interfering compounds;
- sample preservation (temperature control, preservative, and holding time);
- geographic information system (GIS) database information such as sample location, description (manhole identification number), manhole diameter and depth, pipe size and slope, depth of water (at the time of sampling), diurnal flow profile (if metering data are available), and depth of flow variation (if data are available);
- date, time, duration (start/stop) meteorological conditions/observations (e.g., 2.5 cm [1 in.] of rain in the last 6 hours);
- field parameters such as temperature, conductivity, pH, oxidation–reduction potential (ORP), dissolved oxygen, sulfate, sulfide, and so on;
- upstream chemical feed dosing rates and/or operations and maintenance activities (e.g., hydrojetting, subregional reclamation facility sludge disposal); and
- sample/analysis database and reporting.

When properly considered and documented, the resulting data have considerable value.

Whenever several discrete wastewater grab samples are suitable, these may be collected manually using a 2-L plastic pail suspended by a parachute cord for short-term verification sampling and field parameter measurement. For some applications, a more extensive sampling program may be necessary to capture representative data. In these situations, a more extensive or longer sampling period extending beyond normal work hours (or spanning more than 24 hours) may require the use of battery-powered portable automatic samplers. Portable auto samplers hung within a wet well, manhole, or junction structure perform discrete or composite sampling per a programmed schedule or flow-proportional sample draw while logging field parameters from connected instruments. Field parameters can be measured with multiparameter instruments and alkalinity and sulfides may be determined using laboratory kits (e.g., Lamotte).

The parameters listed in Table 5.1 are tests recommended for chemical dosing programs. Table 5.1 values listed are for illustration only. Additional parameters may also need to tested to properly characterize wastewater for odor generation potential such as Na+, Ca^{2+}, Mg^{2+}, K^+, Cl^-, SO_4^{2-}, HCO^{3-}, and CO_3^{2-}, as well as chemical oxygen demand (COD) (dissolved and total). Manhole differential pressure is most conveniently measured with an Acrulog™ logging differential pressure (atmospheric–sewer headspace) sensor placed several feet below cover and offset from manhole cover pick-holes.

For the sample to be representative, it must be collected from near midstream below the surface and above mid-depth. Should the sample be collected from near the center to the bottom of flow, it is likely there will

TABLE 5.1 Typical field sampling program.

		Field Measured Parameters[a]								
Time [24 hr]	Flow Situation	Temp [°C]	Cond [uS/cm]	pH	Alkalinity [mg/L $CaCO_3$]	ORP [mV]	DO [mg/L]	Liquid Sulfide [mg/L]	H_2S [ppmv]	Manhole Differential Pressure (in. H_2O)
4:00	low	26	700	6.5	200	−200	0.05	2.0	150	+0.01
6:30	rising	27	1100	6.3	280	−250	0.5	2.75	200	+0.1
13:00	peak	27.5	900	7.2	220	−50	1.1	0.75	20	+0.15
17:00	avg	28	1000	7.5	230	−75	0.8	1.5	100	−0.01

[a]Data shown represent an illustration. Actual values will vary depending on site specifics.

be a positive bias for suspended solids because, although settleable solids will be more concentrated at the bottom of flow, suspended solids will be drawn toward the center of flow by the velocity gradient. This bias toward high solids will result in an inaccurate assessment of total sulfides and COD.

In the absence of wastewater flow information, a best practice for efficient monitoring at any sampling manhole location is to conduct preliminary vapor-phase hydrogen sulfide monitoring with a continuous logging instrument (e.g., Odalog® or Acrulog™) measuring temperature and sulfur compounds (such as hydrogen sulfide) in advance of collecting liquid samples to provide a record of that sample location's diurnal pattern. When downloaded from the instrument, the logged data provide a record of when vapor-phase concentrations are highest, lowest, and most representative. That information can be used to schedule sampling for other parameters at the appropriate time to collect the planned average sample or a minimum or maximum concentration sample. Generally, wastewater field parameter measurements are taken at the hydrogen sulfide logger deployment and at pickup.

2.2 Air Monitoring

Vapor-phase monitoring with a logging instrument should be completed after conducting a preliminary check with an intrinsically safe 4-gas (oxygen [%O_2], hydrogen sulfide concentration, combustibles [%LEL], and carbon monoxide concentration) personal ambient air monitor to confirm headspace safety status with respect to the aforementioned parameters and to confirm that oxygen concentration is at a safe level (typically 21 ± 3%).

For conveyance system sites (manholes, junction, flow-diversion structures, or siphon structures or pump stations) with chronic emissions and nuisance reports, supplemental monitoring may be justified for investigative purposes. The selection of supplemental vapor-phase monitoring that will be most useful should be established using liquid sampling results and differential pressure data. In addition, sampling tubes need to be regularly checked for condensation that can occur in areas of high humidity that are typically observed in sewers and other collection system facilities (see Figure 5.1 for a typical hydrogen sulfide monitor).

2.3 Advanced Testing and Monitoring

Beyond measuring hydrogen sulfide and field parameters, further liquid-phase and gas-phase characterization may be necessary to plan, design, and manage an odor control project. These more advanced methods include both continuous monitoring and analytical testing for additional parameters to define the composition of the wastewater, headspace atmosphere, and emissions. For the odor control practitioner, continuous monitoring time

FIGURE 5.1 Typical manhole hydrogen sulfide monitor.

series data from liquid or gas phase in any sampling manhole located on an interceptor will observe cyclic patterns superposed on the diurnal flow. These cyclic patterns will recur with frequencies ranging from minutes to hours and result from the superposed effects on flow, pressure, and chemical parameters caused by a confluence of tributary flows, intermittent force main discharge, or flow-paced scalping for an upstream reclamation facility. Suffice to say, continuous and longer term monitoring is necessary to understanding the range and domain of dynamics within a sewer and managing odor generation and emissions.

2.4 Instrumentation

Of the options for continuous monitoring, there are portable instruments available for up to month-long deployments. Additionally, there are fixed installation instruments for permanent monitoring.

For advance liquids monitoring, there are portable instruments and fixed instruments, each with multi-parameter capability. Instruments typically include the following probes:

- temperature and conductivity,
- pH and ORP,
- dissolved oxygen, and
- measured liquid depth and air headspace.

Instruments should be hung from a pipe strap mount in the manhole wall that holds a flexible polyvinyl chloride (PVC) hose of a diameter such that the instrument may be inserted into the end of the PVC hose, with the perforated sensor end protruding from the hose. The instrument is held within the hose with hose clamps, with the screws taped over. A cable lanyard anchored to the pipe mount and inserted through the flexible pipe is attached to the instrument before sliding into the PVC hose. The hose needs to be of such a length so that when hanging, the instrument could not touch the bottom of the sewer.

Fixed monitoring, the multiple parameters identified above can be supplemented by total and dissolved COD and biochemical oxygen demand (BOD), total suspended solids (TSS), bisulfide (HS^-), and hydrogen sulfide. These instruments, including ultraviolet or ultraviolet–visible (UV–VIS) spectroscopy, may be implemented as in-line on sample loops, or in situ on protected perforated pipes. The real difference is that for wastewater spectroscopy, accurate calibration is specific to the water characteristics at that location because of matrix effects on the light transmission.

Advanced vapor monitoring is also divided between portable and fixed locations; however, this has more to do with remote communications options. The portable systems are typically limited to cellular communications, whereas the fixed stations may use any number of cellular and radio options.

3.0 COLLECTION SYSTEM EVALUATION AND DESIGN

Understanding how odor- and corrosion-causing compounds (predominantly hydrogen sulfide) occur in collection systems, and applying that knowledge to the design and operation of wastewater systems, can significantly reduce the impact of these compounds on the public and the wastewater infrastructure. There are some basic collection system design elements and operating methods that can reduce or even eliminate odor and corrosion issues within the wastewater collection system that are discussed below. Also, predictive computer simulations can be used where existing collection systems will be modified, or when new collection systems are planned. These computer simulations can predict the amount of sulfide generated, or estimate the extent of headspace pressurization that could lead to fugitive odor emissions.

3.1 Flow Monitoring and Modeling

Flow monitoring provides valuable data on present-day wastewater flows that lead to better understanding of actual conditions within collections systems, such as wastewater velocities, sewer and force main retention times, differences between peak and minimum flowrates, and frequency of pump cycles.

Whereas flow monitoring provides information for present-day conditions, flow modeling can be useful for prediction of future flows, under different system configurations, or controls. Coupled with actual flow monitoring data, flow modeling can be a powerful tool to understand changes in the system over years or decades and how changes may trigger increases, or decreases, in corrosion or odor generation in the system. One typical example of such change is upgrades to the collection system to allow for the long-term increase in conveyance capacity well in advance of actual increases in wastewater flow.

Upgrades to force mains without an appreciable increase in flowrates can lead to increased corrosion and odor issues in downstream systems because of the sudden increase in retention time, resulting in increased sulfide generation. This is especially true when force mains are sized for a future flow capacity, rather than their initial capacity. By modeling the system, the overall increase and duration of increases in sulfide generation can be estimated. If moderate levels of sulfide generation in the range of 5 to 20 ppmv are predicted only for a few years, this may be considered an acceptable risk; however, if it is predicted to occur over decades, treatment to reduce sulfides may need to be added to the design.

3.2 Hydraulic Effects on Odor Emissions in Gravity Sewer Systems

Hydraulic conditions in the collection system can have large effects on a variety of issues, ranging from the amount of hydrogen sulfide and other odorous compounds stripped from the wastewater to the amount of oxygen introduced to the wastewater stream.

Pipe slope is an important parameter in collection system design for several reasons, including sufficient capacity in the pipe to convey design flows at the required hydraulic grade and at sufficient velocity to minimize solids deposition in the sewer. The effect that pipe slopes can have on collection system odor and sulfide generation are described in the following paragraphs.

Excessive solids buildup at the pipe invert encourages the growth of sulfate-reducing bacteria and leads to increased sulfide generation. Excessive organic buildup can also increase methane production through anaerobic decomposition of deposited organic matter. One potential solution to control excessive solids buildup along the pipe invert in large-size gravity sewers is to add a low-flow channel (or cunette) along the bottom of the pipe to increase the velocity during periods of low flow.

Wastewater velocity and turbulence promotes reaeration, which increases the dissolved oxygen concentration in the wastewater stream; this, in turn, oxidizes existing sulfide and reduces anaerobic conditions in the downstream sections. However, the turbulence in the wastewater may also lead to hydrogen sulfide and other odorous compounds released into the sewer headspace (discussed later in this subsection).

Extreme changes in pipe slope and pipe cross section or changes in direction may cause odors to be released because of turbulence. Excessive turbulence, such as hydraulic jumps, can occur because of the sudden change in pipe slope/velocity. This turbulence can release hydrogen sulfide and other odor-causing compounds into the sewer air headspace.

Venting of the air in the sewer headspace also occurs where a steep slope changes to a flat slope. Conversely, where a low slope changes to steep slope, air inflow can occur. Air accumulation at these transition sections can pressurize the system and result in air venting from manholes and house plumbing stacks. This emission from the sewer headspace can lead to odor complaints by local receptors. This mechanism is discussed further in subsequent sections of this chapter.

Utilities frequently specify a minimum sewer velocity or a minimum pipe slope for various pipe diameters. However, the American Society of Civil Engineers (ASCE) and Water Environment Federation (WEF) now advocate a transition to the tractive force approach for self-cleaning design (ASCE, 2007). Tractive force is the average shear stress exerted by flowing fluid on a conduit. Various names have been used for this force, such as "shear stress", "drag shear", and "tractive tension"; however, it will be referred to as "tractive force" herein. The tractive force design is based on the concept that if a conduit is adequately sized to transport solids, a discrete grit particle will be transported often enough that it does not accumulate on a permanent sediment layer formed along the bottom.

The impact of the tractive force method on design is that the minimum slope required for self-cleaning action in the wastewater stream varies with pipe diameter (Yao, 1976). For example, under full flow conditions, if a minimum velocity of 0.6 m/s (2.0 ft/s) is adequate for a 150-mm (6-in.) diameter sewer, the same self-cleaning effectiveness can only be maintained by a minimum velocity of 0.75 m/s (2.5 ft/s) for a 500-mm (20-in.) diameter sewer (assuming a Manning's roughness coefficient of 0.013 for both). Therefore, if a larger pipe is selected to gain a smaller minimum slope (to meet a local standard), the self-cleaning characteristics are worse than for a smaller-sized pipe used at the same slope (Thistlethwaite, 1972). This poses a challenge for the design of large-diameter sewers because they typically have flat grades. Therefore, under these circumstances, other methods to mitigate or remove sediments may become necessary, such as grit chambers or cleaning/flushing programs.

Maintaining dissolved oxygen concentrations in wastewater minimizes sulfide production and oxidizes any existing sulfide in solution. Surface reaeration can be taken into account in designing new collection systems. The objective is to maintain a minimum dissolved oxygen value of at least 1.0 mg/L and no less than 0.5 mg/L, although this may be difficult to achieve. Proper surface reaeration requires sufficient oxygen concentrations in the sewer atmosphere and adequate interface area between the air and liquid. When pipes flow more than two-thirds full a significant portion of the time, the wastewater-to-air interface is reduced. Larger pipes typically provide a larger air-to-liquid interface, but they must also be designed to greater slopes to maintain the flow velocities. Larger, deeper trunk sewers can inflate the cost of a collection system, so designs frequently compromise relative to velocity and flushing.

Vent holes in manhole covers allow air movement in and out of pipelines and can improve reaeration. However, the vent holes that allow fresh air to enter also permit the egress of foul air when the system is under positive pressure. The direction of airflow is affected by internal factors such as the wastewater flowrate and external influences such as air pressure, temperature, wind speed, and wind direction. Mechanical ventilation has been applied as a means of supplementing the natural airflow in sewers. However, ventilating long stretches of interceptor pipeline is often difficult to accomplish because of high head losses in the pipe spaces combined with potential contrary air movement and differential pressure caused by the flow of wastewater.

Turbulence is typically considered undesirable in collection systems because it induces stripping of sulfide and other odorous compounds. Locations of high turbulence are most at risk for hydrogen sulfide corrosion. However, it should be recognized that turbulence also causes beneficial reaeration. Sewer design objectives commonly include minimizing turbulence as much as possible. Some examples of good practice are submerged discharges from force mains, smooth hydraulic transitions in conduits, vortex-type drop manholes, level controls to submerge weir discharges, and level controls in wet wells to reduce cascading flow.

Turbulence reduction minimizes hydrogen sulfide release into the sewer headspace, but the sulfide remains in solution. If the flow enters a larger interceptor or combines with a greater flow downstream, the sulfide may eventually be diluted and be less likely to come out of solution. However, in situations where the sulfide concentration will remain high and is likely to cause downstream problems (e.g., sensitive receptors, corrosion risks, etc.), it may be more practical to deal with the problem by providing adequate ventilation at the location of the turbulence or artificially inducing turbulence to strip the odor-causing compounds and treating them before they reach the downstream receptors. If the source of turbulence is in a sensitive location or air quality regulations mandate control of emissions, then vapor-phase treatment of the exhausted foul air will be required. Vapor-phase treatment options suited to collection system applications are discussed in Section 8.0 of this chapter.

3.3 Force Main Design Considerations

Force mains can be significant generators of reduced sulfide compounds because full pipes do not provide an opportunity for reaeration. Force mains and submerged pipes can be designed to reduce the formation of sulfide and other odorous compounds by following proper design and operating practices.

One key element of force main design is the velocity of wastewater. As in gravity sewers, prevention of deposits is important. Deposits and debris are more critical in force mains than in gravity sewers because there is extra difficulty involved in removing them if they occur because of the inaccessibility of the force mains and required pump station shutdown. Shear stress analysis is used in force mains to design to control buildup of deposits and thickness of the pipe wall slime layer. As with gravity lines, at least 3.5 N/m^2 (0.07 lb/sq ft) of shear stress is recommended so that the slime layers will largely be sheared off by the flow.

Force mains are typically sized based on a cost analysis of pumping cost versus pipe cost. For cases where excessive sulfide generation is projected, it may be beneficial to design the system for a smaller-diameter force main to maintain higher velocities. This will tend to increase pumping costs, but it will decrease costs for sulfide treatment. Other considerations in force main design and operation include the following:

- Discharged wastewater from force mains will almost always contain some sulfide. To limit local turbulence and odor stripping, this flow should be discharged below the surface, if possible, or with a smooth hydraulic transition to gravity flow.
- Stations to launch and retrieve pigs to clean force mains could be included in the design. Many agencies do not practice pigging because of the potential interruption in pumping and the potential risk of pigs becoming stuck and obstructing the force main for a sustained period of time. Dual force mains would reduce the risks and allow longer shutdowns.
- Minimizing high points in force mains will facilitate the development of pockets of foul air. Venting these pockets by means of air release valves introduces an additional source of foul air to the environment. The best force main profile is a continuous uphill slope so that no gas collects within the pipe.
- Where air release vents are required, the discharge should be designed for minimum odor impact. Venting the gas through a tall stack, well above grade, helps to disperse odors. In sensitive areas, an odor control device could be installed.

Force mains are not prone to corrosion because they are typically flowing full, so hydrogen sulfide cannot form. However, corrosion can occur in force mains if air pockets occur at or near air release vents. Air can be introduced to force mains at pump stations through wet well turbulence and vortexes, pump seals, and joints in suction piping. Air entrained by these mechanisms rises to the pipe crown immediately downstream of the pump station and forms an air pocket that promotes corrosion. The problem can be avoided by proper design and system maintenance to avoid air entrainment. Air-vacuum valves and vacuum breaker valves designed to mitigate transient pressures may also allow air to enter the force main. Lined pipe can be installed in sections where corrosion is expected, but it may be difficult to determine all the locations where corrosion may occur during the design stage.

Force mains designed with sufficient velocity and sufficiently low retention times can still experience problems during low flow periods. This occurs especially during overnight and early morning periods when low pump station wet well turnover is experienced. Wastewater may stay in the force main for many hours before the wet well is sufficiently full for the pump station to turn on. During these low flow periods, sulfide-laden wastewater can be released at the force main discharge manhole, which can lead to corrosion and odor complaints.

Force mains are commonly designed to handle future flows rather than present-day flows. Given the average expected life of a force main, the selected future flowrate may be for a 50-year time horizon. In addition, because the most costly part of installing a force main is the excavation cost, force mains may be oversized to anticipate future conditions (e.g., future flows are greater than predicted). Oversized lines have long retention times during the initial part of their lives until the design flows are reached. In some cases, because of the conservative estimates used, the design flow of the force main may not be reached before the end of its life.

Installing dual force mains is beneficial to address this concern. The first force main is generally smaller (though commonly can be an existing force main that has reached design capacity), with the second force main slightly larger. Once the smaller of the force mains reaches capacity, this force main is temporarily decommissioned and the second force main brought into service. Once the second force main has reached capacity, then both force mains are brought into service. By doing this, the retention time of the wastewater is kept as low as possible under different development timelines. With dual force main designs, there is an opportunity to add a drain (or drains) to allow the "out of service" force main to be emptied if it requires repair or is not expected to operate for an extended period of time.

3.4 Inverted Siphon Design Considerations

Submerged wastewater pipes, such as siphons and surcharged sewers, are conducive to sulfide formation. Inverted siphons are typically short-section sewers that are used to convey wastewater flow under obstructions (such as streams, highways, railway crossings, or other utilities) and then regain elevation after passing the obstruction. However, long inverted siphons are used to eliminate the need for pump stations, as well. These inverted siphons can be many kilometers long, using hydraulic forces to push flows up hills. The inverted siphon is below the hydraulic grade line, so it remains full even during periods of low flow.

Siphons and other submerged lines are potential sources of high sulfide generation because of full pipe flow, which is caused by a slime layer forming on the full circumference of the pipe wall. Siphons are also prone to deposition, which can cause additional sulfide generation in the septic solids. As with force mains, maintaining a high velocity minimizes deposition and sulfide generation problems.

Siphons are also noted for problems with odor release at the inlet structure because the full pipe blocks the path of air flowing downstream with the wastewater. The high turbulence at the siphon inlet strips hydrogen sulfide out of solution, so significant odor may be vented. The inlet structure can be sealed, but it is under high positive pressure and large amounts of air can escape through small leaks or be released upstream of the actual siphon.

One solution is to use an air jumper line to convey airflow from the inlet to the outlet structure. Air jumpers typically follow the path of the siphon line, so provisions must be made to drain condensate from the line. The use of an air jumper merely conveys the hydrogen sulfide-laden air downstream, where it may cause odor and corrosion problems. The addition of chemicals upstream of the siphon can provide some reduction in odor, but even trace amounts of sulfide may cause problems because of high turbulence at the inlet.

Jumpers may also not be practical under certain circumstances, such as at large river crossings. For these circumstances, the most positive solution is to vent the air to a vapor-phase treatment system located near the siphon inlet structure.

3.5 Odor Capture/Containment

Large municipal collection systems are typically not designed to contain fugitive emissions. Municipal systems are intentionally ventilated for safety reasons to prevent the buildup of explosive gases, unsafe concentrations of hydrogen sulfide, or depletion of oxygen. The best method for controlling fugitive organic emissions in municipal systems is source control and pretreatment, which are discussed in detail in Section 5.0, "Source Control". This places responsibility for controlling organic compounds on industrial producers.

Federal air emissions standards may mandate control for industrial sewers and pretreatment systems. Emission suppression techniques that can be applied to enclose industrial sewers include hard-pipe sewers, water seals on process drains and vent pipes, surcharged lines, and negative-pressure sewers (WEF, 1994). In some cases, industries that remove volatile organics from the air using wet scrubbers can cause problems when the spent scrubbing liquid is discharged to the collection system and odors are released.

3.6 Linings and Coatings Selection

Lining and coating systems have advanced significantly in the past decade, providing different options for different conditions. Generally speaking, coatings are applied as liquids or semi-solids to a material surface and then are allowed to chemically cure to a final product, whereas linings are generally manufactured sheets of some type of impervious material that are attached to the material. However, coatings may also refer to products that are applied to the outside of a material and linings as products applied to the inside of a material. Regardless, for the purpose of this section, linings and coatings will be discussed together and interchangeably because the definitions do not impact how these materials can protect wastewater collection infrastructure from corrosion.

Lining and coating systems can generally be divided into two categories: new installations and rehabilitation. New installations will include the lining or coating system as part of the delivered pipe material, whereas rehabilitation systems apply the coating or lining after the initial installation of the pipe. Lining systems in new installations are used where corrosion is of concern, but noncorrosive pipe or structure material is not available. Under these circumstances, linings are provided to achieve corrosion resistance for a parent pipe or structure material that is not corrosion resistant.

One common application is lined reinforced concrete pipe, manholes, wet wells, and diversion structures. Concrete and iron/steel materials are still used throughout sewer systems even when noncorrosive plastic pipes and manholes are provided. Despite the lack of corrosion resistance, concrete and steel are sometimes the only options available because of their structural strength, moldability (i.e., ability to suit site conditions), availability, and good track record. However, in addition to their lack of corrosion resistance, they do have other limitations, such as lack of erosion resistance from high water velocities and solids abrasion, high porosity allowing for bacterial colonization, and potential for poor quality installation.

Lining systems on concrete or iron/steel can be used to provide additional protection to the material to mitigate some of the weaknesses. Lining systems can either be applied after the concrete pipe or structure has been cast and cured or be cast directly into the pipe for added structural support. When cast into concrete directly, risk of delamination and separation from the pipe wall is reduced. If the pipe or structure is installed below the groundwater table, additional precautions must be taken when linings or coatings are being applied to prevent delamination or dislodging the material because of hydrostatic pressure on the backside of the material. Consult with the manufacturer of the lining or coating system to select the proper precautions to control the adverse impacts of groundwater pressure on the lining or coating system. Currently, the common types of pipe and structure coatings are polyurethane and epoxy. Liners include plastics such as PVC, high-density polyethylene (HDPE), and rubber (Nayyar, King, & Crocker, 2000). For metallic pipes, a cement mortar and/or concrete lining can be provided.

Rehabilitation of pipe (or other collection system assets) using lining systems generally occurs when the pipe has reached its end of life and serves as an alternate method to full pipe replacement or as a method to mitigate further damage to the pipe. Pipe lining systems for rehabilitation fall into two categories: structural and nonstructural lining systems. Structural linings systems can provide full structural support, both from internal hydraulic pressures and external soil and traffic loads. Cured-in-place pipe resins or HDPE slip lining are examples of such structural lining systems.

Nonstructural lining systems do not provide full structural support to the various loads, but rather act as a protective barrier to the original pipe material to reduce further damage to the parent pipe material, which is still capable of providing structural support. Nonstructural lining systems may include spray-on type liners and fold-and-form PVC liners. It should be noted that many of the liner technologies are capable of providing both structural and nonstructural lining systems, and, therefore, it should not be assumed or implied that a specific technology falls into one or the other category.

Considerations for selecting liners and coatings include the following:

- reduction in overall inside pipe diameter because of liner thickness;
- smoothness of the new liner (smooth pipes can increase pipe velocity compensating for lost inside pipe diameter);

- cost of liner installation compared to replacing the pipe using other trenchless technologies, such as pipe bursting; and
- construction requirements (such as pipe preparation before installation, cutting service laterals, and remediating delamination and trapped air bubbles).

3.7 Material Selection

Material selection is a key step in any sewer design process—and it can be a complex process. Material selection can depend on numerous things, from the cost of installation, the cost of maintenance, the preference by the owner, frequency of the laterals, corrosivity of the soil, concerns for inflow and infiltration, workability of the material, and construction schedule. Material selection is also not limited to sewer and force main pipes, but also to other components in the infrastructure, such as air release valves, ladder rungs, manholes, hatches, valve components, bolts, and rebar.

The most common agents that cause corrosion are hydrogen sulfide, sulfuric acid, and moisture. However, other types of non-microbially induced corrosion must be assessed, such as oxidation and electrochemical corrosion. There are many different materials that can be used within a wastewater collection system. Refer to ASCE's Manual of Practice No. 60/WEF's Manual of Practice No. FD-5, *Gravity Sanitary Sewer Design and Construction* (ASCE, 2007), for detailed considerations of sewer construction materials.

3.8 Site Selection

Site selection of sanitary sewer collection system elements, such as pump stations, manholes, and pipes, is typically not driven by odor concerns, but rather by wastewater flow requirements, topographical constraints, or cost implications. However, locating odor mitigation facilities must also be considered during the design. Many older wastewater collection and conveyance systems were built before air movement and sewer ventilation guidelines were established. Newer designs generally include analysis of air movement and odor potential in addition to basic hydraulic design parameters that have been well established over time.

When selecting a new wastewater conveyance site, it is prudent to allow space for buffers and/or odor mitigation facilities such as chemical addition or vapor-phase treatment during the planning and land acquisition phases. With pre-existing wastewater conveyance facilities that are odorous, there may be physical space limitations or neighborhood factors that are controlling implementation of odor mitigation plans. In certain situations, odor mitigation facilities can be effectively installed in a different location from the area where the odors are occurring. Examples of this include the following:

- installing chemical addition well upstream of the odorous location;
- installing chemical addition at a wastewater pump station some distance from the force main discharge that is creating the odor; and
- locating vapor-phase treatment on a main trunk line upstream of and away from the pump station by creating air movement counter flow from the wastewater flow, drawing foul air away from the wet well.

Odor treatment (liquid and vapor phase) and sewer ventilation considerations are discussed later in this chapter.

3.9 Pumping Station Building and Ancillary Equipment Requirements

Pumping station wet well design is an important factor in controlling odor. Small wet wells with minimum wall area that allow for adequate wastewater velocity and no deposits are most desirable. It is helpful to avoid any freefall or turbulent conditions at wet wells. Keeping wet well water elevations as close as possible to the inlet sewer water level is a good design and operating practice. This may be difficult to achieve with constant speed pumps.

The volume of wet well storage depends on the peak and average flows. Many packaged pump stations are designed with two pumps, with each pump sized for peak flows. Duty cycles (i.e., the time between successive starts) are typically 15 to 20 minutes, with minimum pumping times of 2 to 5 minutes. This design results in effective wet well retention times of 7 to 20 minutes under average flow conditions. This may lead to excessive wet well retention times under low flow conditions. A maximum wet well retention time of 30 minutes is recommended to minimize sulfide formation (U.S. EPA, 1985). Wet wells should be designed with sloped walls to avoid solids deposition.

Pump station wet wells are often sources of odor release because or air movement in the wet well. Examples of air movement include the following:

- Wet well filling. As the water level in a wet well rises, air is displaced by the increasing water volume. The rate of air displacement is the same as the rate of incoming flow and, therefore, the greater airflow rate occurs during peak flows.
- Air drag. Air tends to move downstream with the wastewater flow and is dragged into the wet well. At this point, the wet well is considered

a "dead end" from a headspace perspective and is, therefore, forced to vent out of the wet well.

- Artificial ventilation. Wet wells can be actively ventilated for the following three different reasons (not including odor control):
  - Fire and explosion safety of electrical equipment. Respective electrical codes in the United States and Canada are typically read in conjunction with National Fire Protection Association (NFPA) 820: Standard for Fire Protection in Wastewater Treatment and Collection Facilities (NFPA 820, 2016). NFPA 820 is widely adopted as the industry standard guideline for ventilation in wastewater systems. NFPA 820 divides the various wastewater spaces into different electrical classifications (based on the electrical code) and assigns a ventilation rate that is required to meet that classification. For example, a wastewater pumping station wet well can either have ventilation fewer than 12 air changes per hour or more than 12 air changes per hour. In the case of fewer than 12 air changes per hour, an electrical classification of Class I, Division 1 is assigned. If ventilation rates equal or exceed 12 air changes per hour, then an electrical classification of Class I, Division 2 is required. These classifications have an impact on the electrical equipment selection. For further detailed discussions and details regarding wastewater collection system ventilation, consult NPFA 820.
  - Occupational health and safety. Although different local authorities have different requirements, in British Columbia, Canada, the provincial Occupational Health and Safety Regulation have specific requirements for confined spaces. Most underground wastewater infrastructure would be considered a confined space because of the enclosed nature of the space and the lack or restricted means of entry/exit. The rate of ventilation can be calculated in different ways, but generally takes into account the number of workers and the size of space being entered.
  - Corrosion management and mitigation. Ventilation in the wet well occurs when hydrogen sulfide is present in the space above 5 ppmv and moisture on the material surface is present. Therefore, by ventilating the space, moisture accumulation on the material surface and hydrogen sulfide concentrations can be lowered. For artificial ventilation of the wet well space, consideration needs to be made for whether fresh air is supplied to the wet well, or foul air is exhausted from the wet well; in some cases, the direction of air flow is explicitly defined in the regulation. If the fresh air is supplied to the wet

well, then foul air may be released from the wet well vent or in the upstream collection system. If foul air is being exhausted from the wet well, then the air from the exhaust vent may become an odor nuisance if not treated.

In colder regions with wastewater temperatures 20 °C (68 °F) or less, the wastewater from a gravity system has lower potential for odor issues than warmer regions with warmer wastewater because it can be "fresh" with higher potential dissolved oxygen concentrations, resulting in lower potential for sulfide generation. Although odors may be present within the vicinity of pump stations receiving "fresh" wastewater, they may not be considered a nuisance because of the lower concentrations of hydrogen sulfide. However, for warmer regions with wastewater temperatures 25 °C (77 °F) or higher, or pump stations that receive industrial wastewater, septage, and discharges from long siphons or force mains, these may have a more significant potential to produce nuisance odors. This may require a fan to draw the air out of the wet well and the use of a vapor-phase treatment system. Refer to Chapters 8, 9, and 10 for information on available vapor-phase air treatment technologies.

3.10 Maintenance and Access Requirements

Regular maintenance and inspection are required to assess the condition of the asset, prevent accumulations of solids that can contribute to odors, and maintain the system on an ongoing basis. For this reason, careful attention must be paid to details such as traffic patterns at manhole locations to allow proper cleaning/access; access size and type (e.g., manhole versus hatch); vehicle access to pump station wet wells and siphon chambers; and wet well access size, type, and location. All of these can aid routine maintenance and cleaning that can help alleviate collection system odors.

Pipe and wet well cleaning is practiced by almost all wastewater agencies to remove debris and deposits in sewers that can cause offensive odors. In some locations, this practice has proven to be an effective means of reducing sulfide generation. The cleaning of small wet wells, sewers, and trunks is easily accomplished, but larger wet wells and pipes may be more difficult because of higher flows. As a result, various methods have been used to clean large wet wells and gravity sewers.

Flushing systems, for example, have been used to increase flow and scour away deposits. This method has limitations because it is difficult to store enough water to provide a significant flush. Gates can be added within sewers to hold the wastewater and create a flushing flow. There is also a large variety of water jetting equipment and vactor trucks available. The types of jet nozzle, pressure, and flowrate vary depending on the type of deposits and the particular application. Some types of water jetting systems use wastewater for the jetting flow, so a clean water supply is not needed.

Regular inspection of collection systems is necessary to assess the condition of the system and determine the need for repair and rehabilitation work and coating/corrosion control measures. Routine inspection activities may include manhole inspections, line inspections, flow monitoring, smoke and dye testing, television (TV) inspections, and private sector inspections.

4.0 COLLECTION SYSTEM AIR MOVEMENT AND VENTILATION

The ventilation standards for sewers and wastewater pump stations are often associated with regulatory authorities such as the Occupational Safety and Health Administration (OSHA) and NFPA. However, these regulatory authorities are primarily focused on other issues, not necessarily odors or odor control. More specifically, OSHA regulations focus on worker safety (i.e., 29 C.F.R. § 1910.146), whereas NFPA regulations focus on requirements for protection against fire and explosion hazards specific to wastewater treatment facilities and their associated collection systems (NFPA 820). Recent studies have focused on the role of sewer air movement and ventilation on odors. The following sections discuss recent developments in collection system air movement and ventilation on odor production and odor control.

4.1 Physical Assessment of Air Movement

The research regarding air movement in collection systems is limited and was thoroughly investigated as part of the Water Research Foundation's (WRF) *Minimization of Odors and Corrosion in Collection Systems, Phase 1* (Apgar et al., 2007). Air movement in sewers is primarily caused by the velocity of the moving water surface and the drag on the air (Pescod and Price, 1981). (See Figure 5.2 for an illustration of idealized air velocity contours within a typical sewer pipe.)

However, the dynamics of sewer air movement are complex, and several other factors impact air movement including wind eduction, temperature differential, barometric pressure differential, and changes in wet well level (Apgar et al., 2007). At pump stations, changes in wet well level can cause significant air movement and may be mitigated by reducing the change in wet well level.

Resistance to the airflow within the collection system causes pressurization of foul air, which forces foul air out of the collection system, projecting

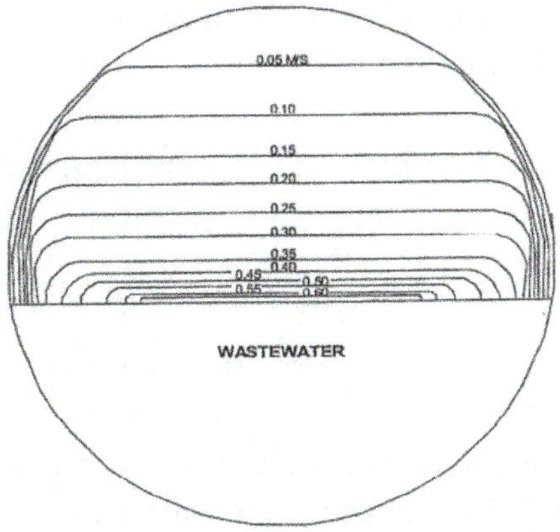

FIGURE 5.2 Idealized airflow velocity contours (Apgar et al., 2007) Reprinted with permission. © WERF and IWA Publishing.

odor-causing and corrosive compounds into the surrounding environment. Often, this is caused by sewers that are surcharged at a certain point, resulting from features such as inverted siphons, sewers that terminate at a wastewater pumping station (WWPS) wet well, and other constrictions of sewer headspaces. The only effective solution to resolve pressure conditions is to mechanically extract air from the headspace and treat it (Apgar et al., 2007).

Pressure conditions and estimates of sewer outgassing must be considered in addition to hydrogen sulfide or other odor causing compound concentrations, since pressure conditions allow the foul air to escape and disperse to surrounding receptors.

4.2 Airflow Measurement

While pressurized air may be forced out of the collection system at some points, other points may be drawing fresh air into the collection system. Physical field verification of manholes, wet wells, siphons, and other sewer features allows for a characterization of positive and negative pressure foul air discharges throughout the collection system. This is an important step in characterizing the collection system air movement before developing any odor control strategy.

Measuring the airflow in sewers may provide useful information for the sizing of an odor control system. However, it is often impractical to collect accurate airflow or velocity readings inside sewer headspaces. It is particularly impractical to collect continuous airflow data to capture diurnal trends.

One method of measuring airflow is using a pulsed tracer gas method as described in WRF's *Collection System Ventilation* (WRF, 2009). The time difference between each pulse tracer release and corresponding downstream concentration peak can be used to determine the average headspace velocity, and thus sewer airflow. In this example, carbon monoxide gas is used as the tracer.

A more rudimentary, but potentially more practical method of airflow measurement is smoke testing and timing the duration of the arrival of smoke to point B from point A. The velocity can be calculated by dividing the distance by the duration. Flow can be calculated if sewer headspace geometry is known. This test should be conducted several times throughout the diurnal flow pattern. Other parameters to measure include sewer diameter, slope, depth of flow, ambient air temperature, wind speed, humidity, barometric pressure, and headspace temperature.

4.3 Air Pressure Measurement

Measuring the differential air pressure at different points in the collection system is another method of understanding airflow dynamics in collection systems. At zero differential pressure, foul air is neither being pushed out nor pulled into the collection system. At positive pressure—even very slight pressures (such as 0.16 cm [0.05 in.] of water column)—foul air is being forced out of the collection system, causing odors. At negative pressure, fresh air is being drawn into the collection system.

Differential air pressure may also be measured after imposing a forced ventilation on the collection system. This can be done as part of a pilot test to verify the sizing and reach of a future odor control system. Additionally, various air extraction points and airflow rates could be pilot tested. Measuring the differential pressure at different airflow extraction rates and different points in the collection system or WRRF would allow for a more informed basis of design for a future odor control system. However, this type of pilot testing could be cost-prohibitive depending on regional costs and requirements.

Differential pressure can be measured using instantaneous meters or data logging devices. When using data logging devices, plan to design a custom-built housing and tubing system that protects the datalogger from hydrogen sulfide, moisture, and vandalism (see Figure 5.3).

4.4 Airflow Modeling

Despite its importance to wastewater odor and corrosion, wastewater collection system ventilation is understood generally at only a crude level among wastewater professionals. Available tools for modeling and analyzing are rudimentary and typically based more on general assumptions rather than fundamental physical principles. Several methods for modeling foul airflow were evaluated in WRF's Collection System Ventilation, and all evaluated

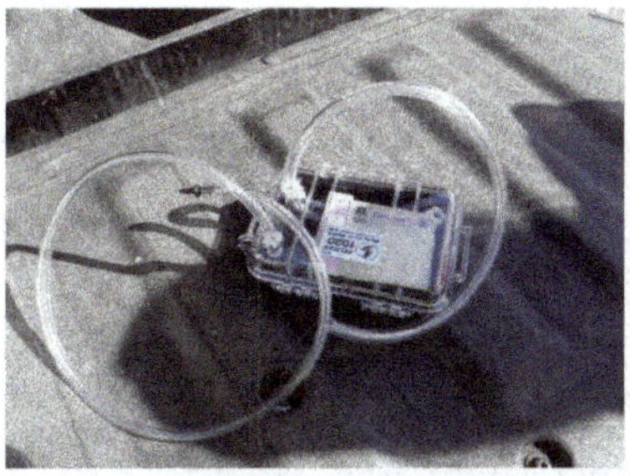

FIGURE 5.3 Typical differential pressure data logger and housing.

approaches exhibited significant errors in estimating sewer airflow (WRF, 2009).

The following equation provides a power fit to the ventilation-hydraulic relationships determined by Pescod and Price (1982):

$$V = 0.397\left(\frac{WV_w}{P_{air}}\right)^{0.7234} \quad (5.1)$$

where V = average headspace air velocity (m/s),
W = wastewater surface width (m),
V_w = wastewater average velocity (m/s), and
P_{air} = pipe dry perimeter (m), which is the pipe circumference minus the wetted perimeter.

Although Equation 5.1 has been shown to be the most accurate widely used sewer airflow model, errors of over 100% are not unlikely. The model is more likely to overpredict headspace velocity, providing a conservative design basis.

Accurate computer models are used, but they are developed on an individual basis, and are not commercially available or user friendly. Models can be developed using conservation of momentum based on a balanced force diagram for each segment of the collection system. It is important to consider forces that are internal to the collection system (liquid drag, pipe friction) and external to the collection system (wind eduction, humidity, barometric pressure).

4.5 Modeling of Hydrogen Sulfide Generation

Modeling hydrogen sulfide generation can be complex and require experience in computer modeling. As a result, a simplified approach to prioritizing further investigation is often used; this includes a mix of GIS sewer data (diameter, slope, location), sewer retention time estimates, and wastewater sample data.

Software exists that can be used to model hydrogen sulfide in sewers, but the programs are not mature in the industry. The following is a list of models that may be used to model hydrogen sulfide generation:

- Innovyze for Infosewer, infoswmm, and infoworks;
- Wastewater Aerobic/Anaerobic Transformation in Sewers Model;
- INTERCEPTOR;
- SPACA;
- ECIS;
- AEROSEPT; and
- U.S. Environmental Protection Agency (U.S. EPA) Sulfide and Corrosion.

Results from hydrogen sulfide modeling should always be field verified. Accurate data can be collected using continuous sewer headspace hydrogen sulfide data loggers at prioritized areas. Collection system hydrogen sulfide values could range from 0 ppmv to over 1000 ppmv.

4.6 Computational Fluid Dynamics Modeling

Computer algorithms for simulating fluid motions in three dimensions are referred to as "computational fluid dynamics" (CFD) models. CFD models are designed to produce simulations of fluid flows influenced by a wide variety of physical processes (e.g., heat conduction, solidification, cavitation, and surface tension). The numerical simulation of fluid behavior has become increasingly used in the wastewater treatment and conveyance field. Commercial computing tools have been developed for the analysis and study of complex fluid flow problems and have become more widely available and user friendly as the practice and computer flow simulators evolve. CFD programs are based on the fundamental laws of mass, momentum, and energy conservation. They are applicable to many situations such as the following:

- Predicting the performance of hydraulic structures;
- Estimating flow distribution, mixing, and air displacement/movement;
- Developing modifications and/or improvements to existing structures; and
- Troubleshooting and testing physical alternatives before design and construction.

CFD is increasingly being used for modeling the complex flow patterns in wastewater collection and conveyance and is an effective tool for gaining insights to flow regimes for different configurations. Although there are limitations to the applications of all CFD models, advancements have been made to address initial limitations such as accurately simulating two-phase (water and solids) and three-phase flow (air, water, and solids) behavior. Currently, CFD models are being used for complex collection system situations involving air/water movement and variable dynamic behavior. These are typically applied on wastewater collection system facilities that have complex air/water interfaces such as tunnel drop structures, flow regulator weir structures, and so on. (See Figure 5.4 for a typical CFD model output showing fluid flow in color and airflow in grey.)

5.0 SOURCE CONTROL

5.1 Constituents Impacting Odor Emissions

Hydrogen sulfide is the most prevalent odorous compound in sanitary sewers. The technical basis for developing numerical limits can include many different considerations. For hydrogen sulfide, the technical basis can be odor emissions from manholes or pump stations, the corrosion rate of concrete pipes and structures, explosivity, or health standard criteria. In developing a numerical standard, it is necessary to define what is to be protected and why that is important. For example, if a model is used to predict concrete corrosion related to hydrogen sulfide in the liquid phase, several items need to be considered, such as the following:

- typical rate of corrosion from hydrogen sulfide,
- acceptable rate of corrosion that occurs because of hydrogen sulfide discharged or generated from industrial sources,
- hydrogen sulfide headspace concentration that results in acceptable concrete corrosion rates, and

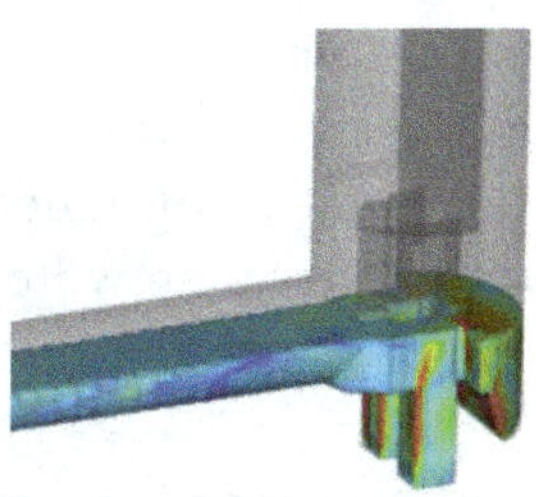

FIGURE 5.4 Typical CFD model output.

- threshold concentration of transition to unacceptable sulfide concentrations to set defensible limits.

The U.S. EPA industrial pretreatment program (IPP) can (and does at many water resource recovery facilities [WRRFs]) limit the introduction of sulfide to the collection system. Some agencies have limited the direct discharge of dissolved sulfide to 0.1 mg/L, whereas others have discharges that have been as high as 0.8 mg/L. However, setting sulfide discharge limits does not address sulfide produced in the collection system. Where sulfate is present in the wastewater and dissolved oxygen concentrations are low, there is always a potential for sulfide production in the collection system.

There are a number of cases where industries, such as food processing, have discharged high-BOD and high-temperature wastewater, which promotes or accelerates the generation of sulfide in the collection system. Municipalities can elect to treat the problem themselves or require the food processing industry to correct the problem. The pretreatment permit is the legal document that IPPs use to control discharges from industrial and commercial entities (U.S. EPA, 1989). To write the individual permit, it may be necessary to define a policy regarding hydrogen sulfide generation in the sewers, odor releases from the sewer, and hydrogen sulfide-related corrosion. Once a comprehensive policy is developed, each pretreatment permit can use the stated policy as the technical basis for the control approach defined in the permit.

One key issue to define in any policy is the acceptable background headspace hydrogen sulfide concentration produced by average strength domestic wastewater. The odor generation potential can be reduced by considering limits on constituents that contribute to odor generation, including BOD (limits on both mass and concentration for average or peak conditions), pH (acceptable range and limits on temporary excursions), sulfide, sulfate, temperature, and dissolved oxygen content. The pretreatment coordinator should use the proper procedures to develop technically based limits.

Other specific organic compounds might be identified as a potential source for odor, a method is available for use to develop a technically based local limit (U.S. EPA, 1992b). This approach can be followed and odor threshold criteria inserted in place of the health effects criteria to develop a liquid-phase limit for a volatile compound. The U.S. EPA procedures use Henry's law, worker health criteria, and the lower explosive limit (LEL) to allow for the screening of organic compounds that may pose explosive or potentially toxic conditions in the wastewater collection system. These procedures are based on the following assumptions:

- The wastewater and sewer headspace temperatures do not deviate significantly from 25 °C (77 °F).

- The airflow rate in the sewer remains constant and negligible.
- Instantaneous attainment of vapor-liquid equilibrium.
- Volatilization rates are not considered.
- The Henry's law constant for each chemical is not affected by other constituents in wastewater.

To convert the LEL or other air criterion to a corresponding wastewater screening level, the criterion is divided by the chemical's Henry's law constant and then multiplied by appropriate conversion factors. The prohibitions of Title 40 of the Code of Federal Regulations (C.F.R.) Part 403 restrict the discharge of pollutants that may create an explosion hazard in the wastewater collection system. The most conservative approach is to use a fraction of the LEL as the indicator for controlling explosive environments in the collection system. The objective is to set a limit that will not result in an explosion or potentially explosive conditions. U.S. EPA has developed a method that converts the LEL into an equivalent liquid-phase concentration. This concentration defines the maximum organic compound discharge that will not create an explosive environment in the sewer. The following equation is used to calculate the explosivity screening concentration:

$$C_{\mathrm{lvl}} = C_{\mathrm{vap}}/H \tag{5.2}$$

Where

C_{lvl} = discharge screening level (mg/L),
C_{vap} = LEL expressed as a vapor-phase concentration (mol/m^3), and
H = Henry's law constant (mol/M^3)/(mg/L).

Organic compounds, when released to the sewer collection system, can produce environments hazardous to collection system workers. To identify industrial discharges that could potentially generate worker health and safety hazards, the city must calculate equivalent liquid-phase concentrations that equate to fume-potential toxicity in the collection system. U.S. EPA has developed procedures to estimate liquid concentrations that produce dangerous vapor concentrations. The gas-phase potential toxicity for the organic chemical can be obtained using Henry's law constants in conjunction with one of the following three criteria:

- American Council of Government Industrial Hygienists (threshold limit value–time weighted average [TLV–TWA]),
- American Council of Government Industrial Hygienists (threshold limit value–short term exposure limit [TLV–STEL]), and

- National Institute of Occupational Safety and Health (immediately dangerous to life and health [IDLH]).

The TLV–TWA is the concentration workers can be exposed to for an 8-hour day and 40-hour work week. The TLV–STEL is the concentration workers can be exposed to for 15 minutes, 4 times per day. The term, *IDLH*, refers to concentrations immediately dangerous to life and health.

The following equation is used to calculate the fume potential toxicity screening concentration:

$$C_{\mathrm{lvl}} = C_{\mathrm{vap}}/H \qquad (5.3)$$

Where

C_{lvl} = discharge screening level (mg/L);
C_{vap} = TLV–TWA, TLV–STEL, IDLH, (mg/m^3); and
H = Henry's law constant (mg/m^3)/(mg/L).

Mixture of compounds can also be an issue. Screening levels calculated from TLV–TWA data represent the potential toxicity of individual compounds. These screening levels do not address the generation of potentially toxic concentrations of gases that are produced from the mixture of chemicals in the wastewater collection system. U.S. EPA has developed a procedure to predict the vapor potential toxicity associated with the discharge of a mixture of volatile organic compounds (VOCs). This procedure requires both actual monitoring data and screening concentrations to determine the specific compound fraction of TLV–TWA. It is assumed that if the sum of the TLV–TWA is greater than 1.0, a potentially toxic condition exists.

Once a material has been designated to receive a local limit, it is necessary to understand, especially for the fume potential toxicity evaluation, the physical and chemical properties of the material to be permitted (Hunter et al., 1995). The important issues are as follows:

- If the material is soluble in water, then the Henry's law relationships apply.
- If the material is insoluble in water, then one of two conditions apply.
- If the specific gravity is greater than 1.1 and the material sinks, there may not be a need to limit the chemical unless there are numerous structures that create enough turbulence to expose the material to the air.
- If the specific gravity is less than 1.1 and the material floats, Henry's law does not apply and the partial pressure of the pure material

should be used to estimate the vapor-phase concentration. In this case, the explosion prohibition would probably be invoked to regulate the discharge of this material to the sewer.

- If the material congeals or precipitates at sewer temperatures, regulation may be a seasonal issue.

Consideration of these issues focuses attention on the key physical and chemical properties of a given material. The physical and chemical characteristics of that material then help determine the correct permitting approach.

5.2 Permits and Ordinances

Most municipalities or utilities have regulations or ordinances that contain prohibitions or numerical limits for selected materials with the express purpose of protecting the collection system. Municipal wastewater treatment authorities or agencies with greater than 19 000-m^3/d (5-mgd) design capacity are required under federal law to have an approved IPP unless designated otherwise by U.S. EPA. The IPP elements are defined in the 40 C.F.R. Part 403. The intent of 40 C.F.R. Part 403 is to provide the municipality or utility with the legal authority to regulate discharges into the collection system for protecting physical facilities and WRRF worker health. These are broad guidelines that can be interpreted to cover many topics. The federal pretreatment program has been in existence for nearly two decades and the legal authority to regulate discharges into the collection system is well established.

The IPP has been successful covering all types of discharges to the collection system. The Effluent Guidelines and Standards, 40 C.F.R. Parts 404–471, are National Pretreatment Standards for Specific Industrial Categories. Municipalities and utilities are also required to establish discharge requirements for local conditions. The implementation of federal standards and local limits has resulted in a dramatic decrease of metals to treatment plants (Hunter & Lodderhose, 1993).

With the focus of National Pollutant Discharge Elimination System permitting shifting to water quality standards, many organic pollutants, specifically the priority pollutants, are being regulated by the IPP. Furthermore, it has been recognized that sewers are dangerous working environments and the regulation of organic chemicals other than priority pollutants is needed. U.S. EPA has developed guidance for the regulation of potentially toxic materials in collection systems and the IPPs also address cross-media pollution. Protection of human health can be used as a basis for regulating the discharge of volatile organic compounds to the collection system.

The IPP contains both general prohibitions and numerical limits for specific substances. The National Pretreatment Standards: Prohibited Discharges are listed in 40 C.F.R. § 403.5. Water resource recovery facilities have broad powers in using prohibitions to protect the collection system. The pertinent federal prohibitions that apply to odor emissions are as follows:

- Pollutants that create a fire or explosion hazard in the WRRF, including, but not limited to, waste streams with a closed-cup flashpoint of less than 60 °C (140 °F) using test methods specified in 40 C.F.R. 261.21.
- Pollutants that will cause corrosive structural damage to the WRRF, but in no case discharges with pH lower than 5.0, unless the works is specifically designed to accommodate such discharges. (This prohibition has two parts. The first part can relate to hydrogen sulfide corrosion or corrosion in the nonwetted parts of the collection system. The second part relates to low pH conditions in the liquid phase. The low pH can be either introduced to or created in the collection system.)
- Any pollutant, including oxygen demanding pollutants (BOD etc.) released in a discharge at a flowrate or concentration that will cause interference with the WRRF. The term *interference* has a very broad connotation, 40 C.F.R. Part 403.3(i).

The provision to prohibit any material that poses a threat to human health is not specifically defined in 40 C.F.R. Part 403; however, it was incorporated to the U.S. EPA Model Sewer Use Ordinance (U.S. EPA, 1992a). Consequently, it does appear in all U.S. EPA-approved IPPs.

U.S. EPA has an approved method for developing technically based local limits (U.S. EPA, 1987). These procedures are geared toward evaluating and developing local limits for materials that have defined water quality or land application standards or defined potentially toxicity threshold limits for aerobic and anaerobic treatment.

5.3 Inspections, Monitoring, and Enforcement

Where industrial waste discharges exist, odor issues may be increased by various characteristics in the discharge. In addition to sulfide, other characteristics of the waste are known to contribute to odor emissions in collections systems. These may include pH, high temperature, high organic loading or excessive oxygen demand, high sulfate, and so on. Many of these discharges can be controlled using inspection and enforcement of actions mandated by federal IPP regulations.

5.4 Pretreatment of Industrial Waste

If the industrial discharge has a high or low pH, high temperature, high organic loading, or excessive oxygen demand, the local municipal authority can require the industry to provide pretreatment before discharge. In the past, IPPs were focused on preventing undesirable compounds from discharging to the wastewater collection system to prevent upsets at the downstream WRRF. More recently, there has been recognition that industrial discharges can be a significant contributor to wastewater collection system odors even when the WRRF treatment performance is not impacted. This sometimes warrants additional controls at the industry such as pH adjustment, heat exchangers to cool the discharge, or chemical/biological treatment to reduce organic loadings. In general, the goal is to establish a mutual solution that is satisfactory to both the municipality and the industry to avoid costly and time-consuming enforcement and litigation proceedings that do not benefit either the municipality or the industry.

For example, in Tucson, Arizona, a Mexican cuisine canning company produces different types of canned beans, peppers, and sauces in batches on a weekly shift schedule. At some point, the canning company changed their production schedule from 7 days a week to a 5-day, 24-hour production, which impacted wastewater collection system odors downstream of the discharge. After the change was made, production stopped every Friday night and resumed every Monday morning.

The canning company's pretreatment system was comprised of pH adjustment chemicals to ensure that they could maintain their discharge between pH 6.5 and 8.5. Because of the nature of the discharge, which comprised lots of fatty acids and natural organic matter, the pH electrodes fouled rapidly, resulting in significant deviations in the pH readings controlling the chemical feed pumps and related uncontrolled pH swings. The local wastewater department feeds magnesium hydroxide slurry into the manhole downstream from their discharge to treat the interceptor downstream for approximately 15 miles. To eliminate the tendency of the discharger to unnecessarily feed acid, the municipality and the industry negotiated to allow a waiver for the pH if they replaced the sulfuric acid and sodium hydroxide caustic pH control with a magnesium hydroxide feed system that more reliably maintains the discharge pH between 7.5 and 9.

The local wastewater department also worked with the discharger to eliminate storage of the untreated wastewater in the equalization tank that was causing pre-fermentation that resulted in downstream fatty acid odors that the pH adjustment chemicals were unable to adequately mitigate (Froehlich, 2018).

6.0 OPERATION AND MAINTENANCE PRACTICES

6.1 Safety Precautions

When entering a manhole or sewer line, it is extremely important to observe the latest OSHA confined-space regulations and other applicable regulations. OSHA's standard for confined spaces is 29 C.F.R. § 1910.146, and it contains the requirements for practices and procedures to protect employees in general industry from the hazards of entering permit spaces (Permit-required confined spaces, 2004). Employers in general industry must evaluate their workplaces to determine if spaces are permit spaces. If a workplace contains permit confined spaces, the employer must inform exposed employees of their existence, location, and the hazards they pose. If employees are prohibited from entering and working in permit confined spaces, employers must take effective measures to prevent them from entering these locations.

If employees are expected to enter permit confined spaces, the employer must develop a site-specific written permit space program and make it available to employees or their representatives. During entry into sewer manholes, vaults, or wet wells, odors may escape or be intentionally discharged via ventilation equipment. Sometimes, it may be necessary to use chemical addition or mobile air treatment equipment connected to portable ventilators to control odor emissions during operation and maintenance (O&M) activities in areas where the public is nearby (Webster, Erstgaard, Parkins, & Koetter, 2018). Figure 5.5 shows a typical mobile odor control unit used to treat foul air exhausted from the wastewater collection system during maintenance or construction activities. Costs for these odor mitigation techniques should be built into annual O&M budgets, if needed.

6.2 Visual and Closed Circuit Television Inspections

When entering a manhole, sewer, or wet well is not feasible, video recording equipment or mirrors can be used. Intrinsically safe or explosion-proof cameras suitable for hazardous areas should always be used for video recording

FIGURE 5.5 Typical mobile odor treatment system (Webster, Erstgaard, Parkins, & Koetter, 2018).

wastewater collection system facilities. If cameras or other video recording equipment are not available, then mirrors can be used. Mirrors can be placed at two adjacent manholes to reflect the interior of the sewer line or a wet well or siphon chamber. Although mirrors can be used to document an immediate problem, they do not document the conditions identified and, therefore, may not be the best choice for planning and design of remediation or repairs.

Closed circuit television (CCTV) inspection will indicate pipe conditions including breaks and leaks, leaking and protruding laterals, root intrusions, and other blockages, including the exact locations of all features and problems. All of these can contribute to odors by

- surcharging, which restricts or blocks natural airflow in the sewer causing unwanted venting of contaminated air;
- deposition of solids or debris in surcharged lines caused by reduced velocities, which can result in septic conditions and increase odors; and
- leakage or spillage caused by the blockage that can be odorous and visually offensive.

The videotapes or photographs provide a visual history of the sewer for future reference and can be stored on a computer network or other portable electronic medium such as a CD or USB flash drive. The CCTV inspections will provide verification that manholes and cleanouts are on proper grade and accessible for future use. Accurate tie information and tape measurements are also used to physically locate manholes and cleanouts in case they are paved over or are otherwise concealed. Inspection records should be updated regularly to include the exact locations of service taps and property lines so that they can be located when maintenance is required in the future.

6.3 Radar, Ultrasonic, and Laser Scan Inspection Techniques

Sonar technology can be useful in inspecting depressed sewers (inverted siphons), where the pipe is continually full of water under pressure. Ultrasonic techniques can be used to assess sewer corrosion and detect the thickness of pipe materials including cast iron, ductile iron, stainless steel, wrought iron, and concrete. Internal and external ultrasonic testing and inspection can be used to determine pipe wall thickness and detect thinning caused by corrosion, which is an indication that odorous gases (hydrogen sulfide) are being generated and concentrated at the location where thinning because of corrosion is detected. If ultrasonic sonar is sent across a confined area such as a sewer pipe filled with liquid, they produce focused ultrasonic energy waves. The energy waves are reflected off the pipe boundary. These ultrasonic techniques can also be used to detect cracks.

Sonic calipers are also available for use in sewer, siphon, and force main inspections. Sonic calipers are capable of taking measurements along every foot of pipe. The sonic caliper is pulled through the pipes from one manhole to the next. It collects data that can be used to calculate the volume of debris underwater, measure the corrosion from the crown of the pipe to the waterline, and determine the percent of deflection at all points around a flexible pipe. The data collected are based on the time it takes a sonic pulse to travel between targets. Debris accumulations are a source of odors and sulfide formation. Locating debris and sediments allows O&M personnel to prioritize cleaning, which, in turn, can help reduce odors.

To assess external pipe corrosion, nonintrusive sophisticated electronic techniques such as three-dimensional laser scanning are available. These are used in place of traditional intrusive methods such as pipe coupons and are able to compensate for vibration that can adversely impact ultrasonic measurements. The three-dimensional laser scanner can accurately profile the damaged surface with repeatable accuracy, replacing typical mechanical attachments with a contactless referencing system. The scanner produces three-dimensional visual images that clearly define the extent and depth of external corrosion.

6.4 Smoke Testing

During smoke testing, collection system O&M staff, or maintenance contractors, blow air and nontoxic smoke into the sanitary sewer system using the manholes in the street as a point of access. As the smoke is added, the location where smoke escapes is visually monitored. The smoke under pressure fills the headspace of the sewer line as well as any connections. It also follows the path to any leak to the ground surface, quickly revealing the source of the problem. The amount of pressure required is small; it only requires enough positive pressure to overcome atmospheric pressure. Smoke testing has traditionally been used to identify sources of inflow and infiltration into sewers. However, it is also an excellent visual method that shows where sewer gases are escaping from the system and, consequently, creating the potential for odors. This allows the collection system O&M staff to quickly pinpoint locations where odors may be escaping and develop repairs and mitigation strategies.

6.5 Cleaning and Housekeeping

Pipe cleaning is practiced by almost all wastewater agencies to remove debris and deposits in sewers. In some locations, this practice has proven to be an effective means of reducing sulfide generation. The cleaning of small sewers and trunks is easily accomplished, but larger pipes may be more difficult because of higher flows. Various methods have been used to clean large gravity sewers. Flushing systems have been used to increase flow and scour away deposits. This method has limitations because it is difficult to store enough water to provide a significant flush. Gates can be added within sewers to hold the wastewater and create a flushing flow. There is a large variety of water jetting equipment available. The type of jet nozzle, pressure, and flowrate vary depending on the type of deposits and the particular application. Some types of water jetting systems use wastewater for the jetting flow, so a clean water supply is not needed.

In some collection systems, a variety of solids discharges occur that may require special attention to contain and control odors. These can include the following conditions:

- Septic tank effluent pumping systems may discharge to local wastewater collection systems, bringing high-solids wastewater directly to the main interceptor systems.
- Septage-receiving stations are sometimes situated on wastewater interceptors at convenient locations such as pump station wet wells. These discharges add slug loads septic solids, which may be highly odorous.
- Waste sludge from an upstream plant may be discharged to an interceptor for transport to a downstream plant for treatment and disposal. These discharges may cause increased anaerobic conditions in the collection system, which will likely result in higher sulfide production.

Discharge of septage and sludge to the collection system should be examined carefully to determine the extent of the effect on the system. In some cases, significant odor effects result from such discharges. If the effect is significant, options must be investigated to reduce or eliminate the problem. Problems with septage in the collection system can be resolved by relocating the receiving station to the treatment facility. Receiving facilities may include storage and sampling of each load and implementation of various pretreatment processes, including grit removal and screening.

6.6 Equipment Maintenance and Exercising

Wastewater conveyance and treatment equipment can only perform effectively if it is properly operated and maintained. If a pumping unit is not operating, it can contribute to odors by causing incoming sewers to surcharge, which, in turn, causes solids to settle and increases the amount of time that the wastewater is held in the collection system. When equipment such as valves or gates are not exercised, they may not open or close when needed, resulting in a spill or diversion of untreated wastewater along with the related release of odors in the area. Intermittently operated odor scrubbers and chemical feed systems should also be exercised and maintained so they function correctly when they are needed to control odors. The treatment system media should be regularly checked to avoid discharging untreated or poorly treated air. Each element in the collection system can contribute to odors if it is not regularly exercised and maintained.

6.7 Protective Coatings and Corrosion Control

A proactive management plan is essential to control odor and corrosion in a collection system. If not controlled, hydrogen sulfide can pose a safety hazard to workers and odorous emissions can cause a severe public nuisance. Odor problems typically serve as indicators of potential corrosion problems. When corrosion is not addressed, concrete pipe and manhole deterioration can progress rapidly to the point of collapse. The money invested in preventive measures is almost always less costly than unplanned emergency repairs.

One objective of a comprehensive odor and corrosion management plan is to conduct regular inspections to determine the condition of the pipelines, manholes, and critical structures. Routine inspection activities may include manhole and line inspections, flow monitoring, smoke and dye testing, and TV inspections. Smoke testing will reveal whether house traps are properly functioning to contain odor. Visual inspections involve walking the sewer reaches and looking for signs of ground subsidence, soggy areas, and odor. Manholes along the route should be inspected with the interior observed from the ground to determine structural integrity, condition of rungs and joints, and infiltration. When visual inspection identifies potential problem areas, more detailed information can be obtained using TV inspection. Television inspections can identify problems such as deformed crowns, cracking concrete, water infiltration, settlement, root penetration, open joints, illegal connections, and damage that occurs because of construction by others.

The inspection routine should also include monitoring of headspace hydrogen sulfide and dissolved sulfide. Where liquid-phase chemical treatment is used, the sampling data establish diurnal and seasonal variations in hydrogen sulfide and track treatment effectiveness as dosages are adjusted. In reaches where corrosion-resistant materials are in place, sulfide treatment may be deemed unnecessary; however, it is still necessary to conduct regular inspections of these reaches to ensure that the corrosion protection remains effective. Leakage at joints in lined pipe or failure of epoxy coatings in manholes can result in rapid destruction of concrete because of high hydrogen sulfide concentrations. Once these areas of liner or joint deterioration have been identified, they can be resealed with epoxy or foam injection. Inspections may also reveal that certain reaches with low flow velocity are subject to solids deposition, which may encourage sulfide formation. In these cases, periodic cleaning should be scheduled to flush solids and reduce sulfide generation.

As noted previously, some agencies inspect approximately 6% of the system per year. This rate of inspection is cost-effective and maintains performance to at least 85% of a new system. However, 17 years are required to complete the full cycle of inspection. Critical areas of the collection system should be identified and inspected more frequently than the average rate. Critical areas may include critical structures such as junction or drop manholes that have high turbulence and are prone to corrosion. Critical sections of sewer lines may include tunnels or road crossings where repair is difficult and collapses would be catastrophic.

In reaches where high sulfide is measured and control is deemed necessary, decisions must be made to select the most suitable treatment. A cost comparison of chemical options can be made by applying local chemical prices to typical dosage rates to obtain a cost per kilogram (or pound) of sulfide treated for each option being considered. However, it is not always an easy task to compare chemical products versus more labor-intensive control methods such as cleaning. Considerable operating experience must be gained with some methods before accurate costs can be determined. Typical dosage rates may vary from system to system, so confirmation data should be collected to verify that effective treatment is obtained. Even within the same collection system, different chemical products may be selected for specific applications. One chemical may be most cost-effective for gravity lines, while another may be better for force mains.

Liquid-phase chemical treatment is typically sufficient to reduce headspace hydrogen sulfide enough to control corrosion, but trace amounts of sulfide may remain and odor emissions will occur at turbulent locations. In areas adjacent to homes or businesses where odor is a concern, ventilation and vapor-phase treatment may be required. A wide range of control devices are available on the market today, and some are better suited to collection system applications than others. Both economic and non-economic factors should be considered in selecting the most suitable odor control system. Once vapor-phase treatment is installed, regular monitoring and maintenance must be performed to ensure that there are no lapses in treatment. Proper collection system maintenance requires a comprehensive management program and constant vigilance to ensure that odor and corrosion are controlled and structural problems are identified and corrected in a timely manner.

6.8 Recordkeeping

Unfortunately, wastewater creates odor nuisances from a variety of collection system sources such as the following:

- sewer manholes,
- siphons,
- force main discharges,
- leaking plumbing,
- plumbing vents, and
- wastewater pumping stations.

It is important to track and measure odors on a regular basis to understand and minimize them. Many large scale urban agencies have dedicated crews available around the clock to respond to odor complaints. However, with the limited resources available in smaller communities, it may not be feasible to provide the same level of response to an odor complaint. For this reason, it is important to set up a simple log or documentation form to record odor incidents. The form should record the following information:

- name and phone number of complainant;
- address or location where the odor was detected (the complainant may not be at home when the odor is noticed);
- date and time when the odor was detected;
- wind direction when the odor was detected;
- weather when the odor was detected (still, windy, foggy, rainy, clear/sunny);
- brief description of the odor; and
- date, time, staff names, and outcome of follow-up investigation(s).

Some agencies promote and accept electronic, web-based complaints if resources are not available to accept and process verbal or phone-based complaints. Chapter 12 contains example odor complaint forms.

It is preferred to have staff follow up with the complaint immediately when it is first recorded. However, this may not always be feasible. Follow-up agencies can have an operator or cleaning crew investigate to verify if there is a blockage or other mechanical problem the next day.

6.9 Budgeting and Financial Management

Agencies or public utilities face the challenging task of collecting, conveying, and treating polluted discharges from a variety of customers. However, they do not always have accurate or timely financial information related to the economic performance of their systems with respect to odors. Effective budgeting and cost accounting systems can help allocate resources and document costs related to odor monitoring, containment, and treatment. It is important to develop plans for an odor control program for the following:

- level of service expectations;
- short- and long-term objectives;
- odor contributors and fair rate apportionment ("polluter pays" principle);
- balance of capital improvements versus O&M;
- detailed cost/revenue categories;
- impacts of odor mitigation on other system components (corrosion, treatment, solids handling, etc.);
- staff resources;
- tracking and monitoring protocols to document results;
- consideration of future life cycle costs;
- impacts on sewer user rates;
- benchmarking against other similar utilities or agencies;
- permitting impacts; and
- land use considerations (reducing odors can open up development opportunities to generate additional revenues).

Effective budgeting and financial management allows agencies and utilities to develop and track trends in spending patterns and better estimate spending in the future (U.S. EPA, 1998).

7.0 LIQUID-PHASE TREATMENT

There are different types of control measures that can be applied to wastewater to treat sulfide and other odorous compounds in the liquid phase before they are emitted. Numerous chemicals are available for controlling hydrogen sulfide odors in wastewater collection systems. It is not necessary to choose one. In many municipalities, multiple chemicals are applied at various locations for various reasons.

For example, in Orange County Sanitation District (OCSD; California), multiple chemicals are used—not only for collection system odor control, but also to assist in reducing odor emissions at its two regional WRRFs. At OCSD, odor treatment begins in the collection system, where foul air emissions from 400 mi of sewer pipes and 15 pump stations could cause nuisance odor complaints. The OCSD odor team mitigates odors and decreases hydrogen sulfide loads to the treatment facilities with chemical additions of ferrous chloride, magnesium hydroxide, calcium nitrate, and/or sodium hydroxide upstream in the collection system. These chemicals are added continuously to the wastewater trunk lines by designated chemical dosing stations, or periodically through chemical batches provided by chemical tankers (Sobhani, Reed, Velasco, Thomas, & Murthy, 2018).

Another form of control is oxygenation. Oxygen injection can be applied to reduce sulfide formation and oxidize existing sulfide. Other approaches use chemicals to halt sulfide production or react with sulfide in the liquid. Some chemicals are sulfide specific, whereas others react with other organic-based compounds. There are also biological products that have been used for sulfide and odor control.

Regardless of the control method or chemical type, it is recommended to pilot test potential options, if possible, to measure the actual effectiveness and understand logistical requirements of implementation and operation. The City and County of Honolulu (CCH) represents an example of successful pilot testing of chemicals for odor control. In 2014, CCH conducted a full-scale pilot test of ferric chloride and calcium nitrate for liquid stream odor control. (See Figure 5.6 for a typical temporary chemical dosing system.)

FIGURE 5.6 Typical temporary chemical dosing system.

One location was dosed with ferric chloride and one location was dosed with calcium nitrate. Various dosage rates were tested over a 3-week testing period. During the testing, the downstream hydrogen sulfide, pH, flow, and temperature were monitored at various locations. Results of the tests suggest that stoichiometric chemical dosages are not always sufficient to achieve targeted hydrogen sulfide reduction. Pilot testing is a useful method to confirm chemical dosages and life cycle cost estimates before constructing full-scale facilities.

Pumps, mixers, piping, and valve material selection varies depending on the chemical chosen. A careful review is recommended for all applications. In some cases, use of inert plastic materials for all pumps, piping, and valves is preferred to allow maximum flexibility to change chemicals in the future. Spill containment materials and requirements also vary from chemical to chemical and must be considered for each application. Many chemicals are hazardous materials. Therefore, use of chemicals must include adequate site-specific health and safety precautions. For indoor storage, chemicals should be stored in a separate room with outside egress. Personal protection equipment, including gloves, goggles, eyewashes, and safety showers for emergencies, should be included in all chemical applications. Gaseous chemicals such as chlorine require additional safety and handling precautions that are more rigorous than liquid chemicals. The following sections discuss the various chemical treatments for odor control in detail.

7.1 Chemical Oxidants

Chemical oxidants chemically attack the odor-causing compounds and destroy them through oxidation–reduction reactions. Although some of the chemicals in this category may contain oxygen as part of their molecular structure, their primary action is to chemically react with the odorous compound in the dissolved form rather than release oxygen for use by bacteria. Chlorine, hypochlorite, hydrogen peroxide, potassium permanganate, and ozone are examples of chemical oxidants.

Chlorine is a relatively inexpensive, powerful oxidant, and the equipment required for its use is inexpensive and widely available. Chlorine is available as pure gas, hypochlorite solution, or hypochlorite granules or tablets. Wastewater pH may be slightly affected by the addition of chlorine solutions. The dissolution of chlorine gas creates an acidic product, whereas hypochlorite solutions are basic. Chlorine reacts with many compounds found in raw domestic wastewater, including hydrogen sulfide. The reactions between chlorine and sulfide in domestic wastewater are as follows:

$$HS^- + 4Cl_2 + 4H_2O \rightarrow SO_4^- + 9H^+ + 8Cl^- \text{ (Acidic pH)} \quad (5.4)$$

$$HS^- + Cl_2 \rightarrow S + H^+ + 2Cl^- \text{ (Basic pH)} \tag{5.5}$$

Equation 5.4 requires 8.9 parts, by weight, of chlorine to oxidize each part of sulfide, where Equation 5.5 requires only 2.2 parts chlorine per part sulfide. Competing side reactions with other non-sulfide compounds may require overfeeding to fully oxidize the sulfide present. Actual practice has shown that, depending on pH and other wastewater characteristics, between 5 and 15 parts, by weight, of chlorine are required for each part sulfide (U.S. EPA, 1985).

Chlorine can also act as a bactericide because it is a strong disinfectant. Depending on the point of application and the dose, it can kill or inactivate many bacteria that cause odors. However, because chlorine is nonselective, it will also kill organisms beneficial to wastewater treatment processes. Care should be exercised when chlorine is added in the collection system near WRRFs.

In some situations, the chlorine dosage required to disinfect the pipe interior is lower than the dosage needed to react with the sulfide being formed. The sulfate-reducing bacteria that produce hydrogen sulfide live on the pipe wall. If the pipe walls are disinfected, it can result in lower overall chemical usage than initial estimates based on sulfide sampling results.

Hypochlorite solutions are safer to use for collection system applications and the equipment is relatively simple to design, operate, and maintain. Sodium hypochlorite is typically available in a 12.5 to 18% solution with caustic added to retard off-gassing. The solution is subject to degradation because of light and heat, so solution strength diminishes during long-term storage.

Hydrogen peroxide is a commonly used oxidant that chemically oxidizes hydrogen sulfide to elemental sulfur or sulfate depending on the pH of wastewater. Hydrogen peroxide reacts with hydrogen sulfide according to the following equations:

$$H_2S + H_2O_2 \rightarrow S + 2H_2O \text{ (pH} < 8.5) \tag{5.6}$$

$$S^- + 4H_2O_2 \rightarrow SO_4^- + 4H_2O \text{ (pH} > 8.5) \tag{5.7}$$

Most wastewater applications are at neutral pH, so the theoretical dosage requirement is 1 part of peroxide per part of sulfide. However, like other oxidant chemicals, peroxide reacts with organic material in the wastewater, so higher dosages are required. For some applications, successful treatment has been reported at a peroxide-to-sulfide ratio of 2:1, but other applications have required ratios as high as 4:1 (Van Durme & Berkenpas, 1989). Peroxide is fast acting, which makes it useful for addition immediately upstream of problem locations. However, it is also quickly consumed and, therefore, multiple injection sites are required to treat long reaches of collection systems.

For collection system applications, peroxide has been successfully applied in both force mains and gravity lines. The decomposition of excess peroxide to water and oxygen increases the dissolved oxygen concentration of wastewater and produces no chemical residue. Peroxide can reduce sulfides to near zero, depending on the dosage ratio, and peroxide reactions yield no byproducts of concern.

Potassium permanganate ($KMnO_4$) is a strong chemical oxidizing agent that reacts with hydrogen sulfide (H_2S) according to the following equations:

$$3H_2S + 2KMnO_4 \rightarrow 3S + 2H_2O + 2KOH + 2MnO_2 \text{ (Acidic pH)} \quad (5.8)$$

$$3H_2S + 8KMnO_4 \rightarrow 3K_2SO_4 + 2H_2O + 2KOH + 8MnO_2 \text{ (Basic pH)} \quad (5.9)$$

In actual practice, several reactions ranging between these two may take place to produce elemental sulfur, sulfate, thionates, dithionates, and manganese sulfide, depending on local wastewater chemistry. For this reason, field studies have indicated that between 6 and 7 parts potassium permanganate are required for each part sulfide to be oxidized (U.S. EPA, 1985).

Potassium permanganate is available in dry crystal, granule, or pellet form and must be mixed with water to approximately a 3 to 4% solution before use. When kept dry and cool, potassium permanganate is relatively stable. However, when contaminated with organics or acids, it can become unstable and decompose, causing potentially hazardous conditions.

7.2 Biological Treatment Methods

Nitrate addition controls dissolved sulfide by two different reaction mechanisms or modes—prevention and removal. In the prevention mode, nitrate is added to fresh wastewater to be used as a substitute source of oxygen. The facultative and obligate anaerobic bacteria, which are responsible for odor and sulfide generation, use dissolved oxygen, nitrate, and sulfate as oxygen sources, in that order of preference. Typically, dissolved oxygen in wastewater is rapidly depleted and there is little nitrate present. Sulfate is typically abundant, so the bacteria reduce sulfate to sulfide, which causes odor and leads to corrosion problems. When nitrate is added to wastewater, the bacteria use it instead of sulfate. This results in the production of nitrogen gas and other nitrogenous compounds rather than sulfide.

In the removal reaction, nitrate can be added to septic wastewater to remove dissolved hydrogen sulfide from wastewater by a biochemical process, which converts the sulfide to sulfate. The nitrate supplies oxygen to bacteria (likely *Thiobacillus denitrificans*) present in wastewater to metabolize hydrogen sulfide and other reduced sulfur compounds.

Nitrate can be used to remove preexisting dissolved sulfide. The product is typically an aqueous solution of calcium nitrate containing 0.42 kg/L (3.5 lb/gal) of nitrate-oxygen. The solution can be tailored to different applications and may contain sodium nitrate, which is more compatible with certain industrial wastewater.

In untreated wastewater, a carbon source (BOD) is consumed via an anoxic denitrification reaction. Assuming a single carbon source (methanol) for simplicity, the reaction is as follows:

$$SO_4^- + 4CH_3OH \rightarrow S^- + 4H_2O + 2CH_4 + 2CO_2 \quad \text{[Untreated reaction]} \tag{5.10}$$

In this process, 0.25 parts of sulfate are used and 0.25 parts of sulfide are produced for every mole of carbon consumed.

Nitrate prevents the formation of sulfide by acting as a preferential oxygen source to sulfate. Again, assuming a single carbon source for simplicity, the reaction is as follows:

$$6NO_3^- + 5CH_3OH \rightarrow 5CO_2 + 3N_2 + 7H_2O + 6OH^- \quad \text{[Prevention reaction]} \tag{5.11}$$

In this process, 1.2 parts of nitrate are used and zero parts of sulfide are produced for every mole of carbon consumed. The nitrate requirement per unit of sulfide prevented is expressed as follows:

$$(1.2 \text{ parts } NO_3^-/\text{mole C})/(0.25 \text{ parts } S^-/\text{mole C}) = 4.8 \text{ parts nitrate per mole sulfide} \tag{5.12}$$

On a mass basis, 9.3 kg of NO_3 (or 7.2 kg of NO_3^-O) are needed per kilogram of sulfide prevented. For a solution containing 0.42 kg/L (3.5 lb/gal) of nitrate oxygen, this translates to 17.2 L/kg sulfide prevented (2.1 gal/lb sulfide prevented). This represents the minimal stoichiometric requirements for prevention given the simplest form of organic matter; the actual dosage rates tend to be higher.

The removal mechanism uses naturally occurring bacteria to biochemically oxidize dissolved sulfide in the presence of nitrate. The dissolved sulfide may be generated upstream of the nitrate application point or may enter the line downstream through a lateral branch. Nitrate addition causes the biochemical oxidation of sulfide according to the following reaction:

$$8NO_3^- + 5H_2S \rightarrow 5CO_2 + 5SO_4^- + 4N_2 + 4H_2O + 2H^+ \quad \text{[Removal reaction]} \quad (5.13)$$

This reaction occurs in the bulk flow and in the outer zone of the slime layer. Nitrate is not added in sufficient quantities to fully saturate the slime layer, so sulfide production continues in the inner zone. The sulfide is removed when it reaches the outer zone or the bulk flow. In this reaction, 1.6 parts of nitrate are used for every mole of sulfide removed. On a mass basis, this requires 2.4 kg of NO_3^-O per kilogram of sulfide removed. For a solution of 0.42 kg/L (3.5 lb/gal) of nitrate oxygen, this requires 5.7 L/kg sulfide removed (0.7 gal/lb sulfide removed) (Hunniford, 1990). (It should be noted that the removal reaction is a biochemical process, so sulfide reduction is not instantaneous and a reaction time of 1 to 2 hours may be required for optimal effectiveness.)

Nitrate has the advantage of being one of the safest of all the sulfide control chemicals to handle. Generally, nitrate salt solutions are considered to be nonhazardous substances and are not included on either federal U.S. EPA or state Conservation Environmental Response, Compensation, and Liability Act (CERCLA) lists. Standard nitrate solutions are also exempt from federal U.S. Department of Transportation placard requirements. There are two potential considerations when using nitrate salts. First, nitrate is commonly supplied as calcium nitrate. During normal operation, there are rarely concerns with precipitation of solids or scaling. However, when calcium nitrate is added at sites with high fats, oils, and grease levels, such as downstream of dairy production facilities, it can lead to calcium salt accumulations that can increase scum concentrations. The scum can clog piping or accumulate in pump station wet wells. In these cases, alternatives such as sodium nitrate should be considered. Second, depending on where the nitrates are being applied, they can interfere with treatment facilities that are practicing biological phosphorus removal by inhibiting the required anaerobic activity within the process reactors.

A bacterial metabolism modifier product that contains a form of anthraquinone has been shown to inhibit the reduction of sulfate to sulfide by sulfate-reducing bacteria (SRB). A key step in the metabolic process occurs in the cytoplasmic membrane of the SRB at the cytochrome c3 location. The chemical inhibits only SRBs by interfering with their respiration system. Anthraquinone interacts with cytochrome c3 and specifically disrupts the sulfate reduction process. When the SRB are unable to carry out sulfate reduction, they switch to an alternate metabolic process (fermentation), which does not produce sulfide.

The anthraquinone product must come in direct contact with the sulfate-reducing bacteria that typically thrive in the slime layers on pipe walls or sediment buildup in gravity lines, wastewater tankage, or lagoons. Therefore, it is effective in treating small-diameter pipes and force mains where the ratio of surface area slime to water volume is fairly high or in wastewater conveyance or treatment processes where septic sediment is a problem. The dosage is dependent on pipe surface area or degree of sediment and not sulfide concentration, so the product is most cost-effective for small lines with high sulfide concentrations, slow-moving gravity lines, or unaerated wastewater tankage and lagoons.

The anthraquinone product can be used by itself when only prevention of sulfide is needed. The two basic methods of application include slug dosing and automated feed. Slug dosing is appropriate for applications such as small diameter–long detention time force mains. When large doses are added, the slime layer/sediment becomes saturated with the product and SRB are inhibited. The SRB remain inhibited until they slough off and are replaced with new growth. Depending on how completely the slime layer can be contacted, sulfide may be inhibited for 10 to 30 days. The supplier recommends a maintenance dosage of approximately 0.24 kg of the product per 1000 m^2 of surface area (0.05 lb/1000 sq ft). The anthraquinone product is environmentally safe and contains no hazardous substances included on federal U.S. EPA or state CERCLA lists.

Bacteria and enzyme products that have been promoted to control odors and reduce sulfide consist of bacterial cultures, enzymes, other bacterial stimulants, or combinations of these. The bacteria are typically touted as proprietary strains of microbes that have been developed in the laboratory. The proprietary bacteria are intended to replace the naturally occurring bacteria in the slime on the pipe walls by competitive exclusion. The proprietary bacteria are said not to generate sulfide, and product literature from one vendor shows reactions that use oxygen or nitrate. As previously noted, the naturally occurring facultative bacteria prefer oxygen, nitrate, and sulfate, in that order. Collection systems typically have little or no dissolved oxygen or nitrate present, so it is unclear how the proprietary bacteria will survive. If dissolved oxygen or nitrate were present, there would not likely be a sulfide problem in the system.

7.3 Iron Salts

Iron and other metals can chemically combine with dissolved sulfide to form relatively insoluble precipitates. The iron salt precipitates are in the form of black or reddish-brown floc that does not deposit in the collection system, but readily settles with other solids at the treatment facility. Iron salts are in widespread use in conveyance systems throughout the United States. Other metal salts such as chromium, copper, and zinc also react with sulfide, but these are regulated heavy metals. High heavy metal concentrations make biosolids hazardous and disposal difficult and costly. Zinc was once used extensively because it would reduce sulfide concentrations to near zero (U.S. EPA, 1985). This practice has been discontinued because zinc is now a regulated heavy metal.

Both ferrous and ferric metal salts can react with dissolved sulfide. The following four types of iron salt solutions are commercially available: ferrous sulfate, ferrous chloride, ferric sulfate, and ferric chloride. The addition of sulfate-based salts has been questioned because sulfate can be reduced to sulfide. However, this is generally not a concern in municipal wastewater systems because sulfate is typically present in excess and sulfide generation is not significantly increased. Sulfate sources for wastewater include drinking water, industrial discharges, and hydrolysis of sulfur-containing organic wastes. Most commercially available iron salts are high-quality products with minimal contaminants, but purchasers should request analytical data from the supplier.

The reaction of ferrous (Fe^{++}) salts with dissolved sulfide is the same for both ferrous solutions because the anionic carrier ion (sulfate or chloride) does not enter into the reaction. A variety of iron-sulfide complexes may be formed, but the simplest reaction is as follows:

$$Fe^{++} + HS^- \rightarrow FeS + H^+ \quad (5.14)$$

For this reaction, the stoichiometric amount of iron required is 1.6 kg Fe^{++}/kg sulfide. Ferric (Fe^{+++}) salts react with dissolved sulfide in much the same way as ferrous salts, according to the following reaction:

$$2Fe^{+++} + 3HS^- \rightarrow Fe_2S_3 + 3H^+ \quad (5.15)$$

In practice, a number of different ferric sulfide species are formed, but the result is a black precipitate of sulfide. The stoichiometric amount of iron required is 1.1 kg Fe^{+++}/kg sulfide.

Ferrous sulfate ($FeSO_4$) is produced in the mining industry as a coproduct with titanium dioxide. Iron and titanium are found in the same ore, and iron is selectively removed by dissolving the ore in sulfuric acid and precipitating it as ferrous sulfate heptahydrate crystals ($FeSO_4 + 7H_2O$). It may also be produced during steel pickling with sulfuric acid (the spent acid is called "pickle liquor"). Solutions produced from crystals can have a pH of 2.0 or greater if free acid is controlled. A pH of 2.0 or greater classifies the material as noncorrosive according to the Resource Conservation and Recovery Act (RCRA). Ferrous sulfate is available in various dry forms (including ferrous sulfate monohydrate), which require mixing with water before use. Dry ferrous sulfate has not commonly been used for sulfide control in collection systems since it requires mixing and handling to prepare the solution for use.

Ferric sulfate ($Fe_2[SO_4]_3$) is produced by reacting natural iron-rich ores with sulfuric acid or oxidizing-spent pickle liquor. Ferric sulfate is commonly used in both water and wastewater as a coagulant and settling aid. The orange-brown liquid is typically a 45 to 50% solution with approximately 10% iron as Fe. The specific gravity is 1.44, so there is approximately 0.14 kg Fe/L (1.2 lb Fe/gal) The pH is less than 1.0, so the material is classified as a corrosion hazard according to RCRA. The crystallization temperature (freezing point) of 60% ferric sulfate solution is −18 °C (0 °F); therefore, insulation is typically not required except in extremely cold climates. Otherwise, the storage and feed requirements are similar to those for ferrous sulfate.

Ferrous chloride ($FeCl_2$) can be produced by pickling steel or processing natural ores with hydrochloric acid. Ferrous chloride has a high solubility with solution strengths of 18 to 28. A solution of 23% contains approximately 10% iron as Fe by weight. The specific gravity is 1.23, so there is 0.12 kg Fe/L (1.0 lb Fe/gal). The light-green colored solutions typically have a pH of less than 1.0; thus, they are classified as a characteristically hazardous material. Ferrous chloride is in widespread use for odor control applications because it is often the least costly iron salt chemical. The crystallization temperature (freezing point) of 37% ferrous chloride solution is −26 °C (−15 °F), so insulation is typically not required except in extremely cold climates.

Ferric chloride is applied as a coagulant in water treatment facilities, so high-purity products are typically available. Liquid iron (III) chloride ($FeCl_3$) is an orange-brown solution that is acidic and corrosive to metals. Solutions range from 37 to 42% $FeCl_3$, by weight, with the standard being 40%, which is 13.8% total iron as Fe. The specific gravity of the standard solution is 1.43, so there is 0.2 kg Fe/L solution (1.6 lb Fe/gal). The crystallization temperature (freezing point) of 40% ferric chloride solution is 10 °C (14 °F) so insulation may be advisable in some extremely cold regions.

Iron salts are fast acting and are often applied just upstream of a treatment facility to remove sulfide before the headworks facilities. Greater benefits in odor and corrosion control are obtained when iron is added upstream in the collection system. Iron salts do not react with organic material in wastewater; therefore, they can be overdosed at one upstream location to treat long reaches of a collection system (i.e., if concentrations are not sufficient to accelerate corrosion). The iron precipitate settles rapidly in a quiescent basin, but, in the collection system, it remains suspended in the flow and does not form deposits. The iron precipitate adds to the overall solids production at the treatment facility, with the volume dependent on the amount of sulfide treated. Even in systems with high sulfide concentrations, the added solids are typically less than 5% of the overall solids.

Iron salts are typically added to the gravity main portion of wastewater systems for easier application. Injection to force mains is not typically practiced because of the requirement for high-pressure metering pumps. For force main treatment, iron salts can be added directly to a pump station wet well or to the gravity main entering the station. When long reaches are to be treated and a large quantity of iron salts is needed, the injection site should have adequate wastewater flow at all times to dilute the chemical and avoid corrosion problems. The chemical piping should direct the solution to the wastewater flow without contacting concrete or metal surfaces in manholes or wet wells.

7.4 Peroxide Regenerated Iron

The use of iron salts for sulfide control is quite common. Enhancements are available that facilitate regeneration and reuse of the reacted iron (ferrous sulfide [FeS]) multiple times throughout the wastewater collection system and on to the treatment facility. The iron regeneration process is called PRI-SC® by US Peroxide (PRI-SC stands for "peroxide regenerated iron for sulfide control"). The application uses hydrogen peroxide (H_2O_2) to oxidize the sulfide from ferrous sulfide (FeS) to elemental sulfur, thereby freeing up the iron as Fe^{2+} and/or Fe^{3+}. The regenerated ferrous or ferric iron is now available to bind additional sulfides as they are generated downstream.

An advantage of the PRI-SC process is that less hydrogen peroxide is required for direct sulfide oxidation from ferrous sulfide than it would take additional iron to bind the same amount of sulfide. The process provides potential cost savings over iron usage alone and has an added benefit of reducing the total amount of iron solids in the WRRF sludge reducing treatment and handling/beneficial reuse costs and helping to control hydrogen sulfide in the digester gas at the WRRF.

The City of San Diego has successfully used PRI-SC for over 10 years. At the Point Loma Wastewater Treatment Plant, the iron from its main influent pump station is regenerated a second time at the headworks to assist with chemically enhanced primary treatment that has been achieving 89% TSS removal and 65% BOD removal in addition to assisting with regulatory compliance for control of digester gas hydrogen sulfide concentrations below 40 ppmv at the Point Loma WRRF (Nunez et al., 2010).

7.5 Oxygenation

The addition of pure oxygen gas to improve the oxygen balance of wastewater and prevent odors is used in the United States and worldwide. Oxygen has a significant advantage over air when added to wastewater because it is 5 times more soluble in water. This means that a smaller volume of gas is required to achieve the same oxygen transfer and the possibility of gas pocket formation is lessened. Reductions in BOD have also been reported following installation of oxygen injection equipment. The oxygen can either be generated on-site for requirements greater than 900 kg O_2/d (2000 lb O_2/d) or purchased commercially and delivered by truck for lesser quantities. The use of aeration (not oxygenation) is not recommended and is not covered in this section.

The equipment required for on-site storage and application includes a specialty steel, double-walled containment pressure vessel for liquid oxygen storage, evaporator–pressurizer, control valve, pressure regulator, flow meter (rotameter), piping, and injectors. This equipment can either be purchased or leased from oxygen suppliers. Liquid oxygen requires a sophisticated containment vessel to maintain the 1730-kPa gauge (250-psi gauge) pressure required to keep it in the liquid state. When the controls call for oxygen delivery, a control valve on the tank opens and releases liquid oxygen to the evaporator. The evaporator is a pressurized, radiator-like device that raises the temperature of liquid oxygen, causing it to vaporize. When the oxygen vaporizes, it creates its own pressure; thus, compressors are not required. Another control valve and regulator at the discharge end of the evaporator meters the vaporized oxygen to the point of application.

Oxygen typically is injected directly to the pump discharge to fully benefit from the pressure and turbulence. Typically, a signal from the pump controls opens the oxygen control valve so that oxygen is only injected when wastewater is flowing. Because of the intimate mixing provided in the pump discharge, many injectors are simple threaded connections, flush mounted with the inside diameter of the pipe. Others are perforated stainless steel lances that protrude into the wastewater flow. Lance injectors that protrude into the bottom half of the pipe have been known to accumulate rags and debris, which decrease efficiency and increase maintenance. Injectors should be located in an easily accessible area for maintenance and repair.

Because of high pressure and turbulence in the force main, wastewater becomes supersaturated with added oxygen. When the force main is not operating, local conditions cause pressure drops, and some of the oxygen may come out of solution. This is a particular problem when vacuum-release valves are used. As the wastewater reaches normal atmospheric pressure inside the pipe, bubbles of oxygen-rich gas develop and coalesce at high points in the main. If air-release valves are present at these locations, a significant amount of oxygen can be lost from the system. Elevated pipeline crossings of streams, rivers, highways, and canals where air- and vacuum-release valves are common should be considered when deciding to use oxygen for force main injection.

El Paso Water Utilities installed a super oxygenation system upstream of their Northwest Wastewater Treatment Plant to control hydrogen sulfide odors and corrosion. Approximately 90% of the plant's 38-ML/d (10-mgd) influent average daily flow is conveyed through a 5.6-km (3.5-mile) dual force main. The super oxygenation system was commissioned in 2015 at the pump station upstream of the dual force main, and average hydrogen sulfide at the dual force main discharge was reduced from 12 ppmv (with calcium nitrate treatment) to 1 ppmv without calcium nitrate treatment (Clidence, 2015). At another location in the Town of Milford, Connecticut, a super oxygenation system was installed at the Gulf Pond WWPS upstream of the West Ave. pump station that discharges approximately 40% of the flow to the 45 000 m^3/d (12 mgd) Housatonic Wastewater Treatment Plant. The Gulf Pond WWPS discharges into dual 18-in. force mains that are 13 200 ft in length. The super oxygenation system was commissioned in 2011. The West Ave. WWPS that receives discharges from the Gulf Pond WWPS had wet well hydrogen sulfide concentrations as high as 900 ppmv before installation of the super oxygenation system. After startup, peak hydrogen sulfide concentrations were reduced to below 10 ppmv (Smith, Bradstreet, & Bowker, 2012). (See Figure 5.7 for a typical super oxygenation system.)

7.6 Adjustment of pH

There are two different approaches to controlling hydrogen sulfide by adjusting the pH of wastewater. One approach involves continuous pH adjustment to hold hydrogen sulfide in solution. The other approach involves using

FIGURE 5.7 Typical super oxygenation system (Clidence, 2015).

intermittent slug doses of caustic to inactivate the slime layer and minimize sulfide generation.

The objective of continuous pH adjustment is to prevent the release of hydrogen sulfide gas by shifting the sulfide equilibrium to hydrosulfide and dissolved sulfide species, which remain in solution. At pH 7, approximately 50% of all sulfide exists as hydrogen sulfide and 50% exists as the dissolved hydrosulfide ion. At a pH of 8, only 10% exists as hydrogen sulfide and 90% as hydrosulfide. A change in pH of 0.5 units can have a significant effect on the amount of hydrogen sulfide emitted to the atmosphere.

In typical collection system applications, magnesium hydroxide $Mg(OH)_2$ is continuously added to raise and buffer the pH of the wastewater to a range of between 8.3 and 8.6. At this elevated pH range, the gas phase hydrogen sulfide equilibrium shifts toward the liquid phase. Additionally, at this pH range, the sulfate-reducing bacteria that produce dissolved sulfides become less productive. However, $Mg(OH)_2$ does not treat the existing dissolved sulfide, but merely holds it in solution. Therefore, dilution of the flow from branch lines may cause a decrease in pH at a downstream location, which would allow hydrogen sulfide to escape from the liquid. There are some formulations available that combine both magnesium hydroxide and nitrate salts. These can help treat the hydrogen sulfide, rather than holding it in solution by elevating the pH of the wastewater.

Magnesium hydroxide dosage is dependent on the wastewater flow and not on the concentration of the dissolved sulfide in the system. Typical dosing rates range from 50 to 100 L of $Mg(OH)_2$ per million liters of wastewater flow. Because the dosing rate is flow dependent, $Mg(OH)_2$ addition becomes economically attractive for systems with dissolved sulfide concentrations greater than 5 to 10 ppmv. Magnesium hydroxide chemical is supplied by one vendor as a nonhazardous 63% slurry, but still requires gloves, goggles, and an approved respirator. U.S. Department of Transportation placards are not required for tank trucks or storage tanks and secondary containment is not required. Storage tanks should be equipped with mixers to supply continuous agitation to prevent settling. Tanks must be heated if subject to freezing. Mild steel or plastic tanks are acceptable. Positive displacement pumps such as progressing cavity, diaphragm, or peristaltic are recommended.

Periodic slug dosing with caustic can effectively remove all sulfide forms. It is not added to shift the equilibrium, but to inactivate or kill the biological slime layer that reduces sulfate to sulfide. Exposure to high pH levels will destroy the slime layer and cause it to slough. The slime layer will immediately begin to re-form, but it may take days or weeks to reach full sulfide production again. The time required for the slime layer to grow back after slug dosing is a function of pH, temperature, and time of contact. For treatment of gravity sewers, the liquid sodium hydroxide solution (caustic) is added directly to manholes upstream of the area to be treated. Slug dosing can also be applied to force mains by adding it at the pump station. However, care must be taken to prevent damaging pumps and valves that might not be designed for elevated pH levels.

The alkali release rate must be controlled at a rate calculated to provide the necessary pH adjustment based on the flow in the area where sulfide controls are required. This prevents the waste of chemical and maximizes contact time. In general, raising the pH to 11 to 12 for a period of 30 to 60 minutes has been demonstrated to inactivate the slime layer for at least several days and possibly up to 2 weeks. Sodium hydroxide may be more desirable than $Mg(OH)_2$ or calcium hydroxide for slug dosing because the rapid reaction and pH increase that occurs is desirable.

An aqueous solution of sodium hydroxide containing 25%, by weight, sodium hydroxide (NaOH) has a crystallization temperature (freezing point) of −18 °C (0 °F). An aqueous solution of sodium hydroxide containing 50%, by weight, sodium hydroxide (NaOH) has a crystallization temperature (freezing point) of 14 °C (58 °F). More dilute caustic solutions should be used in colder climates to avoid freeze protection.

Caustic slugging near treatment facilities can adversely affect pH sensitive treatment processes and effluent discharge limits because of locally elevated pH levels. If equalization facilities are available, they can be used to store the high-pH wastewater for slow release to the facility. If these facilities are not available, acid can be used for neutralization, but this adds to the expense. Some facilities have reported that pH levels as high as 8.5 have not adversely affected the secondary treatment process.

8.0 VAPOR-PHASE TREATMENT

Liquid-phase treatment can significantly reduce sulfide formation and release in collection systems; however, in highly turbulent locations, trace amounts of sulfide can yield hydrogen sulfide concentrations high enough to cause odor and corrosion problems. Headspace ventilation can be used to mitigate corrosion, but if homes or businesses are located nearby, then vapor-phase treatment may be needed to control odor.

As with covered reclamation facility processes, wastewater conveyance processes need to maintain a slightly negative headspace pressure with respect to atmosphere. This ensures the air flows down into the pick-holes in manhole covers and plumbing vents rather than out. Although each location has unique characteristics, as a general rule for manhole covers with pick-holes (perforations), the sewer headspace pressure should be maintained so that it is at least 12 pascals (0.05 in. H_2O) less than atmospheric pressure because this should be enough to resist most fluctuations.

In reaches where corrosion-resistant materials have been used for pipelines and manholes, vapor-phase treatment of problem areas may be necessary in conjunction with liquid-phase treatment of a large flow volume. In such cases, untreated, the hydrogen sulfide concentrations may exceed 100 ppmv; therefore, care must be taken to select a suitable vapor-phase treatment device. It is now understood that even the "corrosion"-resistant materials are susceptible to acid corrosion. Vapor-phase treatment systems are discussed in detail in succeeding chapters. This section provides a general discussion of the applicability of various control devices for collection systems.

Most interceptors less than 1800 mm (72 in.) in diameter have relatively low airflow rates of 140 m^3/min (5000 cfm) or less, so vapor-phase treatment units are typically small in size. For collection system applications situated close to homes, special consideration should be given to noise reduction measures for fans and, in some cases, pumps. Tall stacks are desirable from a dispersion standpoint, but shorter stacks may be necessary to minimize the visual effect in a neighborhood. Because of the proximity to homes, the treatment devices selected must provide high-efficiency odor control, particularly in locations with high inlet hydrogen sulfide.

8.1 Biological

In-ground or above-ground soil or inorganic media biofilters and biotrickling filters with engineered synthetic media have emerged in the last 20 years as the odor control process of choice for conveyance systems because they provide a robust and cost-effective solution. In addition, they avoid the respirable bio-aerosols from carbon beds or the chemical exposure risk from wet scrubbers. In fact, with solar power with battery storage capabilities and the potential to produce recycled water on-site, biofilters could potentially be placed virtually anywhere along the conveyance system—even in areas without access to utilities.

Biofilters and biotrickling filters require lower face velocities than wet scrubbers and carbon units, so they occupy a larger footprint than these other two processes when considering the vessels alone. However, both have lower life cycle costs and are robust and reliable, similar to carbon or chemical scrubber systems. These can require as many or more frequent inspections, monitoring, and maintenance compared to a biofilter or biotrickling filter. Also, when compared to wet chemical scrubbers, one must consider that biofilters will not require space for chemical storage and feed systems or handling of one or more hazardous chemicals, electrochemical pH and

ORP sensors requiring regular calibration will not be required, roadways designed for large chemical delivery trucks will not be required, and truck traffic associated with chemical deliveries to the site will not occur.

Positive features of in-ground soil or inorganic biofilters include their suitability to include landscaping to screen the fan and electrical equipment and the potential to install these items in separate below-grade ventilated vaults so they can be made to be virtually invisible to the public.

Although at-grade emissions from in-ground soil or inorganic media biofilters may not have ideal dispersion characteristics, this means that conventional open-top biofilters may be precluded at some sites where space is not available. The open-top units also have poor dispersion characteristics because of the low velocity ground-level discharge. In some applications with some biofilter media, biofilters may have their own intrinsic earthy odor that may be detected by nearby receptors. Therefore, selecting an experienced vendor who understands and can warranty performance is recommended for sensitive areas.

Above-ground soil or inorganic media biofilters and biotrickling filters can be difficult to screen from view. Therefore, their application requires careful review if visual aesthetics are an issue at the collection system location where they will be installed. (See Chapter 8 for detailed information related to the design of vapor-phase biological air treatment systems.)

8.2 Chemical

Packed tower scrubbers have been effectively used for numerous collection system applications and have demonstrated good performance. They offer the advantage of high-efficiency removal in a compact footprint. Several manufacturers produce low-profile package systems that are ideally suited for installation inside existing pumping stations. If floor space is not available, the units can be installed outdoors, but a building addition is generally recommended to address security concerns.

One of the main disadvantages of wet scrubbers for collection system applications is the use of chemicals. Chemical storage tanks are needed for both caustic and hypochlorite chemicals and containment must be provided. The tanks are typically housed in a building or screened enclosure for security. Although the scrubbers are small, a large amount of site space may be needed as driveway access for chemical delivery trucks.

Smaller delivery trucks may be desired in neighborhoods to minimize the effect of larger vehicles on residential streets; however, smaller vehicles will necessitate more frequent deliveries, which may also be of concern. Wet scrubbers require considerable maintenance for servicing pumps and calibrating probes, which is difficult for operators to attend to in remote locations away from the treatment facility. Wet scrubbers require a potable water supply, which may not be available at some remote sites, for makeup water in the scrubbing solution. (See Chapter 9 for detailed information related to the design of vapor-phase chemical scrubber systems.)

8.3 Adsorption

Activated carbon units are well suited for many collection system applications where the flow-weighted average concentration of hydrogen sulfide is less than 5 to 10 ppmv. Conventional carbon units are slightly larger than wet scrubbers, but they do not require chemicals; therefore, the overall footprint is about the same. A 3.7-m (12-ft) diameter dual-stage unit can handle several hundred cubic meters per minute (m^3/min) (several thousand cubic feet per minute [cfm]), which is a large enough capacity for most collection system applications.

Truck access is necessary for carbon media replacement, but truck traffic is minimal. Carbon units are small enough to be installed indoors at existing pump stations, but they can also be installed outdoors with less concern for security because they have no chemical tanks. Carbon units require much less operator attention than scrubbers, so this makes them desirable for remote locations. Carbon inserts are available for manholes, but the pressure drop required for air to traverse the filter generally renders them less useful than a manhole rain-stopper (seal).

One of the main limitations of activated carbon is rapid media depletion at high hydrogen sulfide concentrations. Virgin carbon media is best suited for low hydrogen sulfide combined with odorous VOCs. Metal-oxide enriched carbon media (where metal-oxides like copper (II) oxide [CuO], magnesium oxide [MgO], zinc oxide [ZnO] are included before activation) is specially tailored to increase hydrogen sulfide capacity, but this limits the VOC removal capability and can be problematic with significant condensation. Caustic impregnated carbon can be chemically regenerated in situ, but many operators prefer replacement to regeneration because of concerns about handling the strong chemical base.

Water-washable (or "water-regenerable") media is available that has about one-half the hydrogen sulfide capacity of caustic impregnated media, but can be safely regenerated using water. When the water-washable media is used in a conventional deep-bed carbon system, the water regeneration is time consuming and the unit may be out of operation for several days. To address this concern, the vendor developed a replaceable canister system. The canisters are arranged in several banks, which are washed sequentially so that only one bank is offline at a time and the unit remains in service. The washing operation is fully automated for ease of operation and maintenance.

The canister system is well suited to collection system applications because of its small size and low maintenance. The compact polypropylene unit can be installed indoors or outdoors. A source of potable water is required for washing. One design concern for collection systems is low pH discharge liquid. At treatment facilities, there is enough flow to dilute this discharge; however, in collection systems, neutralization may be required in upstream reaches to avoid damage to pipelines during periods of low flow. A low cost neutralization system can be provided to address these concerns.

The carbon media in the canister system has been shown to remove some VOCs; however, for applications where high-efficiency VOC removal is required, follow-up treatment with virgin carbon may be necessary. (See Chapter 10 for detailed information related to the design of vapor-phase adsorption air treatment systems.)

8.4 Other

Ionization is a technology that uses ion tubes with alternating current to produce negative and positive ions and form clusters of oxygen molecules. It represents another successful European technology now being applied in other areas such as the United States.

The ionized air disassociates a variety of gaseous compounds and has been shown to effectively treat low concentrations hydrogen sulfide, dimethyl disulfide, ammonia, and a wide range of other odorous compounds. Benign end products such as carbon dioxide, nitrogen, and water vapor are formed. By producing balanced levels of positive and negative ions, the particulates remain suspended and pass through the air return ducts where they are removed by mechanical filtration, so dust collection on surfaces is minimized. The ionization system may not produce unsafe ozone concentrations and has been used in hospitals and offices for air purification (Tulenko, 2001).

Vapor-phase treatment is typically applied to treat exhaust air before discharge. However, some ionization systems treat the supply air that is discharged only into the odorous area. In some other applications, a two-stage approach is used where a polishing ionization reactor is added before the air is discharged outside of the odorous area. Reaction time is obtained within a room, under a cover, or within a wet well.

One method of wet well treatment involves injecting just enough ionized air to form a layer or cap above the foul air. This method requires less ionized air because only the foul air that travels upward during fill events is treated. If fill events are rapid, an in-duct plenum chamber can be installed to allow additional reaction time.

Ionization technology is economical compared to other options. The ionization tubes require cleaning every 2 to 3 months, but this is not time-consuming for smaller systems. For higher volume systems with a large number of tubes, the cleaning process can be staggered so that all of the tubes do not require cleaning at the same time.

Ionization systems are relatively compact and require minimal operator attention, so they may be well suited for collection system applications provided that hydrogen sulfide concentrations are low enough to avoid escaping untreated or unreacted. Pilot testing of ionization systems is recommended to verify performance at locations that have relatively high hydrogen sulfide concentrations (>50 ppmv peak or 10 ppmv average) or for installations that will be located close to neighbors where adequate natural dispersion may not occur.

9.0 REFERENCES

American Society of Civil Engineers. (2007). *Gravity sanitary sewer design and construction*, 2nd ed. (Manual of Practice No. 60). American Society of Civil Engineers.

Apgar, D., Witherspoon, J., Easter, C., Bassrai, S., Dillon, C., Torres, E., Bowker, B., Corsi, R., Davidson, S., Wolstenholme, P., Forbes, B., Quigley, C., Ward, M., Joyce, J., Morton, R., Weiss, J., & Stuet, R. (2007). *Minimization of Odors and Corrosion in Collection Systems Phase 1*. Alexandria, VA: Water Environment Research Federation, and London, UK: IWA Publishing.

Clidence, D. (2015). *ECO2 Case study—El Paso Texas collection system odor control at Frontera Pump Station.*

Hunniford, D. J. (1990, Oct 7–11). *Control of odors and hydrogen sulfide related corrosion in municipal sewage collection systems using a biochemical process: BIOXIDE*. Presented at the 63rd Annual Water Pollution Control Federation Technical Exposition and Conference, Washington, D.C., United States. Alexandria, VA: Water Environment Federation.

Hunter, G. L., Baumgartner, G., Kobylinski, E. A., Esqueda, T., & Pronold, M. (1995, March 5–8). Rational approach to the establishment of local limits for total potentially toxic organic compounds. *Proceedings of Industrial Wastes Multimedia Pollution Control and Prevention Specialty Conference*, Pittsburgh, Pennsylvania, United States. Alexandria, VA: Water Environment Federation.

Hunter, G. L., & Lodderhose, J. (1993, December 5–8). Presented at the AWWA Joint Residuals Conference, Phoenix, Arizona. Denver, CO: American Water Works Association.

National Fire Protection Association. (2016). NFPA 820: Standard for Fire Protection in Wastewater Treatment and Collection Facilities. Quincy, MA: National Fire Protection Association.

Nayyar, M., King, R. C., & Crocker S. (2000). *Piping Handbook*, 7th ed. McGraw-Hill.

Nunez, C., Dornfeld, M., Shankles, K. C., Watson, I., Nguyen, L., & Prellberg, J. (2010). Cost savings and performance improvement of large system iron salt use for integrated sulfide control and chemically enhanced primary treatment by using peroxide regenerated iron technology. *Proceedings of the Water Environment Federation* (pp. 1110–1121). doi: 10.2175/193864710798158517

Permit-required confined spaces, 29 C.F.R. § 1910.146 (2004).

Pescod, M. B., & Price, A. C. (1982). Major factor in sewer ventilation. *Journal of the Water Pollution Control Federation, 54*(4), 385–397.

Smith, C. N., Bradstreet, K. A., & Bowker, R. P. G. (2012). Use of pure oxygen to control odors and corrosion and improve safety in the Milford, Connecticut Sewer System. *Proceedings of the Annual Water Environment Federation Technical Exhibition and Conference*, New Orleans, Louisiana, United States. Alexandria, VA: Water Environment Federation.

Sobhani, R., Reed, B., Velasco, M. P., Thomas, C., & Murthy, U. (2018). Orange County Sanitation District's proactive strategies for preventing odor complaints. *Proceedings of Water Environment Federation Conference on Odors and Air Pollutants*, Portland, Oregon, United States (pp. 45–56). Alexandria, VA: Water Environment Federation.

Thistlethwaite, D. K. E. (Ed.). (1972). *The control of sulfides in sewage systems*. Anne Arbor, MI: Anne Arbor Science.

Tulenko, D. (2001). *Air quality control Bentax™ air purification systems*. National Registry of Environmental Professionals Technical Conference, Arlington, Texas, United States. National Registry of Environmental Professionals and the International Society of Technical and Environmental Professionals.

U.S. Environmental Protection Agency. (1985). *Design manual, odor corrosion control in sanitary sewerage systems and treatment plants* (EPA-625/1-85-018). Washington, D.C.: U.S. Environmental Protection Agency.

U.S. Environmental Protection Agency. (1987). *Guidance manual on the development of local discharge limitations under the pretreatment program*. Washington, D.C.: U.S. Environmental Protection Agency.

U.S. Environmental Protection Agency. (1989). *Industrial user permitting guidance manual*. Washington, D.C.: U.S. Environmental Protection Agency.

U.S. Environmental Protection Agency. (1992a). *Model pretreatment ordinance*. Washington, D.C.: U.S. Environmental Protection Agency, Office of Wastewater Enforcement and Compliance.

U.S. Environmental Protection Agency. (1992b). *Guidance to protect POTW workers from potentially toxic and reactive gases and vapors* (EPA-812). Washington, D.C.: U.S. Environmental Protection Agency.

U.S. Environmental Protection Agency. (1998). *Cost accounting and budgeting for improved wastewater treatment* (B-92-001 [NTIS No. PB92-173-236]). Washington, D.C.: U.S. Environmental Protection Agency.

Van Durme, G. P., & Berkenpas, K. (1989). Comparing sulfide control products. *Operations Forum*, *6*(2), 12.

Water Environment Federation. (1994). Low emissions sewer systems for industry, Project 92-PRE-1. Alexandria, VA: Water Environment Federation.

Water Research Foundation. (2009). *Collection system ventilation*. Alexandria, VA: Water Research Federation.

Webster, N. E., Erstgaard, J., Parkins, P., & Koetter, B. (2018). Portable odor control—A solution to wastewater construction projects odor problems in Albuquerque. *Proceedings of Water Environment Federation Conference on Odors and Air Pollutants,* Portland, Oregon, United States (pp. 498–510). Alexandria, VA: Water Environment Federation.

Yao, K. M. (1976). Functional design of sanitary sewers. *Journal of the Water Pollution Control Federation*, *48*, 1772–1778.

6

Water Resource Recovery Facilities

1.0 ODOR CHARACTERIZATION 140
- 1.1 Background 140
- 1.2 Odor Derivation from Collection Systems 141
- 1.3 Odor Derivation from Water Resource Recovery Facilities 141
- 1.4 Mass Transfer of Odorous Compounds 141
  - *1.4.1 Favorable Mass Transfer Conditions: Turbulence* 142
  - *1.4.2 Favorable Mass Transfer Conditions: pH* 143
  - *1.4.3 Favorable Mass Transfer Conditions: Temperature* 143
  - *1.4.4 Favorable Mass Transfer Conditions: Seasonal and Diurnal Patterns of Wastewater Flows* 144
  - *1.4.5 Favorable Mass Transfer Conditions: Air Pressure* 145

2.0 WASTEWATER ODOROUS COMPOUNDS 145

3.0 ODOR EMISSIONS BY PROCESS UNIT 146
- 3.1 Preliminary Treatment 146
- 3.2 Primary Treatment 151
- 3.3 Biological Treatment 152
- 3.4 Intermediate and Final Settling 154
- 3.5 Chlorine Contact Tanks 154
- 3.6 Stabilization Ponds 154

4.0 PROCESS IMPROVEMENT AND ODOR MITIGATION 155
- 4.1 Odor Prevention 155
- 4.2 Physical System Layout 156

5.0 ODOR CAPTURE AND CONTAINMENT 158
- 5.1 Introduction 158
- 5.2 Odor Cover Alternatives 159
  - *5.2.1 Flat Covers* 162
  - *5.2.2 Barrel Arch Covers* 165
  - *5.2.3 Pitched Covers* 168
  - *5.2.4 Building Extension or Enclosure* 168
    - *5.2.4.1 Building or Penthouse* 169
    - *5.2.4.2 Dome Enclosure* 170
- 5.3 Odor Capture Alternatives for Bar Screens 171
- 5.4 Materials of Construction 175
  - *5.4.1 Concrete* 176
  - *5.4.2 Aluminum* 176

5.4.3 *Fiberglass Reinforced Plastic* 177
5.4.4 *Stainless Steel* 177
5.4.5 *Canvas* 180
5.4.6 *Air-Supported Canvas* 181
5.4.7 *Wood* 181
5.5 Access 181
5.6 Odor Containment 184
5.6.1 *National Fire Protection Association 820 Ventilation Rates* 184
5.6.2 *Standard Ventilation Rates* 186
5.6.2.1 *Nonworker-Accessible Space* 186
5.6.2.2 *Worker-Accessible Space* 189
5.6.2.3 *Additional Considerations* 190
5.6.3 *Ventilation System Layout* 191
5.6.4 *Conveying Odors* 193
5.6.4.1 *Duct Materials of Construction* 194
5.6.4.2 *Shapes* 195
5.6.4.3 *Sizing Criteria* 195
5.6.4.4 *Additional Ductwork Notes* 196
5.6.5 *Fans* 196
5.6.5.1 *Fan Positioning* 197
5.6.5.2 *Fan Balancing* 197
5.6.6 *Dampers* 198
6.0 CASE HISTORIES 199
6.1 Coney Island Water Resource Recovery Facility 199
6.2 North River Water Resource Recovery Facility 201
7.0 REFERENCES 206
8.0 SUGGESTED READING 206

1.0 ODOR CHARACTERIZATION

1.1 Background

Water resource recovery facilities (WRRFs) have historically been recognized by visitors, neighbors, transient passersby, politicians, the media, and so on as sites where nuisance odors are found. Taking into consideration the nature of the wastewater that is discharged by residential, commercial, and industrial facilities, it is clear that these facilities are eager to remove the wastewater from their homes/facilities and forward it to a location where somebody else can take care of it. All the wastewater from these facilities is combined and processed at the WRRF. The "age" of this wastewater, or the amount of time the wastewater remains in the collection system before reaching the facility, is directly related to the size and condition of the collection system network connected to the WRRF, the distance the wastewater must travel, and how it is conveyed (i.e., gravity, force main, or pressure sewer). The "age" of the wastewater is important because the longer it remains in the collection system, the greater the chance that it will be septic (*septic* is defined by the U.S. Environmental Protection Agency [U.S. EPA] as a condition produced by bacteria where all the oxygen supplies are depleted). If severe, bottom deposits and water turn black and give off foul odors; the more septic the wastewater, the higher the odor potential of the wastewater. In general, the more complex the collection system network and the greater the wastewater travel distance, the higher the odor potential of the wastewater entering the WRRF. Initially then, wastewater nuisance odors associated with the headworks are, in large part, a function of the characteristics of the influent wastewater and airspace above the wastewater of the collection system sewers.

1.2 Odor Derivation from Collection Systems

Accordingly, wastewater as received by WRRFs from dischargers through the collection system network is odorous, containing various compounds that have been identified in Chapter 2. The nature and intensity of the odors associated with the wastewater will be based on the wastewater origin, the conditions within the collection system, and the length of time it spends being conveyed to the WRRF. After reaching the WRRF, the odor potential of the wastewater can be reduced, maintained, or made worse by the way it is processed. This chapter will discuss operational and physical methods to mitigate odors from reaching off-site receptors and causing odor complaints.

1.3 Odor Derivation from Water Resource Recovery Facilities

Independent of the characteristics associated with the influent wastewater and collection system headspace air, conditions within the WRRF will also play a large role in defining the characteristics of the odors emitted. The unit operations in service can generate nuisance odors based on how they are operated. For example, maintaining aerobic conditions in the process units generally reduces the formation of reduced sulfur and nitrogen compounds, which generally have very low odor threshold concentrations. Another example is maintaining deep sludge blankets in primary and secondary sedimentation tanks or gravity thickening tanks. WRRF staff often try and balance the depth of sludge in these tanks with the desire to maximize thickening of the solids. Deep sludge blankets can encourage the formation of anaerobic conditions and, subsequently, reduced sulfur and nitrogen compounds. Accordingly, facility operations can have a direct effect on the nuisance odor emissions from a WRRF.

1.4 Mass Transfer of Odorous Compounds

Odorous compounds in the wastewater from the collection system and/or formed within the WRRF need to be released to the atmosphere before they can become a safety or nuisance odor problem to the staff or community.

Mass transfer is the mechanism whereby the odorous compounds are released by the wastewater into the air phase. The mass transfer of odorous compounds in the wastewater into the air occurs because of the gradient conditions that exist, (i.e., higher concentrations in the wastewater than in the air, and the need by nature to achieve equilibrium). The release of odorous compounds will continue until that equilibrium is reached. Ventilation at the wastewater surface will generally help to renew the air/liquid interface with fresh air, encouraging mass transfer to continue even as the wastewater concentrations of the odorous compounds diminish.

Henry's law predicts the maximum theoretical concentration of a compound in the gas phase as

$$P = Hp.$$

Where

P = mole fraction in gas phase,

p = mole fraction of dissolved gas in liquid phase, and

H = Henry's law constant.

In practice, concentrations of odorants such as hydrogen sulfide (H_2S) in the headspace of sewers or enclosed tanks rarely approach equilibrium values predicted by Henry's law because of many changing variables such as waste flowrates and characteristics, headspace ventilation rates, and temperature gradients. Typically, quiescent surfaces such as lagoons and clarifiers have relatively low emission rates per unit area, although increased wind velocities can increase emission rates from such sources. The highest emission rates typically occur under conditions of high turbulence or agitation (sewer drop structures, junction or splitter boxes, weirs, aerated grit chambers, sludge mixing operations, etc.).

1.4.1 Favorable Mass Transfer Conditions: Turbulence

Although the concentration gradient is the prime driving force behind the release of odorous compounds from wastewater, other physical and chemical conditions enhance that release. Wastewater turbulence is a physical condition that significantly increases the surface area that is exposed to the air and thereby over which mass transfer can occur. Elevated concentrations of odorous compounds are often found at surface-turbulent locations across the WRRF. As a result, one of the first steps explored by WRRFs when trying to mitigate nuisance odors is to investigate and identify turbulent areas around the facility and assess whether there are ways to reduce it.

Wastewater turbulence is often found at open drops into tanks/channels, hydraulic channel jumps, weir drops, inlet tank structures, diversion structures, trickling filter wastewater distribution arms, and so on. Reducing the degree of turbulence (e.g., lessening the height of the drop) can help lessen the potential odor release, particularly when the turbulent condition cannot be eliminated. Other physical mass transfer conditions that encourage nuisance odor release from wastewater include channel and tank mixing or agitation and diffused aeration in channels and tanks.

Other parameters that can influence the degree to which odors are released from the wastewater are pH, temperature, wastewater flows, and atmospheric pressure.

1.4.2 Favorable Mass Transfer Conditions: pH

Wastewater pH is a chemical condition that can significantly increase the release of certain compounds. For example, hydrogen sulfide is significantly more soluble in liquid/wastewater at elevated pH ranges, as discussed in earlier chapters. At a pH of 7.0 and less, most of the wastewater sulfides (50% and more) exist in solution as hydrogen sulfide in the gas state and will readily strip out of the liquid to the air; turbulence will further enhance this mass transfer.

1.4.3 Favorable Mass Transfer Conditions: Temperature

Seasonal odor variability is common at WRRFs and is more pronounced where the temperature has greater swings. The microorganisms that are the key driving force in forming odorous compounds have a higher activity level when the wastewater temperature is warmer (around 18 °C and above) and they will start slowing down as temperatures drop below 18 °C (64 °F). Microorganism activity level is a direct function of odor compound formation because the potential for anaerobic/anoxic conditions increases as available oxygen is used up. Under these conditions increases in odorous compound formation will occur. In an opposite scenario, as wastewater temperatures drop so will microorganism activity along with the formation of odorous compounds.

Simultaneously occurring at warmer/colder temperatures is the variability of the solubility of oxygen and the ability of odors to strip from the wastewater. The solubility of oxygen is greater at lower temperatures; for example, at 16 °C and 20 °C (61 °F and 68 °F), the saturation solubility of oxygen at standard sea level in pure water is 9.87 mg/L and 9.09 mg/L, respectively. So, as the microorganism activity begins to decline, the amount of oxygen that could be available is increasing, whereas when activity is at its greatest, the available oxygen is diminishing, thereby supporting the decline into anaerobic/anoxic conditions.

As temperature increases, the partitioning of volatile odorous compounds to the atmosphere increases; as the temperature decreases, the partitioning decreases. Temperature plays a key role in the formation and release of odorous compounds at a WRRF.

When reviewing the mass transfer of odors at WRRFs, consider that the wastewater-focused process units with exposed liquid surfaces to the atmosphere are more susceptible to the effects of temperature than solids-processing unit operations, which have less direct contact with the atmosphere. Accordingly, the liquid operations are more readily affected by air temperature than solids. In general, the influent and headworks areas of a WRRF will experience a decline in odors and hydrogen sulfide more rapidly than solids processing (gravity thickening, storage, dewatering, etc.).

1.4.4 Favorable Mass Transfer Conditions: Seasonal and Diurnal Patterns of Wastewater Flows

A unique feature about nuisance wastewater related odors both in the collection system and at the WRRF is their fluctuating diurnal pattern. WRRF nuisance odors, particularly at the headworks and preliminary treatment processes that are more heavily influenced by the influent wastewater flow, are characterized by diurnal trends that have peaks and valleys that often are closely related to the influent flow. Patterns develop that reflect the individual facilities' collection system operations. Figure 6.1 illustrates a

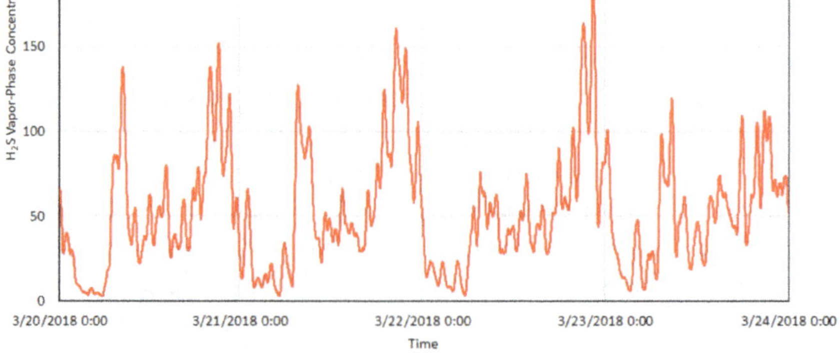

FIGURE 6.1 Diurnal hydrogen sulfide graphical representation.

common pattern observed that is a graphical representation from the continuous hydrogen sulfide monitoring of the air space of an influent wet well. The hydrogen sulfide pattern reflects an increase in the early morning hours when people are getting up and ready to go to work, a second but lesser increase around the midday lunch period, and a third in the early evening.

These three elevated hydrogen sulfide time periods are generally consistent with similar patterns in wastewater flows. Just how well these hydrogen sulfide and wastewater patterns may reflect conditions at a WRRF are a function of the components of the collection system. These diurnal patterns—however they may actually appear at a WRRF—are important to recognize and understand at a WRRF, especially when trying to determine how to implement actions to mitigate nuisance odors. Because of the importance of hydrogen sulfide as an odor indicator, although it is clearly not the only odorous constituent at a WRRF as previously discussed, and because the ability to monitor it on a continuous logging basis, the owner can better understand the more critical concentrations that need to be considered in odor control planning and the time of the day to monitor for additional odorous parameters.

It is these high hydrogen sulfide concentration diurnal patterns discovered through continuous monitoring that generally represent the higher odor loading at the facility, so that additional odor characterization monitoring should follow the trends when establishing an odor sampling plan. In addition, facility operating data should be reviewed and staff consulted to ensure that odor-specific samples are collected during facility conditions that are representative of normal facility operations. (See Chapter 4 for more information on odor source sampling.)

1.4.5 Favorable Mass Transfer Conditions: Air Pressure

The liquid wastewater streams in WRRFs are operated in an atmospheric environment, subject to the standard air pressures at the facility location. High or low air pressures are not part of the liquid treatment process. As a result, there are no practical situations where air pressure would affect the mass transfer of odorous compounds from the liquid to the air phase.

2.0 WASTEWATER ODOROUS COMPOUNDS

Odorous compounds associated with wastewater processing have been defined in earlier chapters. The more egregious of these compounds commonly have sulfur and nitrogen roots; extremely low level (parts per billion and even parts per trillion) detection and recognition odor thresholds; and higher persistence (i.e., they continue to maintain a high level of intensity even as they are being diluted). In addition to the sulfur- and nitrogen-based compounds, nuisance odor compounds can also be associated with carboxylic acids and volatile organic compounds (VOCs).

These odorous compounds can be found throughout WRRFs. Some of these compounds are more closely aligned with the liquid and some more with the solids handling/processing side of the facility; some can be found in both areas. Although any one of these compounds may be found throughout the facility, Tables 6.1, 6.2, and 6.3 list sulfur-based, nitrogen-based, and carboxylic acid compounds, respectively, and identify where in the WRRF they are more likely to be found. Tables 6.1, 6.2, and 6.3 represent general lists of compounds that consider conditions found at average WRRFs, recognizing that exceptions do occur.

3.0 ODOR EMISSIONS BY PROCESS UNIT

3.1 Preliminary Treatment

For purposes of this discussion, preliminary treatment includes influent pumping stations, wet wells, bar screens, grit chambers, preaeration basins, influent channels, distribution boxes, Parshall flumes, and flow equalization basins. The characteristics of air emissions from these processes are highly dependent on the characteristics of the influent wastewater, detention time in the collection system, level of turbulence, and characteristics and volume of process side streams returned upstream of the process and septage received from haulers who can discharge to the facility influent flow.

For facilities that experience high odor emissions from preliminary treatment processes, source monitoring typically shows that hydrogen sulfide dominates the reduced sulfur compounds. Other odorous reduced sulfur compounds are likely to be present, but typically at lower concentrations relative to hydrogen sulfide. In addition, other non-reduced sulfur compounds may also be present. Whether hydrogen sulfide concentrations are greater or not, the common potential compounds that are released from WRRFs (found in Tables 6.1, 6.2, and 6.3) can contribute to odors that are released.

The first location where nuisance odors can be expected is the end of pipe from the collection system. This typically occurs at the influent junction structure or wet well, as the collection system wastewater is initially introduced to the facility and the odors from the headspace of the gravity line reach their terminus and/or the odors from the wastewater force main(s) break to gravity or enter the wet well and release pent-up odors that have remained in the liquid based on the pressure of the system. Turbulence

TABLE 6.1 Sulfur-based compounds at wastewater facilities.

Sulfur-based Odorous Compounds	WW						Solids	WW		Solids								Pounds/ Resv.		WW
	Raw Sewage Pump Station Wet Well	Flow Control Chamber	Screening Facilities	Grit Facilities	Primary Settling Tanks	Aeration Tanks	Return and Waste Activated Sludge Pumping	Final Settling Tanks	Tertiary Filters	Scum Concentration Tanks	Sludge Concentration Tanks	Sludge Holding Tanks	Sludge Thickening Facilities	Digesters	Septage Receiving Area	Sludge Lagoons	Solids Drying Areas	Retention Ponds	Flood Control Reservoirs	Collection System
Hydrogen Sulfide	P	P	P	P	P	P	P	P	NP	P	P	P	P	P	P	P	P	P	NP	P
Carbonyl Sulfide	P	P	P	P	P	P	P	P	NP	P	P	P	P	P	P	P	P	NP	NP	P
Methyl Mercaptan	P	P	P	P	P	NP	P	NP	NP	P	P	P	P	P	P	P	P	NP	NP	P
Dimethyl Mercaptan	P	P	P	P	P	NP	P	NP	NP	P	P	P	P	P	P	P	P	NP	NP	P
Dimethyl Sulfide	P	P	P	P	P	NP	P	NP	NP	P	P	P	P	P	P	P	P	NP	NP	P
Carbon Disulfide	P	P	P	P	P	NP	P	NP	NP	P	P	P	P	P	P	P	P	NP	NP	P
Isopropyl Mercaptan	P	P	P	P	P	NP	P	NP	NP	P	P	P	P	P	P	P	P	NP	NP	P
tert-Butyl Mercaptan	P	P	P	P	P	NP	P	NP	NP	P	P	P	P	P	P	P	P	NP	NP	P
n-propyl Mercaptan	P	P	P	P	P	NP	P	NP	NP	P	P	P	P	P	P	P	P	NP	NP	P
Ethyl methyl Sulfide	P	P	P	P	P	NP	P	NP	NP	P	P	P	P	P	P	P	P	NP	NP	P
Thiophene	NP	NP	NP	NP	NP	NP	NP	NP	NP	NP	NP	NP	NP	NP	NP	NP	NP	NP	NP	NP
Isobutyl Mercaptan	P	P	P	P	P	NP	P	NP	NP	P	P	P	P	P	P	P	P	NP	NP	P
Diethyl Sulfide	P	P	P	P	P	NP	P	NP	NP	P	P	P	P	P	P	P	P	NP	NP	P

(continues)

TABLE 6.1 Sulfur-based compounds at wastewater facilities (*Continued*).

Sulfur-based Odorous Compounds	WW						Solids	WW		Solids								Pounds/ Resv.		WW
	Raw Sewage Pump Station Wet Well	Flow Control Chamber	Screening Facilities	Grit Facilities	Primary Settling Tanks	Aeration Tanks	Return and Waste Activated Sludge Pumping	Final Settling Tanks	Tertiary Filters	Scum Concentration Tanks	Sludge Concentration Tanks	Sludge Holding Tanks	Sludge Thickening Facilities	Digesters	Septage Receiving Area	Sludge Lagoons	Solids Drying Areas	Retention Ponds	Flood Control Reservoirs	Collection System
n-Butyl Mercaptan	P	P	P	P	P	NP	P	NP	NP	P	P	P	P	P	P	P	P	NP	NP	P
Dimethyl Disulfide	P	P	P	P	P	P	P	NP	NP	P	P	P	P	P	P	P	P	NP	NP	P
3-Methylthiophene	NP	NP	NP	NP	NP	NP	NP	NP	NP	NP	NP	NP	NP	NP	NP	NP	NP	NP	NP	NP
Tetrahydrothiophene	NP	NP	NP	NP	NP	NP	NP	NP	NP	NP	NP	NP	NP	NP	NP	NP	NP	NP	NP	NP
2,5-Dimethylthiophene	NP	NP	NP	NP	NP	NP	NP	NP	NP	NP	NP	NP	NP	NP	NP	NP	NP	NP	NP	NP
2-Ethylthiopene	NP	NP	NP	NP	NP	NP	NP	NP	NP	NP	NP	NP	NP	NP	NP	NP	NP	NP	NP	NP
Diethyl Disulfide	P	P	P	P	P	NP	P	NP	NP	P	P	P	P	P	P	P	P	NP	NP	P

Notes:
NP = not present
P = present
WW = wastewater

TABLE 6.2 Amines, nitrogen compounds, and ammonia at wastewater facilities.

	WW						Solids	WW		Solids								Pounds/ Resv.		WW
Amines, Ammonia, and Nitrogen Compounds	**Raw Sewage Pump Station Wet Well**	**Flow Control Chamber**	**Screening Facilities**	**Grit Facilities**	**Primary Settling Tanks**	**Aeration Tanks**	**Return and Waste Activated Sludge Pumping**	**Final Settling Tanks**	**Tertiary Filters**	**Scum Concentration Tanks**	**Sludge Concentration Tanks**	**Sludge Holding Tanks**	**Sludge Thickening Facilities**	**Digesters**	**Septage Receiving Area**	**Sludge Lagoons**	**Solids Drying Areas**	**Retention Ponds**	**Flood Control Reservoirs**	**Collection System**
Dimethlamine	NP	NP	NP	NP	NP	NP	NP	NP	NP	NP	P	P	P	P	NP	P	P	NP	NP	NP
Ethylamine	NP	NP	NP	NP	NP	NP	NP	NP	NP	NP	P	P	P	P	NP	P	P	NP	NP	NP
Trimethlamine	NP	NP	NP	NP	NP	NP	NP	NP	NP	NP	P	P	P	P	NP	P	P	NP	NP	NP
Isopropylamine	NP	NP	NP	NP	NP	NP	NP	NP	NP	NP	P	P	P	P	NP	P	P	NP	NP	NP
t-Butylamine	NP	NP	NP	NP	NP	NP	NP	NP	NP	NP	P	P	P	P	NP	P	P	NP	NP	NP
Propylamine	NP	NP	NP	NP	NP	NP	NP	NP	NP	NP	P	P	P	P	NP	P	P	NP	NP	NP
Diethylamine	NP	NP	NP	NP	NP	NP	NP	NP	NP	NP	P	P	P	P	NP	P	P	NP	NP	NP
s-Butylamine	NP	NP	NP	NP	NP	NP	NP	NP	NP	NP	P	P	P	P	NP	P	P	NP	NP	NP
Isobutylamne	NP	NP	NP	NP	NP	NP	NP	P	NP	NP	P	P	P	P	NP	P	P	NP	NP	NP
Butylamine	NP	NP	NP	NP	NP	NP	NP	NP	NP	NP	P	P	P	P	NP	P	P	NP	NP	NP
Disopropylamine	NP	NP	NP	NP	NP	NP	NP	NP	NP	NP	P	P	P	P	NP	P	P	NP	NP	NP
Triethylamine	NP	NP	NP	NP	NP	NP	NP	NP	NP	NP	P	P	P	P	NP	P	P	NP	NP	NP
Dipropylamine	NP	NP	NP	NP	NP	NP	NP	NP	NP	NP	P	P	P	P	NP	P	P	NP	NP	NP
Ammonia	P	P	NP	NP	NP	NP	NP	P	NP	P	P	P	P	P	P	P		NP	NP	P
Indole	P	P	P	P	P	NP	NP	NP	NP	NP	P	P	P	P	P	P	P	NP	NP	P
Skatoe	P	P	P	P	P	NP	NP	NP	NP	NP	P	P	P	P	P	P	P	NP	NP	P

Notes:
NP = not present
P = present

TABLE 6.3 Carboxylic acids at wastewater facilities.

	WW						Solids	WW		Solids								Pounds/ Resv.		WW
Carboxylic Acids	Raw Sewage Pump Station Wet Well	Flow Control Chamber	Screening Facilities	Grit Facilities	Primary Settling Tanks	Aeration Tanks	Return and Waste Activated Sludge Pumping	Final Settling Tanks	Tertiary Filters	Scum Concentration Tanks	Sludge Concentration Tanks	Sludge Holding Tanks	Sludge Thickening Facilities	Digesters	Septage Receiving Area	Sludge Lagoons	Solids Drying Areas	Retention Ponds	Flood Control Reservoirs	Collection System
Acetic Acid	P	P	P	P	P	P	P	NP	NP	P	P	P	P	P	P	P	P	NP	NP	P
Propionic Acid (Propanoic)	P	P	P	P	P	P	P	NP	NP	P	P	P	P	P	P	P	P	NP	NP	P
Butanoic Acid (Butyric)	P	P	P	P	P	P	P	NP	NP	P	P	P	P	P	P	P	P	NP	NP	P
3-Methylbutanoic Acid (Isovaleric)	P	P	P	P	P	P	P	NP	NP	P	P	P	P	P	P	P	P	NP	NP	P
Pentanoic Acid (Vakeric)	P	P	P	P	P	P	P	NP	NP	P	P	P	P	P	P	P	P	NP	NP	P
Hexanoic Acid (Caproic)	P	P	P	P	P	P	P	NP	NP	P	P	P	P	P	P	P	P	NP	NP	P
Octanoic Acid (Caprylic)	P	P	P	P	P	P	P	NP	NP	P	P	P	P	P	P	P	P	NP	NP	P

Notes:
NP = not present
P = present

experienced at downstream processes such as aerated grit chambers, preaeration basins, aerated or otherwise turbulent influent channels, junction chambers, distribution boxes, and unit process effluent weirs are other processes for which significant release of odorous gases can be expected.

Uncontrolled discharge of septage or other trucked-in waste at the headworks can often cause elevated odor emissions. The increased emissions are dependent on the volume and characteristics of the materials and where and how they are introduced to the facility. In addition, side stream return flows from sludge processing operations can also affect odor emissions at and downstream of the point at which the recycle stream is introduced. The effect of these side streams can be significant if the stream is highly odorous.

3.2 Primary Treatment

Primary clarifiers and sedimentation tanks (clarifiers) can be a significant source of odor emissions, particularly if the influent wastewater contains significant levels of dissolved sulfide and/or sulfides are generated within the clarifiers. Primary clarifiers represent a large odor-emitting surface area, and the character of the odor is considered objectionable. The turbulent effluent launders and, to a lesser extent, the turbulent influent feed wells often represent a significant portion of the total clarifier emissions. Again, this is because of the high turbulence levels that promote the transfer of dissolved odorous gas from the liquid to the atmosphere. The greater the freefall over the effluent weirs, the greater the release of odors. A reduction in the freefall over the effluent weirs can reduce the turbulence and, to some degree, the associated stripping of odors. Odor emissions from the effluent launder can typically represent 50 to 90% of the total clarifier odor emissions.

At one facility in Illinois, hydrogen sulfide emissions from the weir area were estimated to be 95% of total primary clarifier hydrogen sulfide emissions (Ziols, 1997). As a result, some facilities choose to cover only the effluent launder and to treat this air. The volume of air exhausted from the effluent launders is only a fraction of the air volume exhausted from a full clarifier cover, even if the cover is flat. Before choosing this option, the facility should be certain that odors from the quiescent surface do not create offsite odor problems. This can be achieved by monitoring the odor emissions from the quiescent surface and using an odor dispersion model to assess whether nuisance odors can be perceived in the surrounding community.

Operational practices can also affect odor emissions from primary clarifiers. Maintaining high (0.6 m or 2 ft) sludge blanket levels can result in the generation of reduced sulfur compounds within the settled sludge, particularly during the warmer operating months when microbiological activity is high, with a resulting increase in odor emissions from both the quiescent surface and the effluent launder. As a general rule, it is good nuisance odor practice to maintain sludge blanket levels less than 0.45 m (18 in.) in cooler operating conditions and 0.30 m (12 in.) or less during warm operating conditions.

The practice of wasting secondary activated sludge back to the primary clarifiers in an attempt to enhance thickening of the combined sludge will also increase clarifier odor emissions as the substrate starved active microorganisms from the secondary clarifiers are mixed with the substrate-rich primary solids. The subsequent increased activity of the microorganisms will rapidly deplete whatever limited oxygen is available and create anoxic and anaerobic conditions where odorous reduced sulfur and nitrogen compounds are formed. Facilities are warned that these conditions can quickly escalate and create an off-site odor source unless the odors are captured, contained, and controlled. One potential additional side effect of combining waste activated solids with primary solids is the generation of methane when the conditions are right and the combined solids have enough time.

Solids and oil and grease can also build up in the effluent weirs and launders of not only primary settling tanks, but also secondary settling tanks and thickening tanks that have weirs and launders, and contribute to nuisance odors from these process units. Facility operators are encouraged to periodically clean the weirs and launders of any buildup to remove these process areas as potential odor sources.

Other operational practices associated with the buildup of scum (a combination of oil and grease and other lighter-phase materials that float on the surface rather than sink) on the surface of the primary clarifier can create considerable odors when it is allowed to accumulate and remain on the surface. Odorous conditions will worsen in hot weather as the scum heats up and odors are more easily volatilized. Routine continuous removal of the surface scum can prevent these odors. Scum troughs and boxes can also be sources of odor. In many cases, good housekeeping practices such as regular flushing can keep odor emissions to a minimum.

3.3 Biological Treatment

Biological wastewater treatment processes, which are generally aerobic, typically generate odors that are considered less objectionable than those from preliminary and primary treatment systems or sludge-handling processes. Although reduced sulfur compounds may still be components of the emissions (attributed to the wastewater feed from primary treatment or recycle streams), the predominant odor character associated with biological treatment is often described as "earthy" or "musty" as opposed to "sewer" or "rotten egg". Atmospheric hydrogen sulfide levels at the surfaces of biological reactors are typically well under 1 ppmv. However, the strength of the odor as measured in dilutions to threshold and the odor emission rate can vary depending on the type of process, loading rate, and how the process is operated.

Although biological activated sludge tanks emit a more acceptable musty and earthy odor, that odor is continuous and is commonly emitted over a wide-open surface area tank. When neighbors are very close to or at the fence line of a facility, or are immediately downwind of the predominant wind direction, these less obnoxious but more continuous odors over time can become annoying. As a result, it is not uncommon for facilities to have to cover and treat the odors emitted from their biological treatment processes to satisfy the nearby neighborhood concerns over nuisance odors.

The activated sludge process is the most common biological wastewater treatment process in use today. Emissions from activated sludge basins are directly affected by the type of aeration basins. Mechanical surface aerators with the highest turbulence levels typically produce the highest emission rate, followed by coarse-bubble diffusers. Because of the low level of turbulence, fine-bubble diffusers typically have the lowest emission rate assuming all other factors (wastewater characteristics, mixed liquor suspended solids [MLSS], and surface area) are equal.

Activated sludge processes that are lightly loaded, such as extended aeration or oxidation ditch processes, typically have lower odor levels at the surface compared with conventional activated sludge or higher loaded systems. Plug flow activated sludge reactors typically have higher odor emissions at the feed end where some stripping of odorous compounds occurs. This may not be true with step-feed operations because processes operated at high MLSS typically release a stronger musty odor than low-MLSS reactors.

Pure oxygen activated sludge systems typically use covered basins. Vents from such processes have low odor levels and emission rates and rarely represent a concern for off-site detection.

Fixed-film processes such as trickling and roughing filters and rotating biological contactors typically have a higher potential for odor effects than aerobic suspended growth processes. There may be several reasons for this, including the following:

- In trickling filters, the significant turbulence created by the physical drop/application of primary effluent from the distribution arms to the media causes the release of hydrogen sulfide and other odors to the atmosphere.
- There is greater potential for development of localized anaerobic conditions within the media resulting from inadequate oxygen transfer.

- There may be greater turbulence and opportunity for stripping of odorous gases within the media.
- Higher odor emission rates may occur during periods when there are large temperature differences between the air and the wastewater.

With the trend toward biological nutrient removal and the use of selector processes to improve mixed liquor settleability, the use of anoxic and anaerobic suspended-growth processes is more common. These processes are typically mixed with a submersible mixer, but not aerated. Turbulence levels at the surface are typically low, but not quiescent. Anoxic basins are expected to have relatively low odor emission rates because an oxygen source in the form of nitrate is present. However, emission rates will be affected by the quality of the primary effluent and the dissolved sulfide concentration. Anaerobic suspended-growth processes have a higher potential for odor production. Unfortunately, little data are available to support any distinction in odor emission rates.

3.4 Intermediate and Final Settling

Intermediate and final clarifiers or settling tanks typically have low odor emission rates, and the odor is of a faintly musty or fresh character. Exceptions can occur if the solids blankets are allowed to accumulate as operators attempt to thicken the solids in these tanks or as downstream solids processing equipment throughput is limited because of undersized or out-of-service units. For example, concentrations as high as 7.7 ppmv have been recorded at the final settling tank effluent weir overflow in New York City's Newtown Creek WRRF.

3.5 Chlorine Contact Tanks

Chlorine contact tanks have limited odor potential. Occasionally, a faint bleach odor can be detected, but this odor is not typically detected off-site.

3.6 Stabilization Ponds

Stabilization ponds/lagoons are a form of biological treatment typically used for relatively small flowrates in rural areas. They can be aerated or facultative, or a combination of cells. They are typically lightly loaded and generate little odor. However, exceptions occur when facilities allow organic solids to accumulate on the bottom of the pond or when industrial wastewater ponds are highly loaded. Under these conditions, the ponds are likely to have higher odor emission rates and may become anaerobic during certain times of the year.

Odor emissions from facultative stabilization ponds may also increase during a spring turnover. This condition occurs when a strong wind, rapid temperature change, or inflow of a large volume of cold water causes the upper layer of water to be replaced by the lower layer of oxygen-deficient water.

4.0 PROCESS IMPROVEMENT AND ODOR MITIGATION

Before implementing capital improvements to mitigate nuisance odors from the WRRF, the owner is advised to conduct a monitoring investigation to assess where and how strong the odors are. Subsequent odor dispersion modeling using the monitoring data collected with real-time meteorological data can help determine the level of off-site odor impact created by the WRRF and the degree of control necessary at the odor sources to mitigate the off-site impacts. The implementation of required odor mitigation measures can then be planned and staged to address the community's concerns.

Not all required odor control actions need to be capital-cost intensive. In some cases, best management practices are all that are needed to avoid off-site nuisance odor problems. In others, simple modifications or adjustments may provide the necessary relief. In severe odor cases, the capital improvement project that is required will take time to design, bid, build, and commission. As such, the owner should ask, what are some immediate actions that may be implemented by the WRRF in the near term that may not provide the full relief that is expected by the community, but could partially reduce peak odorous emissions as well as demonstrate to the public that WRRF staff are listening? The actions outlined in the following subsections have been demonstrated to reduce, in varying degrees of effectiveness, nuisance odor emissions from the WRRF unit operations.

4.1 Odor Prevention

Earlier chapter discussions noted that WRRF influent odorous compounds and/or conditions that support the formation of nuisance odor compounds can be discharged to the collection system by any one of several sources, including industries, commercial businesses, hospitals, and communities. To control the indiscriminate wastewater discharge from these sources to reduce any potential impact that the discharges may have on the WRRF and the WRRF's ability to process the wastewater or solids and prevent the pass through of harmful compounds to the receiving stream after treatment, WRRFs have local ordinances that will identify discharge restrictions. Many of these local ordinances follow language defined by U.S. EPA in their

General Pretreatment Program regulations, where standard parameters such as five-day biochemical oxygen demand (BOD_5), oil and grease, pH, and others are defined. In addition, the General Pretreatment Program will provide specific guidance for the discharge of several industries that are known to have significant wastewater flows or compounds in their discharge that need to be regulated.

Additional restrictions and discharge limits can be added to the local ordinances to further protect the WRRF, staff, community, and the environment. It is not uncommon for WRRFs to add limits on the following discharge parameters to their collection systems to control the release and/or formation of nuisance odors: dissolved wastewater sulfide, BOD_5 concentrations, sulfate concentrations, pH, wastewater temperature, dissolved oxygen, and oxidation–reduction potential. Restricting the discharge of these parameters is the first step in odor prevention at the WRRF.

It is important to recognize, however, that adding limits to an existing local ordinance will require information supporting the change, discussions with local dischargers that such changes may effect, public review and scrutiny, and legislative action. All of these take time to address opposition and passage of the proposed changes. Although local ordinances provide the legal authority to limit discharges, the WRRF needs the staff to monitor and enforce the local ordinance limits.

4.2 Physical System Layout

Conditions that are favorable to the mass transfer of nuisance odor compounds from the wastewater to the atmosphere have been previously identified. One of the prime conditions was wastewater turbulence, which is the exposure of additional surface area to the air/liquid interface with the air over which mass transfer can occur. Recognizing that reducing turbulence can achieve an associated reduction in nuisance odors, WRRF management may encourage staff to inspect the facility operations to see if there are areas where surface wastewater turbulence could be reduced. For example, wastewater turbulence can typically be found at the following process areas around the WRRF:

- Wet wells—In most cases, turbulence associated with wet wells is the result of the way through which the wastewater is introduced. Inlet feed pipes are often positioned above the mean water line of the wet well, so the wastewater drops from the end of pipe to the wastewater surface below, creating turbulence in the wastewater as it falls and again as it disrupts the surface of the wet well bulk wastewater volume. The degree of turbulence is directly related to the height of the fall. Therefore, the WRRF may want to explore ways to reduce the height of the fall or eliminate the fall all together. Questions to be considered include but are not limited to the following:

  - Can the fall distance be reduced?
  - Can the discharge be redirected to below the water surface?
  - Can a device such as a vortex drop be installed to reduce the fall turbulence?
  - Is there any other approach that would reduce the turbulence created by introducing the wastewater flow to the wet well?

- Influent diversion structures—Influent diversion structures always seem to have some degree of turbulence, whether it is at the influent feed, an overflow weir or drop structure, or passage through a gate structure as the wastewater flow is diverted to the corresponding downstream unit operation. Questions to consider include, can the turbulence be reduced? Can the tail water surface be raised up in an attempt to reduce the wastewater drop?
- Parshall flumes—Unfortunately, Parshall flumes do create some surface turbulence at the throat of the device as the cross section of the channel is reduced. This is part of the Parshall flume operation. However, the inlet and effluent channels should be reviewed to determine whether abrupt channel changes in direction or drops can be avoided.
- Channels—Channels can be turbulent because of external means, mechanical mixing or diffused aeration, or their physical design, such as abrupt changes in direction, drops, turns, hydraulic jumps, and so on. Although these are important concepts to keep in mind when designing new construction, it is generally tough to modify existing facilities without a capital improvement design project.
- Grit tanks—Grit tank turbulence depends more on the style of grit removal. Aerated grit tanks are the most turbulent, especially when compared to vortex-style grit tanks where the wastewater surface is more quiescent. However, the turbulence associated with the removal of grit in an existing process unit will not be changed in the near term and without an impact on the efficiency of grit removal. Nevertheless, effluents from various grit technologies have turbulent overflows that could be reviewed to determine whether a reduction can be achieved by minimizing the potential wastewater drops.
- Weirs—Weirs have been a traditional way of maintaining a specific water surface level in various process units from diversion structures to grit, primary, secondary, tertiary, thickening, and storage tanks. The wastewater that flows over the weir plate, which can have different

styles from plain to V-notch, falls into a receiving launder or channel that directs it to the next downstream process unit. The drop can vary from one to several feet. The smaller the drop, generally the less the turbulence and associated stripping. Thus, reducing the drop distance should be explored to reduce the turbulence. Raising the water surface by backing up wastewater in the receiving launder or channel is one approach to lower the drop, turbulence, and nuisance odors emitted. The options in which the wastewater surface in the receiving launder or channel can be raised will vary, but must be consistent with the existing hydraulic profile of the facility so as not to interfere with facility operations.

- Hydraulic jumps—Hydraulic jumps are typically associated with pipeline or channel flow, where the slope of the upstream segment is steeper in slope than the downstream segment and causes the wastewater to appear to "jump" at this intersection as the faster-moving wastewater from upstream interacts with the slower-moving downstream wastewater and tries to rise up and over the slower-moving downstream flow. The greater the transition (i.e., steeper upstream segment and flatter downstream segment), the greater the jump, turbulence, and nuisance odors emitted. Smoothing out the transition by lessening the difference in the slopes of the two segments is an option to reduce the nuisance odor emissions; however, it may not be a solution that can be implemented in the near term as it typically would require more capital-intensive construction. WRRF staff should monitor the odorous emissions from this process to determine whether the cost for corrective odor reduction is worth the investment.

Any modifications to WRRF process units should examine whether the change will have any effect on normal facility operations. This is particularly important when the facility hydraulic profile is altered in any way.

5.0 ODOR CAPTURE AND CONTAINMENT

5.1 Introduction

The technology to treat nuisance odors and VOCs that are released from municipal and industrial WRRFs is available. The challenge, however, comes in selecting the most cost-effective technology. But the best technology or most costly odor control system can only be effective in reducing emissions if the odors and VOCs reach the odor control device. If the containment and ventilation systems are not able to keep the fugitive odors from escaping, or proper operation and maintenance (O&M) practices do not preserve the integrity of the system over time, then the money and effort spent on the control devices have been wasted.

A critical element of any odorous emissions control system, which is often overlooked, is the thoughtful design of odor containment and area ventilation. These systems dictate the size of the overall odor control system and also prevent emissions from escaping before the control devices can treat them. This section provides useful information on how to contain the emissions at the source, minimize the volume of air to be treated, and ventilate the enclosed area so that the emissions can be transported to the selected control device for treatment. In the past, these tasks have often been taken for granted, when they should be part of the essential first steps in the odorous emissions control process design.

When the first level of odor control actions (including operational adjustments, chemical addition, and air dilution) fails to achieve the goal of reducing off-site neighborhood odor effects and complaints from citizens continue to be received, the second and more comprehensive level of odor control is required. The second level of odor control involves significantly more engineering and construction, and is more cost intensive. The second level consists of

- odor containment by enclosing open channels, tank surfaces, or equipment;
- odor transport by providing negative pressure ventilation under the enclosure;
- odor treatment in the appropriate control device; and
- treated-air discharge with adequate dispersion to minimize odor and visual effects on the surrounding area.

Figure 6.2 illustrates this process.

5.2 Odor Cover Alternatives

Once the sources of odors have been identified through specific odor assessment investigations and the odor control system has been chosen, the odor cover and containment systems need to be selected. There are a full range of available odor cover and containment systems alternatives. The selection of the appropriate cover and containment system depends on several factors, including

- site-specific unit processes that must be controlled,
- area climate,

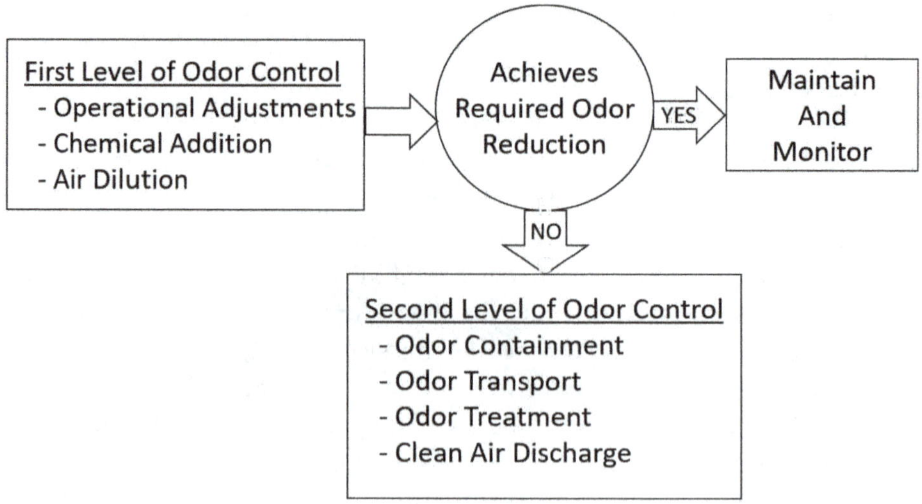

FIGURE 6.2 Progressive odor control process.

- worker safety,
- ease of construction,
- staff needs,
- aesthetics,
- effectiveness,
- durability, and
- cost.

As an example, an approach for designing a cover or enclosure system for primary sedimentation tanks is discussed. This is an area at WRRFs that receives much attention because of the release of nuisance odors that are detected by neighbors. Many facilities around the country have either already covered these tanks or are in the process of doing so.

The evaluation and selection of a functional covering or enclosure system should consider several important factors (see Figure 6.3), including the following:

- Operability and maintainability—Operators must be able to gain access to critical system components for daily inspections and routine maintenance. Critical system components could include
  - sludge- and scum-collector drive equipment;
  - wastewater surface observation for collector system inspection, wastewater and bottom sludge sample collection, wastewater color notation, and degree and character of surface scum accumulation;

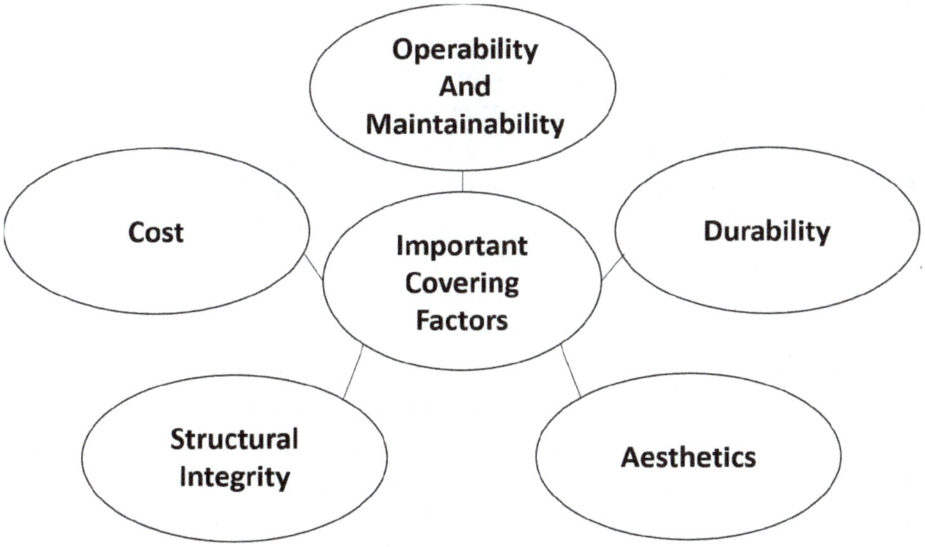

FIGURE 6.3 Important factors in selecting a cover or enclosure system.

  - scum baffles;
  - scum sprays; and
  - effluent weirs.

- Durability—The design should use corrosion-resistant and durable materials of construction to ensure the longevity of the system.
- Architectural conformity of new structures—New structures should be consistent with existing architectural and structural facilities to maintain aesthetic concerns and to reduce attention being drawn to them. "Out of sight" typically means "out of mind" to a neighboring community.
- Structural integrity—The design should include structural considerations and provide necessary modifications as additional support is required.
- Minimization of cost—By designing a system that requires the smallest volume of air necessary for effective odor control, smaller and less costly conveyance and treatment equipment will be needed.

With these important factors in mind, there are different kinds of covers or enclosures available. The first group of alternatives presented in the following three subsections ("Flat Covers", "Barrel Arch Covers", and "Pitched Covers") provide limited-to-no access for the operator. These cover options help to prevent the operator from entering potentially dangerous spaces where their health and safety could be jeopardized.

5.2.1 Flat Covers

Flat covers have become more popular as the cover of choice because their installation minimizes airspace between the cover and water surface. Minimizing that airspace is critical to the ultimate sizing of the transfer duct work, blower capacity, and odor control technology. The smaller the odorous headspace, the smaller the volume of air that needs to be removed and treated. This translates to a smaller capital cost investment to construct the odor control system and reduce the annual operating cost.

Flat covers have been commonly designed for channels; wet wells; tanks (square, rectangular, and circular); distribution boxes; and diversion structures. Figures 6.4 and 6.5 present examples of flat channel covers, and Figure 6.6 presents an example of a tank cover. Depending on the application, it may be necessary for a particular flat cover installation to require internal or external structural support. Figures 6.7 and 6.8 illustrate examples of flat covers with external support for circular and rectangular tanks, respectively. The need for internal or external structural support is determined after comparing specific site conditions with flat cover capabilities. It is most common to design external structural support to avoid the potentially damaging corrosive gases under the cover. Some specific site conditions that would influence the decision for internal or external structural support are

- width and length of process unit opening that needs be covered;
- desired live load (if a person should be able to walk on the cover or a vehicle be allowed to drive on it);

FIGURE 6.4 Flat headworks channel cover.

FIGURE 6.5 Flat influent channel cover.

FIGURE 6.6 Flat aeration tank cover.

FIGURE 6.7 Flat circular tank cover with external truss.

FIGURE 6.8 Flat rectangular tank cover with external support.

- snow load; and
- dead load (if there is a need to build or place anything on top of the cover, typically because of facility space limitations).

The main disadvantage of flat covers is limited operator visibility and accessibility. Limited operator accessibility tends to be the biggest hurdle to overcome. Facility staff have been operating WRRFs for decades by using their senses to judge performance and influent characteristics. For example, facility wastewater influent typically has a consistent color and odor, scum accumulates on the surface of primary sedimentation tanks in a routine fashion, and diffused air creates a recognizable pattern on the surface of aerated channels or tanks. Restricting the operator's visual and olfactory reference hinders the operator's ability to effectively anticipate and react to changing wastewater conditions.

Access hatches are used to address operator accessibility issues. Without access hatches, the flat covers would have to be removed for routine inspections. The likelihood of these covers being properly replaced after each inspection is dependent on the facility, but questionable over time. The use of access hatches provides operations staff with the degree of accessibility that they need to continue to do their job and maintain facility operations, despite the restrictions of flat covers.

However, limited visibility and accessibility is not always a disadvantage. For example, limited visibility and accessibility may be adequate for intermediate channels, wet wells, and tanks that do not have to be inspected on a daily or routine basis. As such, access hatches may not be warranted.

Limited accessibility by installing flat covers can also be an advantage to facility operators. Facility operators would be restricted from entering the confined airspace between the cover and the water–solids surface. Preventing access to these potentially dangerous environments protects the health and safety of the facility operator. The flat covers also help contain the odors at the source and, when the flat covers are located within a building, generally provide for an enhanced building air environment that makes performing O&M duties easier for staff.

There are also situations where certain sections of a process unit are considerably more odorous than others. A good example is the turbulent weir drop area of a primary sedimentation tank, where up to 95% of the odors associated with the primary tank have been attributed (Ziols, 1997). In these instances, instead of placing flat covers over the whole primary tank, consider placing flat covers over the weir area. Depending on the site-specific need for odor control, capturing 75 to 95% of odors may be adequate to minimize off-site odor effects from the primaries. Being able to cover and control the weir areas significantly reduces the volume of odorous air, duct diameter, fan, and odor control device size.

5.2.2 Barrel Arch Covers

Barrel arch covers are appropriate for relatively narrow tanks or channels and are not typically used for extensive wide-open surface areas. Semicircular in nature when viewed down their longitudinal axis, the height of the top of the arch above the water surface is variable. Figure 6.9 presents an example of a barrel arch system. The height selected is often chosen more for aesthetics and visibility of the water surface, rather than for any other function. View windows and inspection/access hatches both at the ends and along the length of the cover provide the operator with an extended and uninterrupted view of the water surface.

Barrel arches can be removed by sections or are equipped with a rollback feature anywhere along their length, but typically at either end for greater access to critical tank areas. For example, critical areas for rectangular sedimentation tanks include the influent (where surface scum can accumulate, and workers need access to remove it) and effluent ends (where effluent weirs need to be maintained). Installing high-intensity lights at access and rollback areas would improve the enclosed tank visibility for both day and nighttime inspections. Figure 6.10 shows an example of a barrel arch with more of the features and flexibility listed above.

FIGURE 6.9 Barrel arch covers made of fiber reinforced plastic (FRP).

FIGURE 6.10 Barrel arch covers made of aluminum and plexiglass.

While providing greater operator access and visibility, the barrel arch cover has a more pronounced exterior profile. The visual effect of the profile and the volume of odorous air needing to be removed and treated can be minimized by reducing the height of the arch above the water surface. However, a balance among exterior profile, operator access and visibility, and volume of odorous air to remove and treat is needed and will dictate the final selection.

Although the theoretical ventilation rate for nonworker space flat and barrel arch covers may be considered equivalent to maintain the design negative air pressure under the cover, actual field conditions dictate that barrel arch covers require a higher ventilation rate. A key difference that supports the need for a higher ventilation rate is the increase in cover surface area and seams associated with the barrel arch cover when compared to the flat cover. Accordingly, the volume of air, and thereby the flow of odorous air, to be treated from barrel arch covers is greater than for flat covers.

However, the relatively low-rise barrel arch cover can also minimize the air volume requiring treatment. Barrel arch covers would be most applicable for long, narrow, distributed areas requiring routine, but not full, operator access. The rollback arch cover option would be applicable for areas requiring regular operator access. The barrel arch cover design does not include provisions to support staff, vehicles, or external structures. This saves on the structural support needed for the covers. However, other loads, such as snow and wind loads, must be considered.

5.2.3 Pitched Covers

Pitched covers are similar to barrel arch covers in that they would be appropriate for long distributed areas with generally narrow width. The visual effect of their exterior profile is also similar. The pitched cover slopes up evenly on both sides to a peak, like the roof on a house. One potential disadvantage of the pitched cover is that it may require, depending on the width of the opening it spans, an internal truss system to support the panel covers. This truss would restrict interior visibility and eliminate the ability to roll back the covers. The covers would then have to be removed by section to allow access within the enclosed area. Therefore, pitched covers would be most applicable to areas requiring limited visibility and minimal worker access. The application of pitched covers at WRRFs is limited.

The second group of cover alternatives presented in the following subsections provide walk-in access to the process unit. To support operator presence, significantly more ventilation air volume is needed.

5.2.4 Building Extension or Enclosure

An alternative to covering an area to be contained is enclosing the area in a building or structure. The premise is that the space enclosed in a building or structure will be occupied. This can entail the extension of existing superstructures or enclosing the area in a new building, enclosure, penthouse, or dome structure. Figures 6.11 and 6.12 illustrate examples of penthouse

FIGURE 6.11 Penthouse enclosure; occupied space.

FIGURE 6.12 Penthouse enclosure; unoccupied space.

enclosures; Figure 6.13 is an example of a geodesic dome. Operators may prefer this alternative because it will provide full tank or section accessibility. Unfortunately, this alternative also requires the maximum amount of air volume to be treated, which is a result of the extended area to be enclosed and higher ventilation requirements for worker space environments.

This alternative is most beneficial for areas that require complete access for frequent maintenance or repairs. This alternative also demands adherence to confined space entry requirements each time the enclosure is entered. Requirements are likely to include training courses, filling out forms, and air monitoring (both before and during entry). Worker health and safety is a concern for buildings, enclosures, penthouses, and dome structures. These concerns can be addressed by modifying the basic structure designs.

5.2.4.1 Building or Penthouse

Because the workers have unrestricted access to the enclosed unit process area, higher ventilation rates are required. In addition, the placement of ventilation air supply and odorous air exhaust registers are critical to ensuring that fresh air being supplied is directed at the workers. The general rule is

FIGURE 6.13 Geodesic dome enclosure.

to design the ventilation system to withdraw odors at the point at which they are released.

For example, in buildings or enclosures, the source of odor release is typically at the wastewater surface, particularly at turbulent locations. To keep the workers in a fresh air environment as they stand above the wastewater surface, the ventilation air supply registers should be located at the enclosure ceiling over work or high maintenance areas, while exhausting odorous air at or near the wastewater surface. This way, the nuisance odors are not being dragged across a worker's face, but rather the worker is always benefiting from the fresh air supply as it flows from the ceiling, across the body, and to the source, where it is removed by the exhaust register.

At most WRRFs where hydrogen sulfide and other reduced sulfur compounds are key nuisance odor compounds, this ventilation scheme of exhausting the odors at the source before they reach worker spaces is advantageous. Hydrogen sulfide gas and other reduced sulfur compounds are heavier than air and tend to accumulate in low-lying areas. With a low-level exhaust, these odorous compounds would be exhausted to an odor control device before it would have a chance to affect the workers or corrode the internal building components.

5.2.4.2 Dome Enclosure

Although buildings and enclosures can be placed over any shape basin or unit process odor source, the dome is generally used to cover circular basins. Circular basins at WRRFs are commonly sedimentation tanks, gravity thickener tanks, or trickling filters. Like the barrel arch, the dome structure has an open interior volume (no trusses are necessary for support) and is variable in its height of rise. The greater the height of rise, the greater the volume of odorous air needing treatment, particularly when occupied. Accordingly, minimizing the height of rise of the dome corresponds to a more economical odor control system.

A standard feature for a dome covering a circular clarifier or sedimentation basin includes a single door, which is provided to allow access to the dome interior to inspect and perform maintenance operations. This means that the dome is a worker space area and, as such, the full volume of air within the dome has to conform to worker safety standards. The volume of air vented from a full dome that needs odor control is significant when compared with flat covers. Modifications to the dome structure now provide options to this standard dome feature so that less air is needed to ventilate the open interior air space. Some options include the following:

- providing a fresh air supply directly over the areas where worker spaces are located when staff is present, and
- avoiding the need for the worker to enter the confined environment of the dome.

Independent accessways are available in a geodesic dome structure. These accessways are separate and independent from the environment of the internal dome. Figure 6.14 depicts a walkway open to the atmosphere, and Figure 6.15 shows an enclosed geodesic dome walkway, separate from the internal odorous air of the dome.

In either option, the necessary ventilation for the enclosed airspace volume under the dome is reduced. Nevertheless, even at reduced ventilation rates, the volume of the dome requires a substantial amount of air to be removed and treated. Table 6.4 presents a summary of advantages and disadvantages of the walk-in cover alternative and the limited-to-no-access cover alternative.

5.3 Odor Capture Alternatives for Bar Screens

The best way to capture odors is at their source. Doing so prevents the odors from contaminating worker spaces and subsequently requiring large volumes of odorous air to be treated. Previous discussions have highlighted how odors can be contained at wastewater sources such as channels, wet wells, tanks, and diversion structures. This subsection addresses how nuisance odors released by process equipment like bar screens can be captured.

FIGURE 6.14 Geodesic dome enclosure with center walkway open to the atmosphere.

FIGURE 6.15 Geodesic dome enclosure with enclosed center walkway.

TABLE 6.4 Comparison of process unit cover alternatives (walk-in vs. limited-to-no access).

Walk-in Cover Alternative	
Advantages	**Disadvantages**
Full access	Requires high air exchange rates for worker access
Full visibility	High volume of air to treat
– Wastewater surface	May be considered confined space
– Mechanical equipment	Larger odor control system
– Weir overflows	High external visibility
– Surface collectors	If confined space, need entry permit
Ease of O&M	Need to assess air environment before entering
Limited difference vs. uncovered	Higher cost
Easy-to-collect samples	
Limited-to-No Access Cover Alternative	
Advantages	**Disadvantages**
Not able to enter confined space	Limited visibility
Worker performs tasks in pleasant environment	Limited accessibility
Minimizes air to treat	More difficult to perform O&M
Smaller odor control system	Need to remove cover panel to access internals
Straightforward installation channels	
Low profile	
Motor drives are outside cover, easy to maintain	

The bar screen is an integral part of most WRRF headworks operations. The challenge posed by containing odors from traditional bar screens is that they are mechanically cleaned, so the bar-rack structure, which captures the unwanted material, extends from below to well above (3 m [9 ft] or more) the water surface and penetrates the concrete access floor on an angle that creates a large opening in the concrete slab. Based on this operation, the concrete floor opening through which the bar screen passes provides a pathway for the odors from the wastewater surface to enter the worker space area. Newer bar screen equipment, especially for finer screens removal, are integral units that operate at the water surface and can be more easily contained.

Typically, odors associated with fresh screening material that is removed from influent wastewater and dragged up the bar rack do not generally reach off-site receptorstravel off-site. However, the opening in the concrete access floor for the bar rack allows wastewater odors that have been traveling with wastewater in the sewer and facility influent channel to escape to the atmosphere (in locations where the bar screen is out in the open) or to the room air (in locations where a building surrounds the bar screen). The key to controlling these higher off-site potential odors associated with the headspace of the wastewater channel is to prevent them from escaping to the atmosphere. The following three approaches provide guidance for controlling odors from bar screens and have been proven effective in field situations:

- **Create a zone of negative air influence around the bar-rack opening.** This approach assumes that the wastewater channel upstream and downstream of the bar screen opening is covered, with the opening at the bar screen allowing odors to escape. Exhaust ventilation pickup points are positioned in covered channels upstream and downstream of the bar screen. The design volume of air for each position is determined by the cross-sectional area of the airspace multiplied by a face velocity that will secure a negative air pressure is maintained across the opening.

  The channel opening is typically relatively large, with a correspondingly large ventilation flow requirement. Modified plastic strip curtains have been used to reduce the opening to sustain the negative air pressure influence and face velocity and reduce ventilation flow. Placed as close to the bar rack as possible without interfering with normal operation at both upstream and downstream channel locations, the polyvinyl chloride (PVC) strip curtains are mounted to the underside of the channel cover or hung off a support bar. The length of the PVC strips is designed to barely reach the water surface at the high-water elevation of the channel.

  The PVC strips will have a gap between the water surface and the bottom of the strip that will vary based on the design of the water surface level in the channel. It is important that the PVC strips do not impede the flow of wastewater through the bar screen. The ventilation is calculated by determining the gap open area and applying a face velocity ranging from 76 to 152 m/min (250 to 500 fpm). The lower face velocity range is applied to bar screen channels that are indoors, whereas the higher range is applied to outdoor or exposed installations where atmospheric wind conditions are involved. The zone of influence created reduces the release of wastewater-related odors at the bar screen and acts similarly to a laboratory hood.

- **Place a cover over the bar screen from the point at which it penetrates the concrete access floor to the top of its rise.** This approach assumes that the wastewater channel upstream and downstream of the bar screen opening is covered. The cover provides minimal clearance between the working bar screen and the cover itself and fits more like a glove.

  These covers are typically painted metal and are provided by the bar screen manufacturer or can be custom built. These covers or enclosures provide access to critical bar screen mechanical areas, so staff can perform routine O&M and observe their operation. Some O&M procedures require that panels be removed for staff access. It is imperative that the panels be replaced afterwards, otherwise the integrity of the odor containment will be lost and odors will be released.
- **Place the bar screen within a building and provide full worker space ventilation.** No special covers are required; however, the volume of air needed to capture and contain the odors is significantly greater than the first two approaches. In this approach, the ventilation layout should be designed to pick up the odors at the source rather than drag them through the worker space of the building. Accordingly, supply registers provide fresh air at a high location, and exhaust registers are located near the floor level.

5.4 Materials of Construction

Selecting the appropriate materials of construction for covers is a key concern for the designer. In general, the materials of construction should be selected to provide durability, ease of maintenance, performance, corrosion resistance, and low cost. Site-specific conditions are important when evaluating the appropriate cover material of construction, particularly regarding the potential corrosive nature of the atmosphere being contained under the cover. Certain cover types may limit the choices of materials to be applied. For example, geodesic dome covers to be placed over circular tanks have not been made of wood. In addition, the material to be selected also depends on the anticipated length of service. Less durable and less expensive materials can provide adequate immediate service for short-term installation, whereas more durable and more costly materials are necessary for long-term installation. The proper selection of the materials of construction should be made to fit the goals, expectations, and needs of the service for which they are intended. The most common materials used at WRRFs for the containment of odors are outlined in the following subsections, with the first three—concrete, aluminum, and FRP—being the most popular:

5.4.1 Concrete

Concrete (precast or cast in place) can support the greatest weight, but also introduces the greatest dead loading as a cover material. Therefore, concrete covers may adversely affect the facility's flexibility by limiting the facility maintenance staff's ability to remove the covering system for significant repairs or to provide access as needed. Furthermore, when used as a flat cover, the additional weight of removable concrete panels makes it more difficult to provide adequate storage for these units while removed. If space is a premium at the facility, flat concrete covers can be designed to allow for other structures to be placed on top. For example, the odor control technology or other facilities can be located on top of the flat concrete covers. Concrete is also subject to corrosion, but may be provided with a protective material to aid in preventing excessive amounts of corrosion. The protective material may be as simple as an additive to the concrete mixture that acts to inhibit corrosion, to applied surface coatings, to liners, and so on. Different approaches are taken for different situations because each material has conditions under which it works best based on field experience. Whenever standard concrete is placed in an enclosed environment with the potential for the presence of corrosive gases, steps need to be taken to protect the concrete. Concrete is typically the highest capital cost investment, especially when installed as a retrofit.

5.4.2 Aluminum

Aluminum covers provide a high-tensile strength with a thin cross-sectional area in a lightweight frame. Some corrosion of this material would be anticipated, but an anodized coating should aid in corrosion prevention. In addition, aluminum covers can also have a factory-applied epoxy coating to provide improved corrosion protection. The lightweight nature and thin cross-sectional area of aluminum make it easier to remove and store during maintenance operations. Although typically less expensive than fiberglass reinforced plastic (FRP) and concrete, aluminum also has the highest resale value, which could result in the risk of theft. Site conditions would, therefore, be an important consideration when selecting aluminum. In addition, the design of an aluminum covering system should consider the incompatibility of aluminum with concrete and other metals (e.g., stainless steel).

Steps should be taken to ensure that these materials do not come in direct contact with each other. If separation is not achieved, the aluminum will disintegrate around the point of contact and the structural integrity of the system could be compromised. Figures 6.16 and 6.17 are examples of aluminum geodesic dome and flat covers, respectively.

5.4.3 Fiberglass Reinforced Plastic

Fiberglass covers provide the greatest resistance to corrosion from sewer gases. When installed outdoors, they typically require periodic maintenance in the form of an ultraviolet-inhibitor coating for durability and longevity protection, particularly when the material is exposed to direct sunlight. Fiberglass reinforced plastic (FRP) covers are also relatively lightweight and typically can be removed by facility staff and stored during maintenance operations. The thickness of the cover members will be dependent on load-bearing requirements of installation. The greater the loading requirements, the deeper the honeycomb web structure (thickness) needs to be. Although lower in strength than concrete and aluminum, its unit cost falls between the two. Figure 6.18, Figure 6.19, and Figure 6.20 present examples of FRP dome, FRP flat, and FRP flat weir area covers, respectively.

5.4.4 Stainless Steel

Stainless steel has almost exclusively been used as a cover material for flat systems, and most typically for covering channels. Stainless steels are resistant to corrosion in many environmental exposures because the metal forms a passive

FIGURE 6.16 Aluminum geodesic dome cover.

FIGURE 6.17 Aluminum flat cover.

oxide film on its surface in the presence of oxygen. When environmental exposure conditions interfere with the formation and/or maintenance of that passive oxide film, stainless steels corrode. Stainless steel has good corrosion resistance and, like aluminum, provides high-tensile strength with a thin cross-sectional area. Unfortunately, it is heavier than aluminum and costs more.

FIGURE 6.18 Fiberglass reinforced plastic (FRP) dome cover.

FIGURE 6.19 FRP flat covers.

Although heavier than aluminum and FRP, stainless steel flat covers are very durable and typically sized to enable facility staff to remove and store them during maintenance operations while avoiding surface damage. Two grades of steel are commonly used: 304 and 316 (American Iron and Steel Institute, 1977). Type 304 serves a wide range of applications and resists sulfuric acids

FIGURE 6.20 FRP weir area covers.

at moderate temperatures and concentrations. Type 316 contains slightly more nickel than Type 304, and 2 to 3% molybdenum, giving it better resistance to corrosion than Type 304, especially in chloride environments that tend to cause pitting. These two grades also have a lower carbon version, 304L and 316L, which provide an increased resistance to corrosion and are among the most commonly used stainless steels in wastewater applications.

5.4.5 Canvas

Canvas style covers provide an alternate means of covering channels, wet wells, tanks (square, rectangular, or circular), and basins. Based on a flat cover design, canvas covers can provide ease of maintenance, typically through zippered access sections; containment of odors; and corrosion resistance (the canvas has an outer coating that resists degradation).

The canvas covers require a support structure when spanning larger openings to prevent sagging and accumulation of rainwater and to allow for carrying snow loads in colder parts of the country. The support structure is positioned below the cover and is, therefore, exposed to potentially corrosive gases and must be capable of resisting corrosion in the headspace of the enclosed area. Zippered sections allow access to enclosed areas; however, because staff cannot walk on the canvas covers, it is not always easy to reach and open the zippered sections. The durability of canvas falls short of aluminum, FRP, concrete, and stainless steel. The cost is higher than would initially be expected, but will still be less than the above materials of construction. Figure 6.21 is an example of a canvas cover over small rectangular tanks.

FIGURE 6.21 Canvas flat covers.

5.4.6 Air-Supported Canvas

A different, seldom seen style of canvas cover is the air-supported system. The positive air-pressure-supported canvas cover is used over large surface areas and is similar to tennis courts or athletic practice facilities used around the country. Only canvas is exposed to the enclosed foul air. Wire bracing is wrapped around the outer canvas to maintain the design dimensions of the cover. There is minimal exchange of ventilation air that makes the interior space non-environmentally friendly. Extreme care must be applied to prevent the release of air from the positive air pressure inside that supports the canvas cover.

5.4.7 Wood

Wood has been used as a cover material, generally for small and narrow openings associated with wet wells, diversion structures, pits, channels, and launders. Generally used as a temporary cover material, the wood cover should not be considered safe to walk on without structural member support beneath the wood. Marine-grade plywood is the wood selection of choice for moist or wet environments associated with wastewater cover applications.

Nevertheless, wood is typically considered a short-term and sometimes temporary cover material rather than a long-term cover solution because the wood does not hold up well under the adverse environment. Over time, the plywood begins to warp and separation of the wood layers around the edges becomes more noticeable. Because of the tendency to warp and separate layers, it is difficult to maintain a good seal with wood. One advantage of applying wood as a cover alternative is the ability of staff to install, and subsequently replace, as necessary.

Each of these materials has been used at WRRFs for the containment of odorous air. Other materials such as plastic have also been used on a small scale with limited success. However, because of its inability to satisfy all the above criteria, it is not used as often for permanent installations.

Table 6.5 presents a comparison of cover materials of construction regarding key decision-making elements including ease of maintenance, durability, corrosion resistance, short- or long-term application, and cost.

5.5 Access

For years, WRRF operators have been trained to rely on their senses when evaluating the condition of the facility operations. Of the five senses, sight and smell are used the most, with hearing a distant third. Operators look for deviations from normal conditions by viewing wastewater for color variations, flow velocity, accumulation of surface material (scum, flotables,

TABLE 6.5 Cover materials of construction comparison.

Cover Material of Construction	Ease of Maintenance[a]	Durability[a]	Corrosion Resistance[a]	Short Term Application[b]	Long Term Application[b]	Cost[c]
Concrete	E	E	P	N	Y	E
Aluminum	E	E	G–E	Y–N	Y	E
FRP	G–E	G–E	E	Y–N	Y	E
Stainless Steel	E	E	G–E	N	Y	E
Canvas	G	P	G–E	Y	Y–N	M
Air supported canvas	P–G	P	E	Y	N	I–M
Wood	P	P–G	P	Y	N	I
Plastic	G	P	E	Y	N	I

[a]Poor – P, Good – G, Excellent – E
[b]Yes – Y, No – N
[c]Inexpensive – I, Moderate – M, Expensive – E

sludge, etc.), excess bubbling on a quiescent or aerated surface, foam accumulation on aeration tanks, and other abnormalities. The odor released at certain facility locations also reveals variations from the standard operating procedure that a trained operator will smell while doing rounds. Placing covers on these once open channels, tanks, wet wells, and so on substantially restricts the operator from evaluating the operating condition of the facility.

The operator is understandably unhappy with the installation of covers over the once open and freely accessible process units, even though the covers may be necessary to contain and control odors. By interfering with the way the operator has performed duties in the past, the additional labor burden needed to complete the same tasks may not be taken well and facility performance could suffer. To accommodate operator concerns and understand access needs, the designer is strongly encouraged to involve the operator early in the conceptual design process. This way, the cover installation will contain the appropriate appurtenances (i.e., accessways) and cover type that best suits the needs of those who will use it every day—the operators. The design intent is to make the cover system more operator-friendly so that routine servicing and O&M has a better chance of being done. The heavier the accessway or more difficult it is for an operator to gain access, the less likely the accessway, if removed, will be placed back in the original position, thereby compromising the design intent of the odor containment.

Cover accessway needs can be best defined by the operators who will use them. In general, cover accessways can be broken down into the following two broad categories:

- Small openings that are large enough to provide operator access, but small enough to restrict personnel entrance to the covered airspace. These cover accessways would allow for
  - visual observation of wastewater surfaces, corrosion to walls, cover and metal supports, aeration patterns, spray headers and nozzles, and surface foam and scum removal;
  - wastewater sampling;
  - cleaning weirs;
  - measuring sludge blanket depths; and
  - O&M procedures.
- Large openings that allow for personnel access to
  - conduct in-tank mechanism maintenance or removal,
  - clean out-of-service units, and
  - perform tank overhauls.

The number of accessways needed is site- and process-unit specific. Discussions with facility operators are the key to quickly defining the number needed.

The type of cover accessway depends on the materials of construction of the cover. Cover vendors typically offer a wide range of accessways that the owner can choose from. Generally, cover accessways need to be sealed to the best extent possible to minimize the fugitive release of odors from around the seams. Traditional gaskets made of reinforced rubber or flexible material (i.e., hypalon and EPDM [ethylene propylene diene monomer rubber]) are currently used to provide a good seal at the external seams.

When accessways are not hinged and an operator must physically remove the accessway cover, the weight and bulk of the cover must allow for easy removal by all staff if management expects the covers to be consistently replaced. Odor capture and containment could be seriously compromised when the covers are not properly replaced.

It is better that the cover accessway is hinged, particularly in an area where the accessway is frequently used by the operator staff to execute their duties. Hinges provide simpler access and better assurances that the cover has a tight seal when closed. In addition, the more frequently a cover is removed when not hinged, the less chance it has of being placed back over the opening and replaced with proper seal. The chances of replacing the cover are much higher if it is attached than if it was loose. In the past, hinges were an additional area of fugitive odor release, but current designs do not have the hinge in contact with the enclosed nuisance odor air.

5.6 Odor Containment

Placing a cover, building, or enclosure around the source of the nuisance odor release is the first step to containing it. Keeping nuisance odors from escaping the cover, building, or enclosure is the second step. The second step requires that negative air pressure be maintained in the airspace under the cover, building, or enclosure. The negative air pressure prevents the release of nuisance odors to the atmosphere by creating an inward flow of air through any cracks, vents, or designed openings. As a result, nuisance odors are captured and contained within the airspace of the cover, building, or enclosure.

A fan is designed to create this negative air pressure in the airspace by venting the odorous air at a rate that is greater than the supply of air to the same airspace cavity. The volume of air vented from the space over time is referred to as the *ventilation rate*. The ventilation rate will vary based on site-specific information such as the type of containment, presence or absence of workers, whether air being diffused to the wastewater process unit is being contained, and so on. The following subsections discuss various approaches to determining the ventilation rate for worker- and nonworker-access applications.

5.6.1 National Fire Protection Association 820 Ventilation Rates

The ventilation rate required to contain odors, reduce corrosion, provide for safe and comfortable worker access, and minimize the volume of air to be treated can be selected based on experience with designing similar systems. Designers typically rely on building codes, which include mechanical, fire protection, electrical codes, and other industry standards as the basis of their ventilation designs. Unfortunately, ventilation rates and methodology are seldom listed in state and local building codes for municipal WRRF-related spaces.

Certain outside groups such as the National Fire Protection Association (NFPA) and Factory Mutual provide information on recommended practice for ventilating specific areas. For example, NFPA has developed a ventilation standard entitled Standard 820 Revision 2020—Standard for Fire Protection in Wastewater Treatment and Collection Facilities (NFPA, 2020). The purpose of NFPA 820 is to provide fire and explosion protection measures at WRRFs. The goals are to protect life and property, to maintain continued operation, and to protect the environment through protection measures that will reduce the risk of fires and explosions. However, the standard does not intend to assist the ventilation designer as to the appropriate ventilation rate for proper worker safety, comfort, or odor control containment. Accordingly, NFPA 820 does not address toxic gas detection and the need and placement of low oxygen, hydrogen sulfide, or carbon monoxide monitoring. NFPA 820 does require that the combustible gas that is or can be present to be detected typically at 10% of Lower Explosive Limit (LEL).

NFPA 820 was first issued as a standard in 1995. The current version is 2016 and the next version is scheduled for release in 2020. It applies to new facilities or modifications to existing facilities. It does not apply to facilities that were built before 1995. However, the authority having jurisdiction (AHJ) can retroactively apply NFPA 820 if they determine the fire or explosion risk does justify applying them. In this case, the AHJ can alter the NFPA 820 requirements where it is deemed impractical to implement if appropriate protection is provided.

The information provided in NFPA 820 is widely accepted as the industry standard of care, and many municipalities and authorities have adopted the recommendations. The standard defines ventilation rates and the corresponding electrical construction classification for most WRRF processes and building structures. For most facilities, the standard lists two options: one with less than a prescribed ventilation rate, but with a strict electrical installation objective (i.e., National Electrical Code [NEC], Class I Division 1 [explosion proof]), and a second whereby a continuous ventilation rate of either 6 or 12 AC/h (air changes per hour) will allow declassification of the electrical requirements to NEC Class I, Division 2, or unclassified altogether. The standard thereby links the fire protection safety of a structure to its ventilation rate.

NFPA 820 has determined that the primary categories for rooms at WRRFs are Class I, Division 1, Group D (NEC defines Group D as containing methane and gasoline vapors), Class I, Division II, Group D, or unclassified. NFPA 820 provides three tables for the three main processes at wastewater treatment facilities and WRRFs. The three processes are collection systems, liquid treatment, and solids treatment areas. Each table provides detail areas for each of the three main categories. Each area is provided the type of fire or explosion hazards present, area classification, and the types of gas detection, ventilation, and fire suppression needed.

NFPA 820 does allow for certain areas to be downgraded. For example, a Class I, Division 1, space can be downgraded to Class I, Division 2. It also allows Class I, Division 2, spaces to be downgraded to nonclassified. NFPA 820 provides the option to use enhanced ventilation to downgrade the hazard designation of certain spaces. Given that gases and vapors are the primary sources of fire and explosion hazards, increased ventilation can reduce or eliminate the danger in the space. NFPA 820 does require both mechanical supply and exhaust ventilation for spaces that are downgraded by ventilation. Ventilation flow switches are required for both supply and exhaust to alarm if either is lost. The alarm is required inside and outside of the space. NFPA 820 also requires that classified areas have negative air pressure when located adjacent to nonclassified areas.

Designers need to be aware of the requirements of NFPA 820 (2020) and should review and become familiar with the three tables provided in the standard for the three main processes at wastewater treatment facilities.

5.6.2 Standard Ventilation Rates

Before developing ventilation rates for covered unit process areas, it is important to understand their accessibility to staff. Ventilation rates are different for the two categories of nonworker- and worker-accessible spaces. A higher ventilation rate is applied to covered unit process areas where staff will be present performing their required duties. As a general rule, the following ventilation rates have traditionally been applied:

- Nonworker-accessible spaces—4 to 6 AC/h and
- Worker-accessible spaces—12 AC/h.

However, today multiple approaches are evaluated to assess independent ventilation rate needs. The limiting or higher ventilation rate determined by these various approaches is then applied to the covered unit process area. The following sections examine the aforementioned two categories: nonworker- and worker-accessible spaces.

5.6.2.1 Nonworker-Accessible Space

Nonworker-accessible spaces at WRRFs, such as airspace beneath flat covered tanks and channels that do not allow worker access, are typically classified NEC Class I, Division 1 (explosion proof), and require that all electrical equipment and controls located within the airspace and within a 0.9-m (3-ft) radius of tanks or channels meet Class I, Division 1, requirements. Ventilation for these types of spaces are used to purge the enclosure and limit corrosion to the covering system, controls, and the concrete or metal structure itself. The second reason to ventilate from below an enclosed space is to develop negative air pressure, relative to the surrounding atmosphere. This negative pressure differential ensures that odors do not escape to the ambient atmosphere and thereby the surrounding community, but are instead contained and conveyed to the appropriate odor control technology for treatment.

The appropriate ventilation flowrate in cubic meters per second (cubic feet per minute) is determined by evaluating as many as three to four approaches and using the highest or more conservative of the four calculated rates. The first approach is calculated by estimating the enclosed volume, assuming the average low water level or empty volume if the area could routinely be empty of wastewater, applying an air change rate, and determining the ventilation rate. As a general rule for nonworker-accessible areas, the minimum air change rate should be 4 AC/h and the common range should be 4 to 6 AC/h.

The minimum air change rate of 4 AC/h is considered the lowest design ventilation rate at which moisture within the air space will be controlled. Lower rates allow buildup of moisture on structural elements in the airspace, such as concrete and metal surfaces. Hydrogen sulfide absorbed by the accumulating moisture is readily converted to sulfuric acid in this environment. The effect of low pH water on concrete and metal surfaces will be negative (i.e., corrosion) unless these surfaces are protected. However, even with surface protection, low pH water can attack and corrode the concrete and metal surfaces over the long term.

When comparing the first ventilation flowrate approach of applying an air change rate to the next three approaches, consider at least calculating the minimum air change rate of 4 AC/h for the space in question to assess which rate should be considered the lower end, where protecting the internal infrastructure is an important concern.

The second approach, which is significantly more difficult to calculate, is the calculation of flowrate across all openings in the cover system at a given pressure differential. An estimation of the cracks and seams between cover plates, valve and gate stem openings, hatches, control ports, inspection ports, and so on must be inventoried to develop an overall opening size (area). This cover porosity (in square meters or square feet), multiplied by a specific velocity (meters per second, or feet per minute) through the opening area, will determine the ventilation system airflow rate at a specific differential pressure. For example, one approach develops a velocity of 390 m/min (1265 fpm) that is required to develop a 0.025-kPa (0.1-in. H_2O) water gauge pressure differential between the covered airspace and atmosphere. The assumption made here is that the pressure differential is equivalent to the velocity pressure. NFPA 820 and other design guidelines state 0.025 to 0.050 kPa (0.1 to 0.2-in. H_2O) water gauge differential as a design standard for odor control systems.

In a more practical approach, the ventilation rate for the unoccupied area under conveyor covers was ultimately based on estimating the leakage area of the gap between the cover and the conveyor belt. This was easy to determine knowing the length of the conveyor. Here, a velocity ranging from 30 to 76 m/min (100 to 250 fpm) was applied to the leakage area of the gap. The lower velocity range, 30 m/min (100 fpm), was applied to the conveyor belt at indoor locations, whereas the upper velocity range, 76 m/min (250 fpm), was applied to outdoor conveyor locations that were subject to ambient winds. Applying these velocities over the leakage area provided ventilation rates for the various conveyor locations and proved to be very effective after they were installed.

The third approach considers the ventilation required to reduce high concentrations of specific compounds. High concentrations of compounds such as hydrogen sulfide can have multiple effects, including the following:

- Corrosion to concrete and metal surfaces. Corrosion to concrete and metal surfaces is an important consideration when protection of these surfaces with proven coatings or liners is not planned.
- Inability of the odor control technology to achieve the desired exhaust concentrations. The various odor control technologies all have a desired range of hydrogen sulfide influent concentrations where they work best. Exceeding these design influent concentrations may not interfere with the design removal rates, but can affect the exhaust concentrations, which may be regulated by the state via an operating permit. To use a preferred and proven odor control technology in this situation, an increase in the ventilation rate could be implemented to avoid a different odor control technology or a second stage odor control device.
- Limiting access for routine sampling, observation, and maintenance operation.

The ventilation rate increase is based on site-specific information.

The fourth method used to determine the design ventilation rate for nonworker-accessible spaces is provided by vendors of flat cover systems who recommend the application of a standard flowrate per unit surface area ($m^3/(m^2{\cdot}s)$ or cfm/sq ft). These standard values will vary between vendors, and they do not always take site conditions into consideration. It is highly recommended that, when considering using standard flowrates per unit surface area, that the owner and consultant work closely with the vendor, compare the calculated ventilation rates with the first three approaches presented above, and seek performance guarantees from the vendor. In addition, smoke tests are recommended during system performance testing to confirm containment.

When assigning the design ventilation rate for nonworker-accessible spaces, the values for each of the three to four (the fourth approach being applied to flat cover systems) approaches presented are evaluated and compared. Typically, to be conservative, the highest ventilation rate of the three is selected as the design ventilation rate. Remember that site-specific conditions, regulations, and preferences need to be considered when making the final decision.

The above ventilation flowrate discussion for nonworker-accessible spaces assumes that the liquid or solid surface beneath the cover is nonaerated. Higher ventilation flowrates are needed to account for air diffused through the liquid and released at the surface. The easiest and proven approach to estimate the adjusted ventilation rate is to first calculate the ventilation flowrate as if no diffused air were present, following the procedures discussed in previous paragraphs. Next, determine the flowrate of air being injected or diffused to the liquid or solids matrix. If the air flowrate varies, use the maximum flowrate to be conservative. Add this injected or diffused air flowrate to the nonaerated calculated ventilation rate. The sum of the two flowrates represents the design ventilation rate for the covered and contained airspace.

Alternative approaches have been practiced. For example, using a unitized air flowrate per square meter (or per square foot) (approximately 0.015 to 0.020 $m^3/(m^2{\cdot}s)$ or [3 to 4 cfm/sq ft]) of area to be covered has been applied in the past. The flowrate as used has been independent of airspace between the cover and water surface and the amount of diffused air applied. Accordingly, caution should be followed when using a unitized flowrate per covered surface area. The final design ventilation rate should be verified.

5.6.2.2 Worker-Accessible Space

Worker-occupied buildings and structures are typically ventilated at higher air change rates, in accordance with NFPA 820 guidelines. Supply air is typically filtered, tempered to 12.8 °C (55 °F), and mechanically supplied to areas generally occupied by workers. Typical wastewater odors are some form of reduced sulfur compound, typically hydrogen sulfide, and are heavier than air. Therefore, the exhaust side of the ventilation system should draw air from low-lying dead spaces, where hydrogen sulfide tends to accumulate. Another more aggressive form of exhaust ventilation would be to cover the odorous process and exhaust from within this enclosure or space (i.e., covers located over influent channels within a headworks building or installation of a hood over a grit dumpster). The common theme is to contain the odors and capture them at their source.

Domed, covered tank ventilation systems are like those of a formal building structure. Fresh air is supplied to worker access areas and exhaust is drawn off from the most odorous location. For a dome-covered primary settling tank, the most odorous location is generally at the overflow weirs. Figure 6.22 illustrates a summary of the ventilation rate evaluation approaches for both worker- and nonworker-accessible spaces.

The tank ventilation rates are dependent on the effectiveness of the cover or enclosure system design to be tight (i.e., without leaks). Accordingly, the final tank ventilation rate selection is also dependent on the type of cover or enclosure system because some are more leak resistant. Table 6.4 presents a comparison of process unit cover alternatives, walk-in versus limited-to-no access.

Many worker-accessible spaces are not occupied by staff for significant portions of the day, and yet the ventilation rate for the space is continuously maintained as if it were occupied. Owners and engineers have been evaluating ventilation alternatives that provide an adequate and safe environment for workers when they enter and inhabit a space and reduced ventilation when it is not occupied. Both ventilation rates must satisfy NFPA 820, building codes, and local requirements and consider the industry standard of care practices. Dual-speed fans or fans with variable frequency drives have been used for this service. The energy saved applying this ventilation approach, depending on the application, would help the facility be more energy efficient. Design of this ventilation alternative requires consultation with operators to assess occupancy needs and frequencies to determine appropriateness and standard operating procedures for its implementation.

5.6.2.3 Additional Considerations

Not all basins to be odor controlled (whether occupied or not) are quiescent; some are aerated, and some have fluctuating liquid levels. The ventilation rate for aerated basins is the sum of the appropriate ventilation rate and the design aeration rate. The ventilation rate for fluctuating liquid level basins is

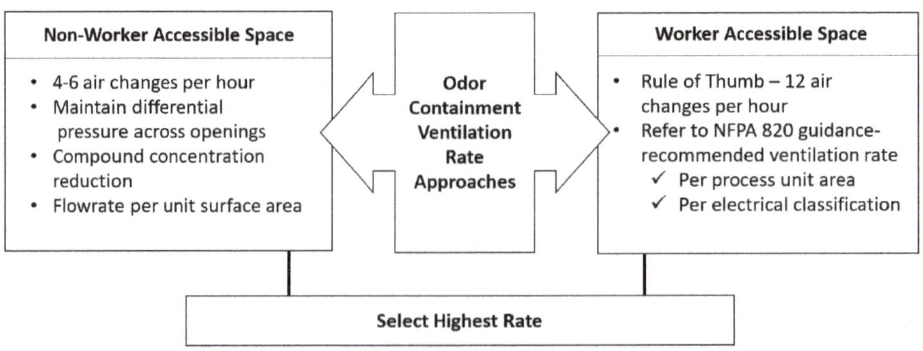

FIGURE 6.22 Summary of the ventilation rate evaluation approaches for worker- and nonworker-accessible spaces.

the sum of the appropriate ventilation rate and the maximum liquid fill rate of the basin. The ventilation rate for aerated fluctuating-level basins is the sum of all three values. This summed ventilation rate provides the capacity to contain the supplied diffused air or air displaced by basin filling while maintaining a good negative pressure under the cover.

5.6.3 Ventilation System Layout

Historically, ventilation systems have been designed in accordance with principles consistent with heating and cooling occupied spaces instead of with odor control in mind. The main objective in designing the ventilation system layout for an occupied space is to provide fresh makeup air in worker space areas in such a way that the worker is bathed in fresh air. Designing for this objective gives the worker a more comfortable environment to work in, minimizes worker safety related concerns, and improves worker attitude and, thereby, productivity.

For worker space areas, forced (fan) air supply and (fan) exhaust are recommended, with the exhaust air flowrate selected to exceed the supply air flowrate by no more than 10%. The higher exhaust air flowrate provides for negative air pressure in the building or enclosure. NFPA 820 stipulates forced air supply and exhaust for most worker-space applications. This approach provides for better control of the worker environment and of negative air pressure in the enclosed space. Passive air supply (i.e., open louvers) and forced air exhaust provide unidirectional airflow and account for stagnant air zones for those areas of the enclosure not between the passive louvers and active exhaust registers. Odors and corrosive compounds accumulate in these stagnant air zones and create a poor working environment for staff and possible damage to the enclosure infrastructure.

Most common odor control fans are base-mounted centrifugal fans with backward inclined wheels. The backward inclined wheels ensure that the motor will not overload under varied system pressure occurrences. The base-mounting arrangement allows for ease of maintenance on the bearings, belt, and motor. All fan housings should be drained to limit the buildup of condensed liquids. Other fan arrangements include base-mounted radial bladed fans and in-line centrifugal fans.

The ventilation system layout for nonworker cover or enclosure areas should include exhaust points, but no forced air supply. Rather than a forced air supply, a passive makeup air device (such as an intake box, louver, or gooseneck pipe) is recommended and positioned strategically away from the exhaust to achieve the desired sweep direction of airflow motion. To avoid the release of odors through the makeup air device, it should be counterbalanced to close if positive pressure builds up in the enclosed headspace. An assumption is made that the final cover or enclosure structure is constructed with high attention to detail regarding containing the odors (i.e., covers are designed with minimal members and gaskets to minimize leaks, and all other gaps between openings are covered).

In some instances, the makeup air device will not be needed because other sources of supply air exist. For example, when covering facility influent channels, air in the headspace of the sewer commonly provides makeup air. For other facility locations, such as loose-fitting covers, openings (that are not addressed) around bar screens, slide gates, and valve structures, makeup air will be supplied through these existing cracks and openings and, therefore, makeup air devices are unnecessary.

Special consideration should be given to ventilating the influent channel of the treatment facility based on physical characteristics of the influent sewer. For those facilities that do not have a common headspace between the influent sewer and plant influent channel, ventilation determinations should follow the approaches previously presented. The types of systems that fall in this category include, but are not limited to, facilities that

- receive their influent wastewater through force mains,
- receive their influent wastewater through surcharged lines that discharge to the influent channel below the water surface,
- use a submerged weir to control the flow of influent wastewater,
- control sewer odors in the collection system by extracting odors and treating them before they reach the facility, or
- install a barrier between the sewer and influent channel.

For those facilities where the influent sewer has a common headspace with the influent channel, additional ventilation volume is necessary to account for the positive pressure of the sewer airspace volume. To estimate the additional ventilation volume, a review of sewer hydraulics is necessary to determine the critical flow regime that will create the worst-case positive pressure conditions.

For worker space areas, the ventilation system layout is site specific. However, there are a few design criteria rules to be considered, including the following:

- Position ventilation system exhaust points at or near the source of odors. Avoid fouling worker space areas by using alternate exhaust locations.
- Understand the odorous components of the controlled airstream. For example, hydrogen sulfide is heavier than air. To efficiently capture

the hydrogen sulfide odor, the exhaust points should be low, near the floor level or water surface.

- The most prevalent odor at WRRFs is hydrogen sulfide. As a result, the makeup air supply is located high, near the roof or ceiling of the enclosure, whereas the odorous exhaust is low, at the floor level where the source of odors is typically found. This positioning also supports the concept of keeping the workers immersed in a flow of fresh air.
- Avoid positioning supply and exhaust registers at equal elevations and on the same enclosure wall because this can support short-circuiting of the ventilation system. Short-circuiting will occur even though the correct ventilation rates are used.
- Forced air supply and exhaust are recommended (NFPA 820 stipulates forced air supply and exhaust for most applications) for worker space areas. Forced air supply and exhaust provide the best ventilation coverage of enclosures. Passive louvered air intakes for supply air, for example, are too unidirectional and allow for short-circuiting and dead zones (areas with no apparent ventilation or air motion) where odors and harmful compounds can accumulate.
- Design the ventilation system layout to avoid dead zones in worker space areas. This condition can be avoided by following the NFPA 820 requirements regarding forced air supply and exhaust.
- Equip the makeup air supply and exhaust registers with volume dampers to control the rate of airflow. Volume dampers are useful when field-flow balancing the ventilation system.
- The makeup air supply should be somewhat less than exhaust to create negative air pressure within the enclosure.
- These criteria will help in the design of an effective ventilation system that also provides for a comfortable working environment.

5.6.4 Conveying Odors

The ability to transport the captured and contained odors from the source to the odor control equipment rests upon the design of a duct network and fan. The fan provides the driving force to create the movement of odorous air through the ductwork. It is just as important to design the ductwork to minimize leaks as it is to reduce leaks from covers and enclosures. Consider each element in the odor control system (i.e., covers, containment, conveying, and treating) as a link in the odor control chain; the chain will only be as strong as the weakest link. As a result, insist that attention to detail be provided for the design of each element.

5.6.4.1 Duct Materials of Construction

Duct materials of construction are similar to corrosion-resistant materials used in covers and enclosures, including

- FRP,
- stainless steel,
- aluminum, and
- PVC.

For corrosive odor environments, FRP and Type 316L stainless steel are the preferred materials. Filament-wound or contact-molded vinyl ester FRP and Type 316L stainless steel have a greater resistance to corrosion from H_2S and chlorides than aluminum and 304L stainless steel and have become the standard for most municipal wastewater odor control installations. For milder odor corrosive environments, aluminum and Type 304L stainless steel are available.

The most commonly used stainless steels in wastewater applications are Types 304L and 316L because they both perform well in acidic odor control environments where sulfuric acid concentrations are likely under 10%. However, when selecting stainless steel alloys for use in odor control, the engineer must carefully consider the chloride ion concentration in the bulk wastewater as well as the probability of chloride concentration effects in the gas phase. Types 304L and 316L will begin to experience pitting and crevice corrosion when bulk wastewater chloride concentrations are near 300 mg/L and 1000 mg/L, respectively.

Odor control airflow conditions tend to exacerbate chloride concentration effects. Therefore, when wastewater chloride levels are even reasonably high (150 to 200 mg/L), Types 304L and 316L stainless steel would not be the preferred selection for duct materials of construction. Duplex stainless steel alloys, Types 2205 and 2507, are growing in popularity because of their higher strength, better chloride pitting resistance, and superior stress corrosion resistance when compared to Type 316L.

Although PVC has been used for small flow installations, it is limited to duct diameters of less than 45.72 cm (18 in.). There are concerns over its ability to maintain its shape over time in hostile and corrosive environments.

Below ground, the choice of materials of construction narrows. The two most commonly used are as follows:

- High-density polyethylene (HDPE)—Check with the manufacturers to be sure that the HDPE being considered is rated for below-ground use.

- FRP—This is used for conditions where truck and vehicular traffic may affect the heavy stress on the buried duct (i.e., if a duct runs under a heavily traveled roadway at the facility).

5.6.4.2 Shapes

Duct shapes used for odor control are typically round or rectangular. In some cases, the selection of duct shape depends on the material of construction. For example, aluminum duct construction is almost exclusively rectangular. In other cases, tight-fitting installations may dictate the shape because of limited space to route the duct.

The round shape is more product efficient (i.e., for the same cross-sectional area, the round shape has a smaller circumference than a rectangular duct has perimeter). As a result, less material is needed. For FRP, the material with the highest corrosion resistance to odors encountered in WRRFs, it is easier to make a round duct or pipe using filament-wound techniques that provide a high degree of quality control and minimize the need for hand layup.

5.6.4.3 Sizing Criteria

Duct-sizing criteria are based on the odorous air flowrate and pressure maintained within the network. A good design range for odor control metric duct is 610 to 914 m/min (2000 to 3000 fpm), within exceeding duct-pressure drop losses of 0.062 kPa (0.25 in. H_20) water gauge per 30.48 m (100 linear ft) of ductwork. Velocities can exceed 914 m/min (3000 fpm); however, the designer must be prepared to enhance the duct and fan to account for

- increase in pressure losses throughout the duct network,
- modifications or redesign of the fan to transport the same volume of air at greater static pressure requirements,
- potential to generate noise at higher velocities, and
- potential need for reinforcing the duct (increasing the thickness) to accommodate the increased pressure.

Pressure-drop calculations should be performed systemwide in accordance with the standards of the American Society of Heating, Refrigerating and Air-Conditioning Engineers (ASHRAE) *Fundamentals Handbook* (American Society of Heating, Refrigerating and Air-Conditioning Engineers, 2001), or Sheet Metal and Air Conditioning Contractors' National Association (SMACNA) *HVAC Systems Duct Design Manual* (Sheet Metal and Air Conditioning Contractors' National Association, 1990).

5.6.4.4 Additional Ductwork Notes

The following are miscellaneous additional notes regarding the design of odor control ductwork:

- Accommodations must be made within ductwork to account for thermal expansion and contraction. How severe this process will be depends on the extremes of temperature experienced at the site and whether ductwork is inside or outside. Expansion joints are used to address thermal expansion and contraction.
- Ductwork must be sloped to a drain at low points of the system. Expansion joints should be located at high points in the duct run because they tend to inhibit the flow of water in the duct. The odorous airflow collected at the odor source is typically moist. As temperatures decrease along the ductwork, condensation will occur, and water will accumulate in the ductwork. Stainless steel and FRP, commonly used in outdoor situations, behave somewhat differently. Stainless steel is an excellent conductor of temperature and condenses the moisture out of the airstream during the cooler months. FRP is more insulated and less of a conductor. Nevertheless, condensation will occur, but to a lesser degree.
- Ductwork drains should include a water seal trap that allows the condensate to drain and does not allow ambient air into the system nor odorous air out. Consider the local weather conditions to determine whether heat tracing is required to keep the drain line and trap from freezing in cold months.
- Placing the ductwork upstream of the fan provides enhanced odor containment. The fan in the downstream position creates a vacuum (negative air pressure) along the length of the ductwork. Accordingly, any leaks within the ductwork will pull air in through the leak, thereby preventing any release of odors.

5.6.5 Fans

Fans represent the heart of the odor control system by creating negative air pressure at the source, providing containment, delivering the odors to the control technology, and discharging the treated air to the atmosphere. Because of the corrosive nature of the odorous airstream, resistant materials are required. The most common material for odor control fans is FRP, with a graphite impregnated wheel needed to ground any static charge buildup experienced during operation. Stainless steel wheels may be evaluated for applications in very warm climates where FRP may be affected, if there may be particulate matter in the airstream that could scratch or damage the FRP, or for high fan revolutions per minute applications.

5.6.5.1 Fan Positioning

Positioning the fan within the ductwork network and odor control technology is at the discretion of the designer. However, some advantages to the push or pull positioning are dictated by site constraints and equipment being used, as follows:

- Transporting odorous air in ductwork benefits from positioning the fan at the end of the system. Negative pressure is maintained throughout the length of the ductwork, and odors are precluded from escaping leaks, cracks, or small openings.
- Certain odor control technologies force air through the device. For example, fans associated with activated carbon and biofilters are almost always found pushing through individual media beds to achieve treatment.
- Wet scrubbing odor control technology can be operated as effectively in either a push or pull arrangement. Although the fan materials of construction are selected to resist the corrosive nature of the odorous airstream, placing the fan downstream of wet scrubbing in a pull-through or induced draft arrangement would eliminate the corrosive effect of the untreated odorous airstream on the fan. Downstream positioning would also prevent any odor from escaping the fan housing, duct connections, and drive shaft that sometimes occurs with an upstream fan arrangement in front of the wet scrubber.
- Upstream fan positioning could minimize ductwork on single- and multiple-stage wet scrubbing systems. Pushing odorous air through a single- or multiple-stage system allows for treated exhaust to be discharged out the top of the unit. Pulling the air through an induced draft arrangement requires ductwork from the top of the unit be directed to a fan located somewhere below, increasing the length of ductwork.

The designer must weigh advantages and disadvantages of either approach before selecting the appropriate fan arrangement.

5.6.5.2 Fan Balancing

The airflow within the odor ductwork system should be balanced to ensure the design amount of air is being drawn from its intended location. Testing, adjusting, and balancing (TAB) services are the way to confirm that design airflows are meeting compliancy. These TAB services are commonly contracted to an independent licensed balancing firm.

The TAB services should be completed before the installation is accepted by the owner. Discrepancies must be addressed and fixed immediately, or at least before the system is accepted. The TAB services should also be used after the system has been operating to ascertain if design flow levels have been maintained and any time modifications are made to the ductwork, exhaust pickup points, or damper settings.

Provisions should be made upstream of each balancing damper to drill two holes large enough to insert a velocity probe that would be required to test the airflow within the duct. The duct hole should also be large enough to pass a flexible line with video capability to allow for viewing the condition and position of the balancing damper. The position of the holes should be at a right angle to each other (e.g., at the top of the duct and at the side of the duct, 90° from the top hole). This allows velocity measurements to be monitored from the top of the duct down to the invert of the duct, and from side to side. A cap should also be provided to fill the hole after the probe is removed.

5.6.6 Dampers

The use of dampers within any ducted ventilation system is typically required. The dampers are commonly used for balancing the airflow quantities and isolation of equipment from the system. The same damper is not recommended for both services. Dampers on the supply air systems should follow the guidelines of ASHRAE or Sheet Metal and Air Conditioning Contractor's National Association (SMACNA).

The following discussion focuses on dampers for foul air exhaust systems, starting from the exhaust point and traveling to the fan and control device. Dampers should be located on each duct connection or at each intake register. These dampers allow each point source to be balanced from one another. In exhaust register installation, a high-quality volume damper should be installed in lieu of the volume damper that may be purchased as an option with the register because they are not typically of sufficient quality to act as a reliable and repeatable control device. Secondly, the branch ductwork should have its own balancing damper to balance the branches independently from the trunk duct. Permanent test ports should be installed downstream of the dampers to allow for airflow measurements to be taken.

Each fan should have an isolation damper upstream and downstream of it. Inlet vane dampers may be added upstream of the fan because they act as a highly efficient means of system flow control.

Volume dampers can be manufactured using the same materials as the ductwork systems (i.e., FRP, stainless steel, aluminum, and plastic). Most commonly installed dampers are butterfly type for round duct installations and airfoil opposed blade type for rectangular installations.

6.0 CASE HISTORIES

The New York City Department of Environmental Protection owns and operates 14 WRRFs in the five boroughs of New York City. Cover and odor control designs for primary sedimentation tanks at two of the facilities (North River and Coney Island) were prepared and construction was completed to reduce the off-site odor effect from these unit process sources.

6.1 Coney Island Water Resource Recovery Facility

The Coney Island Wastewater Treatment Plant is surrounded by high-density residential neighborhoods. In addition, it has a Boys Club park at its fence line and a four-lane road that divides the facility in two. Figure 6.23 shows an aerial view of the plant. At Coney Island, the long and narrow rectangular primary sedimentation tanks, aerated influent channels, and effluent launders are key odor sources. A total of over 20 cover alternatives were initially considered, including all flat covers, barrel arches, enclosures, and combinations, as well as different construction materials. After discussions with facility staff regarding their needs for safety and access to the process,

FIGURE 6.23 Aerial view—Coney Island WRRF, New York.

the list of alternatives was reduced to three. Table 6.6 presents the three screened alternatives.

To assist in evaluating differences between cover types and materials, Table 6.7 presents the unit costs at the time of evaluation for the various alternatives.

It was clear after meeting with facility staff that the overriding concerns were not solely related to minimizing the flow of air to be treated, but to provide ease of access to perform routine and significant O&M procedures. Table 6.8 presents the airflow and comparative cost figures for the three alternatives. (Alternative 1 was divided into two sub-options, where the materials of construction were varied between aluminum and FRP.) The final selected alternative, Alternative 2, is clearly not the least expensive. However, for the facility, it was the most practical.

TABLE 6.6 Coney Island WRRF—primary sedimentation tank covering alternatives summary.

Alternative Number	Alternative Description	Alternative Advantages
1	Flat covers on influent channel.	Easy operator access and visibility at either end of the tanks.
	Movable barrel arch covers on influent and effluent tank ends.	Relatively low air treatment volume (approximately 623 m^3/min [22 000 cfm]).
	Flat covers on remaining tank areas.	
	Access hatches on effluent launders.	
2	Flat covers on influent channel.	Easy operator access and visibility throughout the entire tank.
	Barrel arch covers over the entire tanks with movable sections on both ends and at center of tank.	Moderate air treatment volume (approximately 850 m^3/min [30 000 cfm]).
	Access hatches on effluent launders.	
3	Flat covers on influent channel.	Full operator accessibility at the effluent end.
	Barrel arch over the entire tanks with movable sections at influent end and center of tank.	Easy operator visibility throughout the entire tank.
	Extend skimmings enclosure building approximately 6.1 m (20 ft).	
	Access hatches on effluent launders.	

TABLE 6.7 Comparison of alternative cover and materials of construction costs.

Cover Type	Cover Material	Cover Unit Costs	
		$/m^2$	$/ft^2$[a]
Flat covers	Aluminum	269	25
	Fiberglass reinforced plastic	323	30
	Precast concrete (25.4 cm [10 in.])	38	
	Cast-in-place concrete	398	37
Barrel arch covers	Aluminum	431	40
	Aluminum, movable	861	80
	FRP	48	
	FRP, movable	1100	
Enclosures	Domes, aluminum		
	15.24-m (50-ft) span standard	20	
	15.24-m (50-ft) span low rise	25	
	36.576- to 48.768-m (120- to 160-ft) span standard	161	15
	Domes, FRP		
	15.24-m (50-ft) span	215	20
Buildings	—		
Limited enclosures	—		

[a]Costs received from vendor quotes in 2001. Sq ft × 0.0929 = m^2.

An important lesson is to never rule out the opinions of staff that have to work with the system after it is installed. In addition, site constraints and conditions can also be a deciding factor in the final selection of covers or enclosures. Figures 6.24, 6.25, and 6.26 illustrate, respectively, an overall view of the barrel arch covers, weighted damper, and access door at the end of each barrel arch, and an access hatch near mid-tank with a high-intensity lamp available for easier viewing.

6.2 North River Water Resource Recovery Facility

The North River WRRF is also surrounded by the public, both residential and transient. The North River WRRF has a 11.3 ha (28-ac) state park as a rooftop neighbor where 3000 to 20 000 people may visit on a given day.

TABLE 6.8 Coney Island WRRF—primary sedimentation tank covering costs.

Number	Alternative Description	Cover Area, Approximate Square Meters (square feet)	Cover Unit Cost Per Area $\$/m^2$ ($\$/ft^2$)	Ventilation Flow Rate, Approximate m^3/min (cfm)	Cover Cost (Subtotal)
1a	Aluminum flat and barrel arch covers	5220 (56 200)	270 (25)	1250 (44 000)	$1 405 000
		950 (10 200)	430 (40)		$408 000
					$1 813 000
1b	Fiberglass flat and barrel arch covers	5220 (56 200)	320 (30)	1250 (44 000)	$1 686 000
		950 (10 200)	520 (48)		$489 000
					$2 175 600
2	Flat fiberglass covers on aerated influent channel and weir overflow channels	560 (6000)	320 (30)		$180 700
	Fiberglass barrel arch covers over entire PSTs	5460 (58 800)	520 (48)	1700 (60 000)	$2 823 400
					$3 004 100
3	Extend skimmings enclosure 6.096 m (20 ft)	300 (3200)	3800 (350)		$1 134 000
	Flat aluminum covers on the aerated influent channels and weir overflow channels	560 (6000)	270 (25)		$150 600
	Barrel arch aluminum covers on remaining primary settling tanks	4660 (50 200)	430 (40)	2350 (82 500)	$2 007 040
					$3 291 640

FIGURE 6.24 Primary sedimentation basins barrel arch covers—Coney Island WRRF, New York.

FIGURE 6.25 Primary sedimentation basins barrel arch cover (end view with weighted damper [left] and access door [right])—Coney Island WRRF, New York.

FIGURE 6.26 Primary sedimentation basins barrel arch cover: Tank access point—Coney Island WRRF, New York.

To its immediate east is a six-lane commuting highway. To the east of the highway is a high-density residential neighborhood. Figure 6.27 shows an aerial view of the plant, park, and community. Odor control has been a focal point in the design and operation of the facility. At the North River WRRF, a similar cover system evaluation was conducted for the long and narrow rectangular primary sedimentation tanks. At the multilevel North River facility, the primary tanks are sandwiched between storage rooms on the level below and a well-used state park on the level above. However, around the primary tanks, there are no side walls. In fact, two sides of the primary tanks are directly adjacent to open arches that rim the perimeter of the facility. Six alternatives were developed subsequent to discussions with facility staff, as follows:

- Alternative 1—enclosure of the entire primary tank area,
- Alternative 2—enclosure of individual primary tank areas,
- Alternative 3—flat covers,
- Alternative 4—arched covers,
- Alternative 5—pitched covers, and
- Alternative 6—combination of alternatives 2 and 3.

FIGURE 6.27 Aerial view—North River WRRF, New York.

Because of the sensitivity of the park area immediately above the tanks, the proximity of a highly used six-lane commuter access highway on the east side of the facility, and concerns from the community beyond the highway, the cover system had to provide complete odor control at all times. This included when the tanks were taken out of service and during maintenance procedures. This requirement precluded some low-flow alternatives such as all flat, barrel arch, and pitched arch covers. On the other hand, complete enclosures over individual tanks or the entire primary tank area amounted to an enormous volume of air requiring treatment (i.e., 9910 m^3/min [350 000 cfm] and 15 000 m^3/min [530 000 cfm], respectively).

The facility selected alternative 6, a combination of limited enclosures at the influent and effluent areas of each tank and flat covers for the quiescent tank surface areas. Under normal operating conditions, where 12 continuous AC were provided in the enclosures and 6 AC under the flat covers, a flowrate of 2125 m^3/min (75 000 cfm) would be necessary, which is considerably less than the enclosure alternatives.

For staff at both facilities, it was clear that worker access in key areas was required. The key areas were the influent and effluent ends of the tank, where operator accessibility is frequent and of great importance.

Odor containment and ventilation play a crucial role in the effectiveness of any odor control system. Clearly, an evaluation of available alternatives and design criteria are required to select the most functional and least costly system. However, facility staff are also a factor because they will be the primary users of the facility after it is built. Thus, staff input lends a more practical approach to the selection.

7.0 REFERENCES

American Iron and Steel Institute. (1977). *Design guidelines for the selection and use of stainless steel—A designer's handbook series.* Washington, D.C.: Committee of Stainless Steel Producers.

American Society of Heating, Refrigerating and Air-Conditioning Engineers. (2001). *Fundamentals handbook.* Atlanta, GA: American Society of Heating, Refrigerating and Air-Conditioning Engineers.

National Fire Protection Association. (2020). Standard 820—Standard for fire protection in wastewater treatment and collection facilities. Quincy, MA: National Fire Protection Association.

Sheet Metal and Air Conditioning Contractors' National Association. (1990). *HVAC systems duct design manual* (3rd ed.). Chantilly, VA: Sheet Metal and Air Conditioning Contractors' National Association.

Ziols, G. E. (1997). Odor control in an urban environment. *Proceedings of the Water Environment Federation's Control of Odors and VOC Emissions Specialty Conference*, Houston, Texas, United States (pp. 3–12). Alexandria, VA: Water Environment Federation.

8.0 SUGGESTED READING

American Conference of Governmental Industrial Hygienists. (2001). *Industrial ventilation, manual of recommended practice* (24th ed.). Cincinnati, OH: American Conference of Governmental Industrial Hygienists.

American Society of Heating, Refrigerating and Air-Conditioning Engineers. (2003). *HVAC applications handbook.* Atlanta, GA: American Society of Heating, Refrigerating and Air-Conditioning Engineers.

Water Environment Federation. (2004). *Control of odors and emissions from wastewater treatment plants* (WEF Manual of Practice No. 25, 1st edition). Alexandria, VA: Water Environment Federation.

7

Solids Handling

1.0 BACKGROUND: WHY BIOSOLIDS ODOR AND VOLATILE ORGANIC COMPOUND EMISSIONS MATTER 207
- 1.1 Vector Attraction Reduction and Pathogen Kill 209
- 1.2 Federal and State Regulation Restrictions Versus Community Odor Impacts 209

2.0 ODOR SOURCE CHARACTERIZATION IN SOLIDS HANDLING 210
- 2.1 Odor Generation Mechanisms and Odorous Compound Summary 210
- 2.2 Thickening 211
- 2.3 Digestion 211
- 2.4 Thermal Hydrolysis 212
- 2.5 Dewatering 212
- 2.6 Lime Stabilization 213
- 2.7 Composting 213
- 2.8 Cake Storage 213
- 2.9 Odor Data Summary for Various Types of Biosolids 213

3.0 CONTROL METHODS AND TECHNOLOGY OPTIONS 215
- 3.1 Solids Process Configurations to Minimize Odors 217
- 3.2 Liquid-Phase Sludge Treatment Summary 218
- 3.3 Odor Capture, Containment, and Ventilation 219
- 3.4 Gas-Phase Treatment Summary 221

4.0 REFERENCES 223

1.0 BACKGROUND: WHY BIOSOLIDS ODOR AND VOLATILE ORGANIC COMPOUND EMISSIONS MATTER

Odors associated with water resource recovery facilities (WRRFs) are an area of growing concern (Water Environment Federation, 1995). Often, among the more significant odors at WRRFs are sludge- and biosolids-related processes, including the dewatering of liquid sludge and biosolids to make biosolids cake and the various sludge processes used to stabilize the biosolids. This can also include off-site recycling of biosolids at land application sites.

Figure 7.1 is a biological oxidation–reduction potential (ORP) scale (Apgar et al. 2007). Although this figure was developed for a study of odor and corrosion in collection systems, the concepts also apply to the potential for odor generation in liquid biosolids and stored biosolids cake. Typically, fresh biosolids with positive ORP, characteristic of aerobic systems, have less tendency to generate offensive odors than anaerobic systems with low or negative ORP. Fresh sludge can, however, rapidly deteriorate to anaerobic septic conditions, which often results in odor problems (U.S. EPA, 2012).

Anaerobic (low ORP) conditions typically produce reduced-sulfur and nitrogen-based compounds that can have very low odor detection thresholds and tend to create the potential for negative odor effects. Some of the more typical odorous compounds associated with anaerobic conditions are hydrogen sulfide, methyl mercaptan, dimethyl sulfide, dimethyl disulfide,

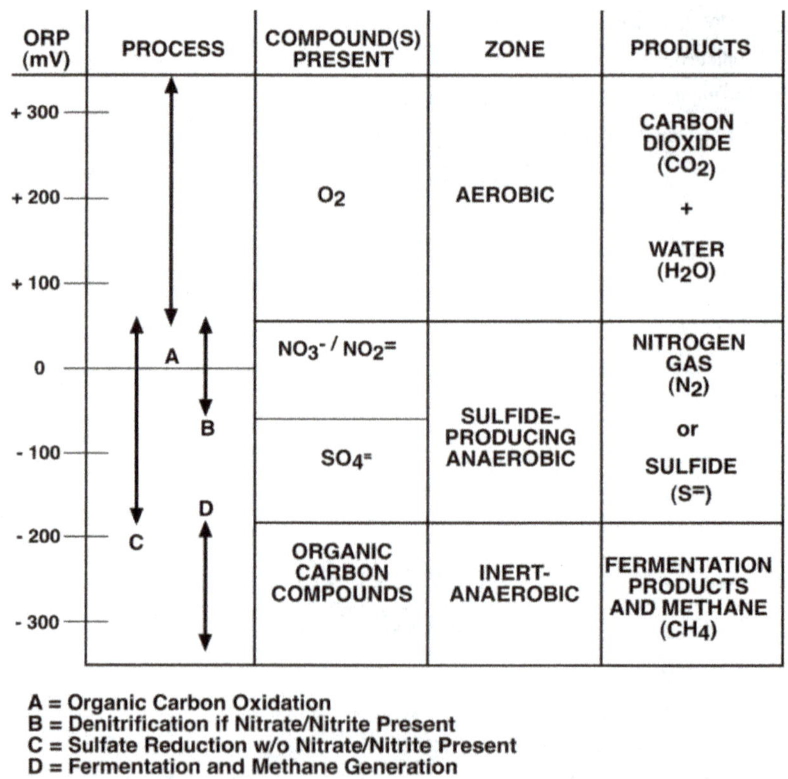

FIGURE 7.1 Biological ORP scale (Apgar et al., 2007).

and carbon disulfide. Ammonia and amine-based odors also may occur in digested biosolids cake and lime stabilization processes. Other volatile organic compounds (VOCs) may be present, but they typically contribute less to odor effects.

Of particular concern for managing stabilized biosolids is the perception that if odors are detectable, then a biosolids product represents a public health and safety issue. Although this is not typically the case, the perception can be problematic enough to be a concern.

1.1 Vector Attraction Reduction and Pathogen Kill

Vector attraction-reduction requirements have been established under the U.S. Environmental Protection Agency's Part 503 regulations, with the intent that reducing attraction to flies, rodents, birds, and other vectors will reduce the potential for the spread of infectious disease. There are various methods for producing a stabilized cake product. Digestion methods have been established to achieve vector reduction that are based largely on achieving a minimum volatile solids destruction of 38% or more. For other stabilization processes, the Part 503 regulations rely on time and temperature under either elevated heat or elevated pH conditions, or both. Finally, heat drying by either solar or artificial means are also used.

All of these methods attempt to provide pathogen kill and reduce vector attraction. Arguably, they would also provide odor reduction. This, however, is not always the case. In fact, a Water Research Foundation (WRF) (2004b) study that looked at 11 WRRFs with anaerobic digestion indicated that even WRRFs with effective volatile solids destruction and relatively long sludge retention times in digesters could still be relatively odorous. No specific odor reduction or odor emission criteria exist in the federal regulations for how a biosolids cake is processed. States may include odor or VOC emissions control, but federal regulations do not directly dictate odor treatment.

1.2 Federal and State Regulation Restrictions Versus Community Odor Impacts

Emissions from WRRFs are governed by a variety of federal and state regulations, including sections of the 1990 Clean Air Act (42 U.S.C. § 7401) applicable to the release of VOCs and hazardous air pollutants (HAPs). Typically, regulation of VOCs and HAPs under the Clean Air Act have little or no direct requirement for odor reduction. Odor effects are more typically controlled by states or communities and may be limited to state or municipality codes that limit activities that prevent people from enjoying their property or recreational environments such as bike paths or athletic fields (e.g., nuisance odor regulations).

Potentially far more important for biosolids is the public's perception that the presence of malodors indicate potential harm to human health. In their study on the health effects of biosolids odors, WRF (2004a) completed an extensive investigation that concluded that odor sensations do not cause illness. Although there is no clear evidence of a direct link between odors and health, this is a much more difficult and sensitive topic to address.

2.0 ODOR SOURCE CHARACTERIZATION IN SOLIDS HANDLING

This section highlights the nature of odors and how differences in biosolids sources might differ in the resulting odors.

2.1 Odor Generation Mechanisms and Odorous Compound Summary

Odors from biosolids processes typically comprise a relatively complex mixture of reduced sulfur compounds (RSCs) such as hydrogen sulfide, methyl mercaptan, dimethyl sulfide, dimethyl disulfide, and carbon disulfide. Ammonia or amine-based odorants may also be present. If scum or fats, oils, and grease (FOG) processes are added to the cake, then fatty acid odors might also be of concern.

Anaerobic (septic) conditions typically increase odors because of greater generation of hydrogen sulfide (H_2S) and various reduced-sulfur organic species. Figure 7.2 is based on work completed under a WRF study (Adams et al., 2008), which proposes a variety of ways that odorants can be generated on the basis of the breakdown of protein compounds.

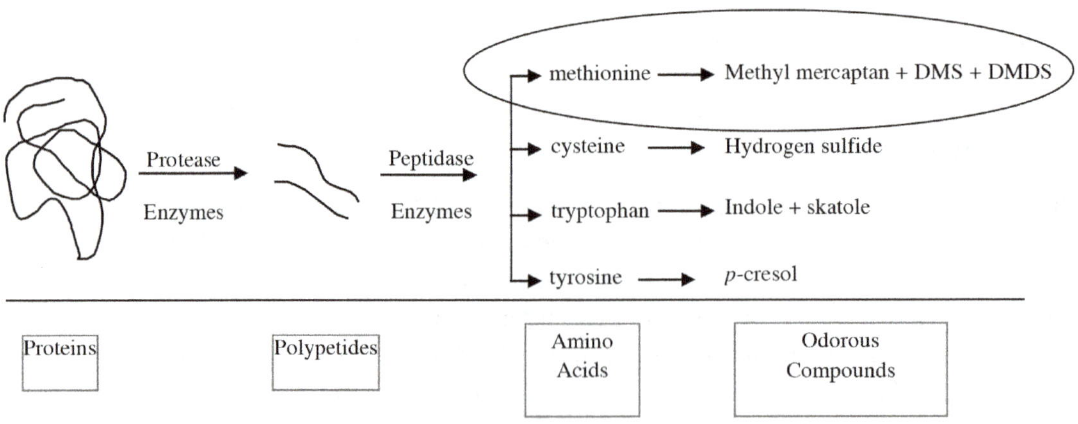

FIGURE 7.2 Odorant production pathways (Adams et al., 2008).

2.2 Thickening

Thickening in general is intended to capture free water, thereby concentrating the solids content in liquid biosolids and reducing volumes that need be further processed during dewatering. Sludge thickening can be accomplished by a variety of methods including gravity thickening, dissolved air floatation, gravity belt filters, rotary drum thickeners, and thickening centrifuges. The types of odors emitted from sludge thickening will be highly dependent on the nature of the sludge; for example, primary sludge odors differ from waste activated sludge (WAS) odors, odors from septic sludges differ significantly from those from aerobic sludges, and sludges affected by FOG will tend to have different odors as well.

Fresh aerobic WAS tends to have an earthy or musty odor that is less intense and less offensive than primary sludge. If, however, WAS is stored and becomes septic, then malodor can become an issue. Blended sludges (primary sludge plus WAS) can be even more odorous because the biological activity of the WAS combined with liquid primary sludge, a food source, tends to cause the sludge to become septic quickly. Septic conditions generate high levels of hydrogen sulfide and related RSC species.

Some of the thickening systems are inherently enclosed and self-contained to one degree or another. Open thickening processes, such as gravity belt thickeners or gravity thickening tanks, will tend to have more fugitive odor emissions.

2.3 Digestion

Digestion is provided to help stabilize biosolids, which in turn reduces vector attraction potential, reduces potential pathogens and provides volatile solids destruction. Digested-cake odor concentrations will depend on whether digestion was aerobic or anaerobic. Odors from aerobic digestion processes and of aerobically digested dewatered cake depend greatly on the content of dissolved oxygen during the process. Fresh, aerobically digested cake tends to have musty odors typical of aeration basins, but is typically stronger than aeration basins.

Anaerobic digestion processes can lower the odor of the final biosolids product, but can also begin to create new odorants of concern, such as ammonia or amines. Anaerobic digestion occurs in an enclosed vessel, forcing anaerobic conditions. The headspace of the digester will typically contain high concentrations of hydrogen sulfide in the range of 0.01 to 1% (WEF, 1995). Other RSCs, ammonia, and potentially lower levels of VOCs will be present. Care must be taken to capture and control this highly odorous (and potentially explosive) digester off-gas by providing flares or other treatment methods should any gases be released.

Autothermal thermophilic aerobic digestion (ATAD) is a high-rate digestion approach using elevated temperatures. It typically includes a fill-and-draw batch cycle and an aerated period that can be odorous. This is particularly the case if the biochemical oxygen demand of the batch feed overwhelms the aeration systems. The ATAD odors can be complex, with very high hydrogen sulfide, mercaptan, dimethyl sulfide, ammonia, and amine-based odors.

2.4 Thermal Hydrolysis

Newer thermal hydrolysis predigestion processes are also being introduced to the U.S. market, such as the "Cambi" process. The thermal hydrolysis process (THP) consists of a high-temperature, high-pressure steam pretreatment of solids to create thermally hydrolyzed solids (THS) before treatment in a mesophilic anaerobic digestion process.

In general, THP produces liquid sludge with a high tendency for ammonia odors, but the dewatered cake is less odorous than other traditional digestion processes.

2.5 Dewatering

Dewatering is provided to further remove free water and further concentrate biosolids. It is typically the last volume reduction process before biosolids disposal. Dewatering is typically completed by means of mechanical dewatering equipment such as belt filter presses, centrifuges, rotary presses, or plate and frame presses. Alternatively, open drying beds are used in warm arid environments.

The types of odors emitted from dewatering are highly dependent on the character of the raw sludge. Septic feedstock tends to result in high hydrogen sulfide and RSCs such as methyl mercaptan, dimethyl sulfide, dimethyl disulfide, carbonyl sulfide, and carbon disulfide. Digested sludge may also release relatively high levels of ammonia and amines. Amines (particularly tri-methylamine) can be aggravated by the type of polymer being used, particularly if the pH is alkaline. Amine-based polymers can cause a fishy smell in the cake.

Recent work indicates that, for anaerobically digested biosolids (Apgar et al., 2007), higher levels of bioavailable (labile) protein in biosolids cake can create more odors over time. Different dewatering practices can affect bioavailable protein differently; some dewatering practices tend to increase odors in the biosolids cake. For instance, with high solids centrifuges, post-dewatering odors can be higher. Additionally, volatile sulfur compounds, or RSCs, are typically the most significant sources of odors in digested biosolids.

2.6 Lime Stabilization

The addition of lime can provide elevated temperatures and pH sufficient to meet sludge stabilization requirements. Lime stabilization typically will cause the release of elevated concentrations of ammonia and amine odors during and shortly after the mixing process. Otherwise, well-stabilized limed cake has reduced long-term odors.

2.7 Composting

Composting provides biological stabilization of biosolids by means of elevated temperatures over time. Literature and field experience reveal that composting facilities emit a wide variety of potentially odorous compounds, including hydrogen sulfide, RSCs, ammonia, amines tri-methyl, amines, skatole, volatile fatty acids, ketones, and other compounds such as terpenes, alcohols, and aldehydes (WERF, 2003). Experience also reveals that odor release is time dependent in the composting process.

In composting, the non-hydrogen sulfide compounds are typically of greater concern. The level of odor emission from composting is also highly dependent on the composting cycle, with higher odor levels being emitted during the initial weeks of the process.

2.8 Cake Storage

Cake storage in open bins can also be odorous. Cake storage facilities typically have covered canopies to keep rain off the cake. Odors depend on the nature of the cake solids (digested or undigested, for instance). Cake is typically most odorous when initially placed on a storage pad with odor levels dropping over time. Disturbance of a cake pile by front end loaders can then produce new odors as the dried cake cover is disturbed and lower levels of the pile that have gone anaerobic are exposed.

2.9 Odor Data Summary for Various Types of Biosolids

Table 7.1 is excerpted from a study summarizing typically reported odor levels (WERF, 2003). Many additional summaries listing the wide variety of odor emissions from composting are also available (Epstein, 2007). Odor data in Table 7.1 are provided both as specific, odor-causing compounds and in terms of odor units (OUs) or dilutions-to-threshold (D/T), a measurement that is based on odor panel analysis methods. These measures are a way to quantify broad spectrum odors based on the number of dilutions of odor-free air required to dilute an odor sample to a point that an odor panel just barely detects a difference between the diluted sample and an odor-free blank (based on American Society for Testing and Materials 369-04).

TABLE 7.1 Odor source characterization (data pulled from WERF, 2003; WERF, 2004a).

Source	Odorous Compounds: Compound Name	Odorous Compounds: Concentration (ppmv)[a]	Panel Analysis (D/T)[b]	Reference
Digester vents	H_2S	0.91–19.5		Witherspoon et al., 1999
Digester gas	H_2S	58–80	29 750	CH2M HILL, 1997
		0.1 to 1%		Water Environment Federation, 1995
	Carbonyl sulfide	7–15		CH2M HILL, 1997
Biosolids tank	H_2S	100	24 150	Bowker, 1999
Biosolids gravity thickener	H_2S	50–200		Burrowes et al., 2001
Biosolids dewatering and storage	H_2S	0.24 (avg.)	24 OU/m^3 (avg.)	Koe and Yang, 1999
		1.5 (peak)	34 OU/m^3 (peak)	
	NH_3	0.33 (avg.)		
Biosolids dewatering	H_2S	3	3800–4600	Burrowes et al., 2001
	Mercaptans	0.005		
	Carbonyl sulfide	0.030–0.035		
Biosolids dewatering	H_2S	34–42		
	Mercaptans	0.0003		
	Methyl sulfide	0.002–0.007		
	Dimethyl disulfide	0.0055		
Biosolids holding tank	H_2S	111 (avg.)	69 200–81 400	Jubinville, Stallings, & Bowen, 1997
		20 to 300 (range)		
Digested biosolids cake	N/A	N/A	340–650	Sereno et al., 1993

[a]Values are expressed as parts per million volume (ppmv), with the exception of data from the Water Environment Federation (1995) expressed in percentage value (%).

[b]Values are expressed as dilutions-to-threshold (D/T), with the exception of data from Koe and Yang (1999) that are expressed as OU/m^3.

The data summarized in Table 7.1 suggest that hydrogen sulfide and RSCs dominate. It should be mentioned that digestion-related odors can be much higher for strong industrial digester or upset digestion processes. This may or may not prove true in a particular application, especially if the biosolids processing trains, such as digestion, are not operating properly. As already mentioned, hydrogen sulfide concentration is sometimes reported in the range of 0.01 to 1% (approximately 100 ppm to 10 000 ppm). Hydrogen sulfide levels are often in the range of 100 to 300 ppm.

Tables 7.2 through 7.4 provide additional odor emission summaries of anaerobically digested cake that had been processed to achieve 38% VOC destruction (Easter et al., 2008). These data are from a municipal WRRF in southeastern Virginia using traditional anaerobic digestion.

This summary also allows a comparison between digested cake comparing belt filter press and centrifuge dewatered digested cake. Centrifuge cake stored over time appeared to be more odorous than belt press cake. Odors in both cases were primarily hydrogen sulfide, methyl mercaptan, dimethyl sulfide, carbonyl sulfide, carbon disulfide, dimethyl disulfide, tri-methyl amines, and ammonia. Out of 17 volatile acids analyzed, acetic acid was the only one always detected in the range of 16 to 130 ppb. Lower levels of propionic acid, isobutyric acid, methylbutanioc acid, isovaleric acid, and isocaproic acids were observed in some of the samples.

3.0 CONTROL METHODS AND TECHNOLOGY OPTIONS

This section highlights a comparison of various odor mitigation technologies with a focus on how they may or may not apply to a given biosolids odor source. Other chapters of this manual provide greater detail on the various liquid- and gas-phase technologies themselves.

TABLE 7.2 Odor detection threshold concentrations for anaerobically digested biosolids from a municipal WRRF in southeastern Virginia (data from Easter et al., 2008)

Sample age	Sample Description	Detection Threshold	Hedonic Tone	Principal Odor Descriptors
One-day	Belt press pile	810	−4.4	Petroleum, grease, oil, offensive
One-day	Centrifuge pile	820	−6.2	Offensive, sour
One-week	Belt press filter pile	1300	−4.2	Tar, chemical, rancid
One-week	Centrifuge pile	3600	−5.6	Putrid, ammonia, tar, chemical, offensive, urine

TABLE 7.3 Anaerobically digested WRRF biosolids reduced sulfur odor species data from a municipal WRRF in southeastern Virginia (concentrations in parts per billion by volume). (data from Easter et al., 2008).

	One-Day Old[a]		One-Week Old Pad Storage[a]		
Reduced Sulfur Species	**Belt Filter Press**	**Centrifuge Cake**	**Belt Press**	**Belt Press (mixed pile[b])**	**Centrifuge**
Hydrogen sulfide	10.00	7.50	27.00	26.00	14.00
Carbonyl sulfide	52.00	17.00	14.00	35.00	11.00
Methyl mercaptan	22.00	33.00	23.00	18.00	49.00
Ethyl mercaptan	—	—	—	—	—
Dimethyl sulfide	62.00	410.00	—	—	25.00
Carbon disulfide	85.00	48.00	9.1	28.00	16.00
Isopropyl mercaptan	—	—	—	6.00	—
tert-butyl Mercaptan	—	—	—	—	—
n-Propyl mercaptan	—	—	—	—	—
Ethyl methyl sulfide	—	—	—	—	—
Thiophene	—	—	—	—	—
Isobutyl mercaptan	—	—	—	—	—
Diethyl sulfide	—	—	—	—	—
n-Butyl mercaptan	—	—	—	—	—
Dimethyl disulfide	2.60	8.80	20.00	3.00	78.00
3-Methylthiophene	—	—	—	—	—
Tetrahydrothiophene	—	—	—	—	—
2,5-Dimethylthiophene	—	—	—	—	—
2-Ethylthiophene	—	—	—	—	—
Diethyl disulfide	—	—	—	—	—
Total	233.60	524.30	84.00	116.00	193

[a]All blanks were sampled and evaluated, but reported below the detection limit of the laboratory test procedure.

[b]Mixed pile reflects a disturbed, broken-pile surface from removing the sunbaked coating.

TABLE 7.4 Amine-based compounds detected from anaerobically digested cake biosolids at a municipal WRRF in southeastern Virginia (data from Easter et al., 2008).

	Day One		One-Week Pad Storage	
Compound	Belt Press, No Lime	Centrifuge, No Lime	Belt Press, No Lime	Centrifuge, No Lime
Dimethylamine	3.80	3.50	—	—
Ethylamine	—	—	—	—
Trimethylamine	5.90	40.00	3.70	11.00
Isopropylamine	—	—	—	—
t-Butylamine	—	—	—	—
Propylamine	—	—	—	—
Diethylamine	—	—	—	—
s-Butylamine	—	—	—	—
Isobutylamine	—	1.30	—	—
Butylamine	—	—	—	—
Diisopropylamine	—	—	—	—
Triethylamine	—	—	—	—
Dipropylamine	—	—	—	—
Total amines	9.70	44.80	3.70	11.00

3.1 Solids Process Configurations to Minimize Odors

Solids processes that are enclosed allow less fugitive odor emissions and are easier to contain and capture the odors for treatment. Otherwise, treating the liquid sludge may be required to control odors such as sulfides.

Solids processes with long detention times can also be more problematic because the sludge has more time to go anaerobic. Longer periods of septicity will tend to cause more hydrogen sulfide and RSC odors.

Treatment of odors can either be achieved by having a more effective biosolids stabilization process, treating the odors in the liquid sludge stream, or capturing and scrubbing the headspace odors. Combinations of these options may be needed in more sensitive applications.

3.2 Liquid-Phase Sludge Treatment Summary

Hydrogen sulfide is often the predominant odor-causing compound. For this reason, liquid-phase treatment often focuses on preventing or removing hydrogen sulfide. Liquid-phase control is a method of odor prevention by which chemicals are introduced to the wastewater or liquid sludge stream to react with or prevent the formation of dissolved sulfides, thereby preventing the release of hydrogen sulfide into the gas phase. Various chemicals can be suitable for liquid-phase odor control. Table 7.5 groups each class of chemical according to the odor control method. These are discussed in more detail in other sections of this manual.

TABLE 7.5 Liquid-phase odor control methods. (U.S. EPA, 2012).

Method	Description
Oxygenation and aeration	Ambient air or pure oxygen is injected to oxidize sulfides or to maintain an aerobic biofilm, thereby preventing sulfide formation. This approach is seldom used in sludge processing except for aerobic digestion applications.
Chemical oxidation	Chemical oxidizing agents are added to oxidize dissolved sulfide to sulfate, which remains in solution. This can include oxidants such as sodium hypochlorite, sodium chlorite, peroxide, or permanganate. Often, the dosing requirements for sludge applications can result in very high operating costs.
Sulfur precipitation	Metal salts are added to form a metal sulfide precipitate with the dissolved sulfide, thus removing it from the waste stream. Iron salts are often used in liquid sludge treatment applications.
Nitrate addition	Nitrate salts are added to the waste stream, acting as the preferred electron acceptor for anaerobic bacteria and preventing sulfide formation from sulfate. Nitrates can also assist in biological oxidation of sulfide using the oxygen present in nitrate. This approach is more typical in collection system wastewater applications, but it has also been used on liquid sludge in limited applications.
pH adjustment	The pH of the waste stream is changed to inhibit growth of hydrogen sulfide-producing bacteria or to drive volatile H_2S to soluble ionic forms such as HS^-. The most extreme version of this is pH shocking at higher pH values to temporarily shut down slime layer activity and sulfide formation. These approaches are more typical for wastewater applications and are not typically used with sludge.

Liquid-phase dosing in biosolids applications most typically includes dosing with either an oxidant such as permanganate or peroxide or with a precipitant such as iron salts. These can control the odors before they are released and either become an odor impact problem or create poor working conditions in enclosed work spaces such as dewatering buildings or truck unloading bays. Cost and effectiveness may require bench-scale testing to determine the required chemical dose and full-scale pilot testing to verify effectiveness.

3.3 Odor Capture, Containment, and Ventilation

Gas-phase treatment options dictate that odors are effectively contained and captured so they can be treated. Odors that cannot be controlled effectively by either process controls or liquid-phase approaches typically must be contained, captured, and treated. Treatment technologies such as wet scrubbers, biotechnologies, carbon adsorption, thermal destruction, ionization, and other approaches are used to treat the captured air. In some instances, stack dispersion might even be sufficient.

Odorous air is typically contained by enclosing odor sources in containment rooms by providing hoods or equipment enclosures, or by covering the source if it is a process basin. Enclosures, including any exposed concrete on covered basins, must be resistant to corrosion. This can necessitate facilities to consider anticorrosion coatings, which is often recommended as good practice.

Equipment enclosures can be either unique designs or prefabricated systems provided by equipment manufacturers. Enclosures are typically made of fiberglass, stainless steel, or aluminum. Some dewatering equipment is inherently self-contained such as centrifuges or rotary presses. Other equipment require engineered enclosures.

Figure 7.3 shows a belt filter press with an integral enclosure system. The cover includes hinged access hatches along the side to allow operators visual access when needed.

Ventilation rates are potentially governed by the following criteria:

- National Fire Protection Association (NFPA) standards for fire protection in wastewater applications;
- the effective capture of odorous-air emissions (rates specified in industrial ventilation guidance manuals and by field experience); and
- worker comfort and safety.

NFPA 820, Standard for Fire Protection in Wastewater Treatment and Collection Facilities (NFPA, 2016), provides recommended standards in

FIGURE 7.3 Belt filter press with integral enclosure system.

air changes per hour for ventilation of wastewater processes. Although the standards in NFPA 820 are not always adopted as required codes, it is generally recommended that they are followed as a normal standard of practice. The ventilation rates in NFPA 820 are intended to protect against potential explosion hazards from volatile compounds that can be emitted from wastewater or sludge. NFPA does not provide ventilation guidance related directly to odor control issues. However, if the NFPA ventilation requirements govern, they might establish the volume of odorous air that must be captured and treated.

In some instances, other local guidance documents might dictate rates of ventilation and capture of odorous air. Examples of these guidance documents include 10 States Standards (Water Supply Committee of the Great Lakes, 2003) and TR-16, titled *Guides for the Design of Wastewater Treatment Works*, published by the New England Interstate Water Pollution Control Commission.

The recommended ventilation rates for effectively capturing and containing odors are a mix of guidance from codes, industrial ventilation guidelines, and field experiences. Ventilation rates should be high enough to capture odors and to keep the contained environment either acceptable for workers or relatively noncorrosive for materials, such as concrete. From this perspective, high ventilation rates are better. On the other hand, it is desirable to minimize ventilation rates to keep the costs of the fan and scrubber system low. The costs include both the initial capital investment in hardware and operation of the system over time. The actual ventilation rate must strike a balance between these two opposing goals.

3.4 Gas-Phase Treatment Summary

Table 7.6 provides a summary comparison of gas-phase odor treatment technologies along with advantages and disadvantage of each. Each of these are covered in more detail in other chapters of this manual.

TABLE 7.6 Summary of odorous air treatment alternatives (U.S. EPA, 2012).

Technique	Frequency Of Use	Cost Factors	Advantages	Disadvantages
Packed tower wet scrubbers	High	Moderate capital, high operation and maintenance (O&M) costs.	Effective and reliable; long track record; small footprint. Can be used on intermittently operated odor sources, which is often the case in biosolids applications.	High chemical consumption, high O&M.
Activated carbon absorbers	High	Cost-effectiveness depends on carbon replacement or regeneration frequency.	Simple, few moving parts; effective; several media options. May be suitable to biosolids applications, depending on odor compound concentrations. Might be advantageous as a polishing step.	Best for relatively dilute airstreams to ensure long carbon life.

TABLE 7.6 Summary of odorous air treatment alternatives (U.S. EPA, 2012). (*Continued*)

Technique	Frequency of Use	Cost Factors	Advantages	Disadvantages
Inground biofilters	High	Low to moderate capital; low O&M.	Simple; low O&M; effective; no chemicals.	Large footprint; design criteria varies; some failures from short-circuiting, overloading. Not well suited to intermittent odor sources, which is sometimes the case in biosolids applications. Older bark, mulch, or wood chip oriented biofilter media has to be replaced on an approximate 3 to 5 year cycle.
Pre-engineered biofilters	Medium	Moderate to high capital; low O&M.	Low O&M, no chemicals; longer media life and smaller footprint than in-ground systems. Can be very effective on biosolids cake applications that run 24/7.	Higher capital costs than inground biofilters. Not well suited to intermittent odor sources, which is sometimes the case in biosolids applications.
Bioscrubbers, biotrickling filters	Medium	Moderate capital; low O&M.	Smaller footprint than biofilters; high H_2S loadings possible; little or no chemicals.	Limited track record in United States; little long-term performance data for non-H_2S applications. Not well suited to intermittent odor sources, which is sometimes the case in biosolids applications.

TABLE 7.6 Summary of odorous air treatment alternatives (U.S. EPA, 2012). (*Continued*)

Technique	Frequency of Use	Cost Factors	Advantages	Disadvantages
Thermal oxidizers	Low	Very high capital and O&M (energy).	Effective for wide spectrum of odors and VOCs. Sometimes required for very complex odorous air streams.	Only economical for high-strength, difficult-to-treat airstreams.
Ionization Systems	Low	High initial capital costs, lower operational costs.	Some systems have proven effective on complex biosolids odors.	Systems are typically for smaller applications (51 m/s [10 000 cfm] or less). Limited vendors with proven experience.

4.0 REFERENCES

Adams, G. A., J. Witherspoon, Z. Erdal, B. Forbes, D. McEwen, R. Hargreaves, M. Higgins, and J. Novak,. (2008). Identifying and Controlling Odor in the Municipal Wastewater Environment, Phase III: Biosolids Processing Modifications for Cake Odor Reduction (Report no. 03-CTS-9T). Alexandria, VA: *Water Environment Research Foundation and London*, UK: IWA Publishing.

Apgar, D., J. Witherspoon, C. Easter, S. Bassrai, C. Dillon, E. Torres, B. Bowker, R. Corsi , S. Davidson, P. Wolstenholme, B. Forbes, C. Quigley, M. Ward, J. Joyce, R. Morton, J. Weiss, and R. Stuet. (2007). Minimization of Odors and Corrosion in Collection Systems. Alexandria, VA: *Water Environment Research Federation and Water Environment Federation*, and London, UK: IWA Publishing.

Burrowes, P. A., Witherspoon, J., Quigley, C., & Easter, C. (2001). Biofilters-wastewater collection and treatment odour experiences of a sustainable technology in North America. *Proceedings of the First International Water Association International Conference on Odour and VOCs*. Sydney, Australia: University of New South Wales.

CH2M Hill. (1997). Department major interceptor odor and hydrogen sulfide monitoring project. Internal Report. (61b). Anaerobic Digestor Sampling.

Clean Air Act, 40 C.F.R. §§ 50–88. (1963).

Clean Air Act, 42 U.S.C. §§ 7401–7671. (1990).

Easter, C., Williams, T., Witherspoon, J., & Feltner, M. (2008). Odor emissions from anaerobically digested municipal waste water treatment plant biosolids during storage and land application: Impacts of lime dosing and the type of dewatering process used (Belt presses versus centrifuges). *Proceedings of the Water Environment Federation/Air & Waste Management Association Odor and VOC Emissions Conference*, Phoenix, Arizona, United States, April 6–9. Alexandria, VA: Water Environment Federation.

Epstein, E. (2007). *The science of composting*. Boca Raton, FL: CRC Press.

Jubinville, R., Stallings, R., & Bowen, W. (1997). Implementation of a comprehensive odor control program at the New London, CT. Water Pollution Control Facility. *Journal of the New England Water Environment Association, 31*(1), 40–48.

Koe, L., & Yang, F. (1999). The effectiveness of the cover and treatment scheme for controlling odorous emissions at a municipal wastewater treatment facility. *Proceedings of the Chartered Institution of Water Environment Management and International Association on Water Quality Joint International Conference on Control and Prevention of Odors in the Water Industry*. London, UK: IWA Publishing.

Melbourne Water. (1989). *Hydrogen sulphide control manual.* Victoria, Australia: Melbourne Water.

National Fire Protection Association. (2016). NFPA 820, standard for fire protection in wastewater treatment and collection facilities. Quincy, MA: National Fire Protection Association.

Sereno, D. J., McGinley, C., Harrison, D., & Haug, R. (1993). Dewatered sludge storage emissions control using multistage wet scrubbing. *Water Environment Research, 65*(1), 66–72.

U.S. Environmental Protection Agency. (1985). *EPA design manual for odor and corrosion control in sanitary sewerage systems and treatment plants* (EPA/625/1-85/018). Washington, D.C.: U.S. Environmental Protection Agency.

U.S. Environmental Protection Agency. (2012). *Solids process design and management textbook*. Washington, D.C.: U.S. Environmental Protection Agency.

Water Environment Federation and American Society of Civil Engineers. (1995). *Odor control in wastewater treatment plants*, Water Environment Federation Manual of Practice No. 22 and ASCE Manuals and Reports on Engineering Practice No. 82. Alexandria, VA: Water Environment Federation and American Society of Civil Engineers.

Water Environment Federation. (2004). *Control of odors and emissions from wastewater treatment plants*, Manual of Practice No. 25. Alexandria, VA: Water Environment Federation.

Water Environment Research Foundation. (2003). *Identifying and controlling odor in the municipal wastewater environment, Phase II: Literature search and review* (Report no. 00HHE5A). Alexandria, VA: Water Research Foundation.

Water Environment Research Foundation. (2004). *Health effects of biosolids odors: A literature review and analysis* (Report no. 00HHE5C). Alexandria, VA: Water Research Foundation.

Water Supply Committee of the Great Lakes—Upper Mississippi River Board of State and Provincial Public Health and Environmental Managers. (2003). *Recommended standards for water works* (aka "10 States Standards"). Albany, NY: Health Research Inc.

Witherspoon, J.; Sidhu,S.; Castelberry, J.; Coleman, L.; Reynolds, K; Card, T.; Daigger, G. (1999) case study-odor emissions estimates and control strategies using models and sampling for east bay municipal utility districts. *Proceedings of the Chartered Institution of Water Environment Management and International Association on water Quality Joint International Conference on Control and Prevention of Odors in the Water Industry, London, England, United Kingdom.* IWA Publishing: London, England, United Kingdom.

8

Biological Odor Control

1.0 INTRODUCTION 228
2.0 BIOLOGICAL ODOR TREATMENT FUNDAMENTALS 229
- 2.1 The Treatment Process 229
- 2.2 Biochemical Processes 230
- 2.3 Critical Operating Parameters 232
  - *2.3.1 Moisture* 232
  - *2.3.2 pH* 233
  - *2.3.3 Nutrients* 234
  - *2.3.4 Temperature* 234

3.0 BIOFILTERS 235
- 3.1 Biofilter Reactor Configurations 235
  - *3.1.1 In-Ground Biofilters* 235
  - *3.1.2 In-Vessel Biofilters* 236
  - *3.1.3 Modular Biofilters* 237
- 3.2 Biofilter Media Types 237
  - *3.2.1 Organic Biofilter Media* 238
  - *3.2.2 Inert Biofilter Media* 239
  - *3.2.3 Biofilter Media Additives and Coatings* 240
- 3.3 General Biofilter Design Guidelines 240
  - *3.3.1 Pressure Drop* 240
  - *3.3.2 Gas Contact Time* 241
  - *3.3.3 Air Distribution* 241
    - *3.3.3.1 Open Air Plenum* 242
    - *3.3.3.2 Media-Supported Air Plenum* 242
    - *3.3.3.3 Prevention of Preferential Airflows* 242
  - *3.3.4 Moisture Control* 243
    - *3.3.4.1 Pre-Humidification* 243
    - *3.3.4.2 Surface Irrigation* 243
    - *3.3.4.3 Subsurface Drip Irrigation* 244

4.0 BIOSCRUBBERS AND BIOTRICKLING FILTERS 244
- 4.1 Defining Bioscrubbers and Biotrickling Filters 244
- 4.2 Bioscrubber and Biotrickling Filter Reactor Configurations 245
  - *4.2.1 Recirculating and Once-Through Operating Mode* 247
  - *4.2.2 Single-Stage and Multi-Stage Treatment* 248
  - *4.2.3 Cocurrent or Countercurrent* 248

4.3 Media Types 249
4.4 General Design Guidelines 250
4.4.1 Pressure Drop 250
4.4.2 Gas Contact Time 251
4.4.3 Gas Distribution 252
5.0 DESIGN AND OPERATIONAL CONSIDERATIONS 252
5.1 General Advantages and Limitations of Biological Odor Treatment Technologies 252
5.2 Best Practices and General Considerations 255
5.2.1 Startup 255
5.2.2 Water Supply Sources 255
5.2.3 Drain Water Considerations 256
5.2.4 Instrumentation and Controls 256
5.2.5 Operations and Maintenance Considerations 256
5.3 Performance Parameters 257
5.4 Reliability 257
5.4.1. Defining Robustness and Resilience 257
5.4.2. Examples of Upset Conditions 258
5.4.2.1 Water Supply Failure 258
5.4.2.2 Fan Failure 258
5.4.2.3 Power Outage 259
5.4.2.4 Nutrient Dosing Failure 259
5.4.2.5 High Chlorine in Supply Water 259
5.4.2.6 Hypochlorite in Inlet Air 259
5.4.2.7 Toxics in Inlet Air 260
5.4.2.8 Low Air Inlet Concentrations 260
5.4.2.9 Sudden Air Inlet Peak Concentrations 260
5.4.2.10 High Inlet Concentration for Extended Period 261
5.4.2.11 Extreme Weather Conditions 261
5.4.3 Opportunities to Improve Reliability 262
5.4.3.1 Sustaining Biological Activity 262
5.4.3.2 Increasing the Buffering Capacity 263
6.0 PROCESS INTEGRATED BIOTECHNOLOGY 264
6.1 Introduction 264
6.2 Activated Sludge Recycling 268
6.2.1 Introduction 268
6.2.2 Critical Design and Operating Parameters 270
6.3 Activated Sludge Diffusion 271
6.3.1 Introduction 271
6.3.2 Critical Design and Operating Parameters 272
6.4 Oxidized Ammonia Recycling 274
6.4.1 Introduction 274
6.4.2 Critical Design and Operating Parameters 274
7.0 REFERENCES 276

1.0 INTRODUCTION

This chapter discusses biological odor control using different vapor-phase treatments and process-integrated biotechnologies. The treatment of foul air emissions using biological methods has gained acceptance over the last two decades as a cost-effective and sustainable alternative to conventional techniques such as chemical scrubbing, physical-chemical adsorption, and thermal oxidation. Biological odor control is considered a "green" technology with a relative low carbon footprint and may help reduce the required operational budget. In addition, biological odor control is generally safe because it is operated at ambient pressure and temperatures, does not involve a concentration step with risk of smoldering, and operates without the requirement for chemicals and their associated risks with storage and handling.

Biological odor control has become more mature over the last two decades (Kennes & Veiga, 2001; Kennes & Veiga, 2013; Shareefdeen & Singh, 2005) and is now commonly applied in the wastewater industry (Estrada et al., 2011; Kraakman et al., 2014). Design limitations or operational problems associated with conventional biofilters, like acidification, media dehydration, biofilter media replacement, and a relatively large footprint, can now be better addressed with technologies like bioscrubbing and biotrickling. In addition, process-integrated biotechnological methods for the prevention of odorant formation and/or odorant release from wastewater can now be considered as attractive options when developing facility-wide odor management strategies.

2.0 BIOLOGICAL ODOR TREATMENT FUNDAMENTALS

2.1 The Treatment Process

Biological processes in general can take many forms and are almost infinitely varied and diverse; most life forms have unique biological and biochemical features. To understand biological odor control, it is not necessary to have a full understanding of biological processes. Biological odor treatment eliminates or transforms odorous compounds through the action of microorganisms, for which odorous compounds serve as an energy and/or carbon source. The microbial degradation processes involved are generally oxidative, meaning the addition of oxygen to a compound accompanying a loss of electrons, and they produce mostly carbon dioxide, water, sulfate, and nitrate.

Microorganisms, primarily bacteria, are the catalyst of this process. Biodegradation of odorous compounds takes place by mixed microbial communities thriving in a complex ecosystem. This ecosystem composition is mainly a function of startup cultures, inlet gas stream composition, and the prevailing environment inside the bioreactor. The microorganisms are mostly embedded in a biofilm, a slimy film that adheres to a surface and protects the microorganisms from extreme or changing conditions and enhances communication among them. The mixed microflora living in the biofilm consist of pollutant degraders, competitors, and predators, which typically have significantly different properties from free-floating bacteria of the same species such as an increased resistance to extreme contaminant concentrations.

The biological degradation of the odorous compounds generally takes place in the biofilm, an aqueous environment, because bacteria need an environment containing a relatively high-water content. While odorous air passes through the odor treatment system, odor compounds are typically first dissolved into the moisture in and on the media. Two processes are occurring simultaneously in biological odor treatment systems. This first step is the mass transfer to the biofilm, which is dependent on the odor compound's solubility in water and its Henry's law gas constant. This air–water partition coefficient reflects the compound's affinity for the gas phase versus liquid phase. Once the odorous compounds are transferred to the microorganisms, the second step, biological degradation of the compound, can occur. These processes of mass transfer and biological degradation occur almost instantaneously within the bioreactor, typically operating under either mass-transfer limited or biokinetic limited conditions.

Volatile odorous compounds and the oxygen required for the biological degradation must be transferred from the gas phase to the liquid phase. Most biological odor treatment systems operate under mass-transfer limited conditions and, therefore, the mass-transfer processes play an essential role. The mass transfer is dependent on the odorous compound, bioreactor type, packing materials, as well as different operating conditions (Dorado et al., 2009; Kim & Deshusses, 2007; Kraakman, Roche Rios, & van Loosdrecht, 2011). Mass-transfer takes place through both diffusion, the random Brownian motion of individual compounds in a medium, and by advection, in which compounds are transported by the larger-scale motion of currents in the medium.

Higher molecular weight organic compounds are less soluble than smaller molecular weight organic compounds. Insoluble or practically insoluble odorous compounds cannot be treated well in a typical biological odor treatment system because the odorous compounds usually must be dissolved in an aqueous matrix to be biologically degraded. However, biological odor control processes have successfully demonstrated the ability to remove all the common and typical odorous compounds found in municipal wastewater facilities and many hazardous, potentially toxic, and industrial-related compounds.

2.2 Biochemical Processes

The biological odor control technologies discussed in this chapter operate mainly through the basic biochemical processes of autotrophic and heterotrophic microbial conversions.

Autotrophic (meaning self-feeding) organisms are typically used for hydrogen sulfide removal, where hydrogen sulfide is used as an energy source and carbon dioxide as a carbon source for building and maintaining cell material. Autotrophic microorganisms are a group of organisms that use inorganic compounds for their energy and carbon sources. Autotrophic bacteria typically have a thin cell wall to allow relatively free movement of inorganic chemicals in and out of their cell. They also prefer to live as single cells in suspended culture or as thin biofilms on surfaces so that maximum cellular surface area is exposed to collect food and nutrients. Therefore, autotrophic bacteria typically do not clump together or form thick masses of cells like heterotrophic microorganisms.

Autotrophic bacteria are also efficient nutrient scavengers and can efficiently recycle nutrients from dead cells around them. Because autotrophic bacteria do not get as much energy from the conversion of inorganic chemicals as heterotrophs, they must convert more chemical compounds at a faster rate. Autotrophic bacteria can, therefore, quickly assimilate and convert large quantities of inorganic compounds such as hydrogen sulfide.

Heterotrophic refers to a class of microorganisms that obtain energy and carbon for cell growth from organic compounds. Heterotrophic organisms cannot use carbon dioxide as a carbon source for cell synthesis and must break the carbon-carbon bonds of organic sources to assimilate carbon. There are many more types of heterotrophic organisms than autotrophic organisms, and they are the primary organisms responsible for the degradation of volatile organic compounds in biological odor control systems.

Heterotrophic organisms are also required for the degradation of the highly intense odorous organic compounds such as mercaptans, methyl- and ethyl-sulfides, and amines. Because heterotrophic bacteria feed on relatively high molecular weight organic compounds, they must have a method of capturing these compounds. To help this, their cell walls are covered with a sticky, excreted polysaccharide material called "zooglea", which simply means animal glue. Large organic compounds stick to this material, and the cell excretes extracellular hydrolytic enzymes that break down the large organic molecule into smaller, more soluble parts. Once broken down, compounds can enter the cell wall and become acted on by intracellular enzymes for further degradation and energy extraction. Because of the sticky cell walls of heterotrophs, they tend to clump together and form relatively thick, slimy biofilms.

In addition to autotrophic and heterotrophic conversions, there exists the process of biological uptake, which is different from food consumption, carbon, and energy production. In biological uptake, the compound is used as a secondary component in the process of cell function and new cell synthesis. An example of biological uptake is nutrient consumption by bacteria, discussed more in the next section. Uptake processes also believed to contribute to the fast uptake and removal of hydrogen sulfide gas in aeration basins are activated sludge diffusion (ASD) and activated sludge recycling (ASR), discussed later in this chapter.

2.3 Critical Operating Parameters

The primary design objective of all vapor-phase biological odor treatment technologies is to build a house that is ideally suited for the specific types of microorganisms and provide supply of suitable food, oxygen, and nutrients and maintain moisture, pH, and temperature under optimal living conditions. Operating conditions like moisture, pH, nutrient concentration, and temperature are basic design parameters directly affecting performance.

2.3.1 Moisture

Biological odor treatment system must be maintained in an optimal moist condition to sustain biological activity. A moist surface of the packing media must be maintained, meaning a water activity of at least 0.9, which requires a moisture content of 40 to 60% on a dry weight basis for most biological filter media types. Too-dry conditions may result in inactive biomass, whereas too-wet conditions may result in anaerobic zones, an increased pressure drop, preferential airflow distribution, and premature media decomposition.

The inlet air is typically not saturated with water (i.e., the relative humidity or RH is less than 100%), which leads to water evaporation from the biological filter media, thus drying it out. Moreover, even when the inlet air is fully water saturated (RH = 100%), the small amount of heat generated from the exothermic biological activity will still slowly dry out the media. Therefore, irrigation of water to a biological odor treatment system media is always required regardless of the moisture content of the inlet air.

In addition to preventing too-dry conditions, the supply of water is important because water also serves as a supply medium for nutrients for the microorganisms and removes degradation products (e.g., sulfuric acid). Biological systems treating high concentrations of hydrogen sulfide, such as bioscrubbers or biotrickling filters, require extra water to prevent the system from becoming too acidic and limiting the biological activity.

Although enough water supply is critical, minimizing the amount of irrigation water, on the other hand, may have benefits when treating hydrophobic contaminants. Minimizing irrigation water rates can reduce the thickness of the water barrier for mass transfer of the odorous compounds from the gas phase into the biofilm.

2.3.2 pH

The pH in biological odor treatment systems needs to be maintained around neutral (pH 5 to 8) for acceptable microbial activity, unless specific microorganisms (extremophiles) are used (e.g., hydrogen sulfide can effectively be converted in an environment with a pH as low as 2). The autotrophic organisms for hydrogen sulfide removal require different environmental conditions than heterotrophic organisms and are, therefore, preferably separated in a biological odor treatment system.

Odorous air from wastewater collection systems and water resource recovery facilities (WRRFs) contains many different odorous compounds, among them mixtures of volatile inorganic and organic reduced sulfur and nitrogen compounds. The different odorous compounds can be best treated by a combination of organisms requiring different environmental conditions, among them pH being the most critical. Therefore, by using different zones or media layers optimized for the different required microorganisms, a broader range of odorous compounds can be removed in the biological odor treatment reactor more effectively.

For example, although multiple microorganisms can degrade highly odorous sulfur compounds such as mercaptans or dimethylsulphides, the treatment of an airstream containing mixtures of reduced sulfur compounds remains challenging for two main reasons. First, the energy yielding process of hydrogen sulfide oxidation is higher and thus preferred over the oxidation of other reduced sulfur compounds (Myung Cha, Suk Cha, & Lee, 1999). Secondly, the degradation of many of these sulfur compounds is only possible at neutral pH, and a degradation product from sulfur compounds is sulfuric acid, which reduces pH.

It is most common to degrade the hydrogen sulfide first because hydrogen sulfide has a higher water solubility (low Henry's constant) compared to most other organic odor compounds. Removal of hydrogen sulfide in the initial part of the treatment reactor prevents acidification of the biological filter media in the following part and provides optimal conditions for the heterotrophic organisms' growth and survival.

In addition to maintaining different pH zones in a biological odor treatment system, the pH within these zones should be kept relatively stable, as sudden pH fluctuations of more than 2 to 3 pH units generally lead to reduced microbial activity that can be reestablished once pH returns to optimal levels. To maintain the pH inside the odor treatment system, buffering materials such as limestone and dolomite are typically added to biofilter media. The pH in biotrickling filters treating hydrogen sulfide is typically controlled by bed irrigation to remove degradation-product sulfuric acid;

neutralization by means of dosing sodium hydroxide in a recirculating liquid flow can also be used when water supply is limited.

2.3.3 Nutrients

The odors entering a biological odor control system form the main carbon and energy sources for microbial activity, whereas oxygen for the predominately aerobic conversions is found in the air. In addition, macro-nutrients (nitrogen, phosphorus, potassium, magnesium, cadmium, and sulfur) and micronutrients (metals, vitamins) are essential for maintaining biological activity.

Some nutrients may naturally be present in the odorous airstream or may be present in the irrigation water. Nutrient requirements can also be fulfilled by the biofilter media materials. Natural materials such as composts are well known to contain various nutrients, but nutrients can also be incorporated to a coating on inert media materials.

Irrespective of the media material used or the origin of the odorous air, the addition of nutrients via the irrigation water is typically necessary to sustain a satisfactory microbial degradation activity over the life of a biological system. The most commonly used nutrient solutions include KH_2PO_4, NaH_3PO_4, KNO_3, $(NH_4)_2SO_4$, NH_4Cl, NH_4HCO_3, $CaCl_2$, $MgSO_4$, $MnSO_4$, $FeSO_4$, Na_2MoO_4, and vitamins. Secondary effluent water contains all the nutrients required, typically in enough concentrations, and its use as water supply would eliminate the constant supply and dosing of nutrient solutions. Liquid sidestreams from processes like anaerobic digestion or biosolids dewatering (centrate) may also be considered as a nutrient source.

2.3.4 Temperature

The temperature inside a biological odor treatment system is often dictated by the temperature of the inlet foul air extracted and not manipulated because of the costs associated with modifying the temperature. In addition, the nature of the microbial ecosystem enables it to slowly adapt to the temperature range inside the biological odor treatment system.

Although increased temperatures typically double biodegradation rates with every 10 degrees Celsius over the 10 to 30 °C range, higher biodegradation rates are less relevant because most biological odor treatment systems are mass-transfer limited. Temperature influences the mass-transfer processes because the mass-transfer parameters solubility and diffusion (increases with temperature) and Henry's constant (decreases with temperature) are somewhat temperature sensitive, but in general not as significant as other biodegradation parameters (Barton et al., 1999). Most biofilters for odor control are mass-transfer limited and, therefore, gain limited advantage of increased temperatures.

A biological odor control system can operate efficiently at low temperatures (5 to 10 °C), even for hydrogen sulfide removal. Biofiltration at high temperature (>35 °C) is also possible, but comes with new challenges, such as higher water evaporation rates, less favorable conditions for mass transfer, and reduced microbial diversity.

3.0 BIOFILTERS

3.1 Biofilter Reactor Configurations

A biofilter system can have different reactor configurations depending on their application, size, location, and operational requirements. Regardless their configuration, all properly designed biofilters should have the following components:

- Air extraction system (air ducting with balancing dampers and fan),
- Pre-humidification system,
- Inlet air distribution system,
- Biofilter media,
- Water irrigation system, and
- Instrumentation for process control and monitoring.

All biofilter configurations can be built with or without a roof or a canopy to limit impact from unwanted external factors such as weather. Wind may impact sprinkler irrigation systems, excessive rain may impact media moisture content, and sun may make weeds grow (brought in by wind and birds) or elevate the drying out of the media. Enclosed biofilter systems have an added benefit of being able to be connected to a dispersion stack for improved dilution of the residual odors. In the following section, the three basic biofilter configurations are discussed: in-ground, in-vessel, and modular biofilters.

3.1.1 In-Ground Biofilters

In-ground biofilters do not use a structural material to contain the biofilter main components and have the air-distribution system built in the ground below grade. They are custom made and can be cost-effective, but are not enclosed, making them more vulnerable to external factors such as weather. They require the least construction materials, but do require a protection material (e.g., acid-resistant geotechnical membrane) under the biofilter system to collect any drain water and protect the surrounding soil from potentially extreme low-pH leachate. Figure 8.1 presents an illustration of a typical custom-made in-ground biofilter.

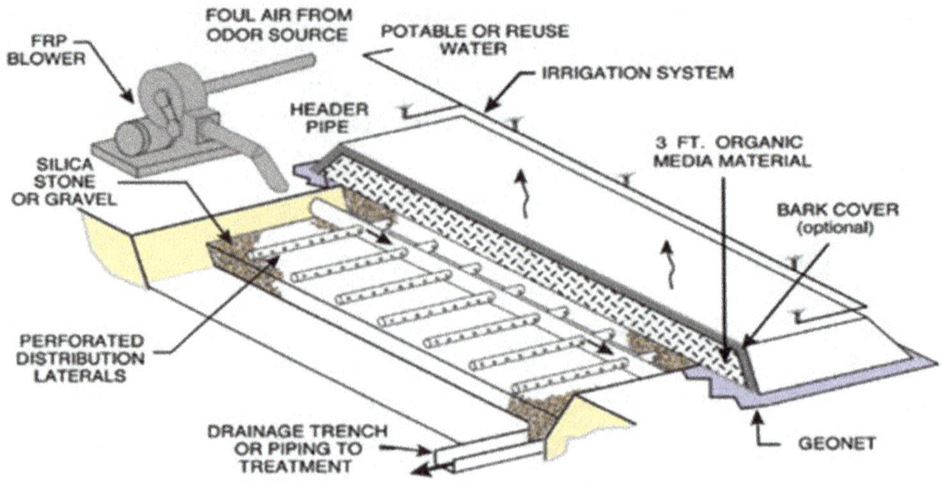

FIGURE 8.1 Typical custom-made in-ground biofilter system for 15 000 m^3/h.

3.1.2 In-Vessel Biofilters

In-vessel biofilters use a structural material to contain main biofilter components, including the air-distribution system (Figure 8.2). Although concrete is typically used, hardwood has also been used in some cases. In-vessel biofilters are custom made, especially for large airflows, and are typically more durable than in-ground biofilters if their construction material is protected

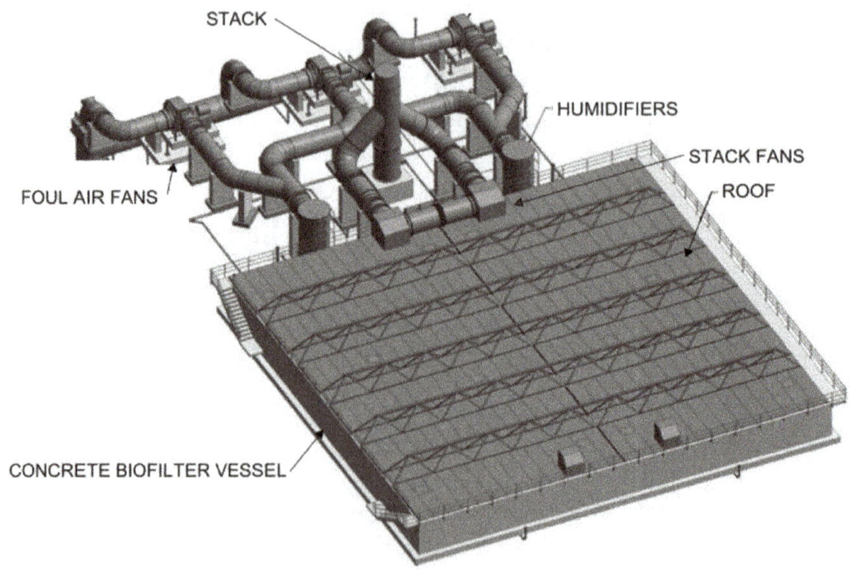

FIGURE 8.2 Typical custom-made in-vessel biofilter system for 150 000 m^3/h (courtesy of Sacramento Regional County Sanitation District and Jacobs Engineering).

against potentially extreme low-pH leachate (e.g., acid-resistant coating, liner, or geotechnical membrane). In-vessel biofilters can be made enclosed more easily than in-ground biofilters, with extra benefits discussed above.

3.1.3 Modular Biofilters

Modular biofilters are in-vessel prefabricated units made for easy transportation and installation. The structural material of the units comes in many forms such as fiberglass, high-density polyethylene, or other plastic material, but lined concrete boxes as well as modular shipping containers can be used as well. Modular biofilter systems tend to be more cost-effective for relatively small airflows and come in various sizes, from small manhole inserts treating a few cubic meters per hour to midsize, still transportable units treating several thousand cubic meters per hour. Figure 8.3 presents an example of a typical modular biofilter system.

3.2 Biofilter Media Types

Biofilter performance is highly dependent on the nature of the biofilter media, also called "filter bed", "support material", "carrier material", or "medium". Biofilters have been applied for several decades, although they were initially used for odor control at WRRFs and composting facilities by means of soil or compost as biofilter media. Other media have been developed to reduce biofilter sizes (affecting capital costs and footprint). New media have also been developed to improve physical stability (affecting

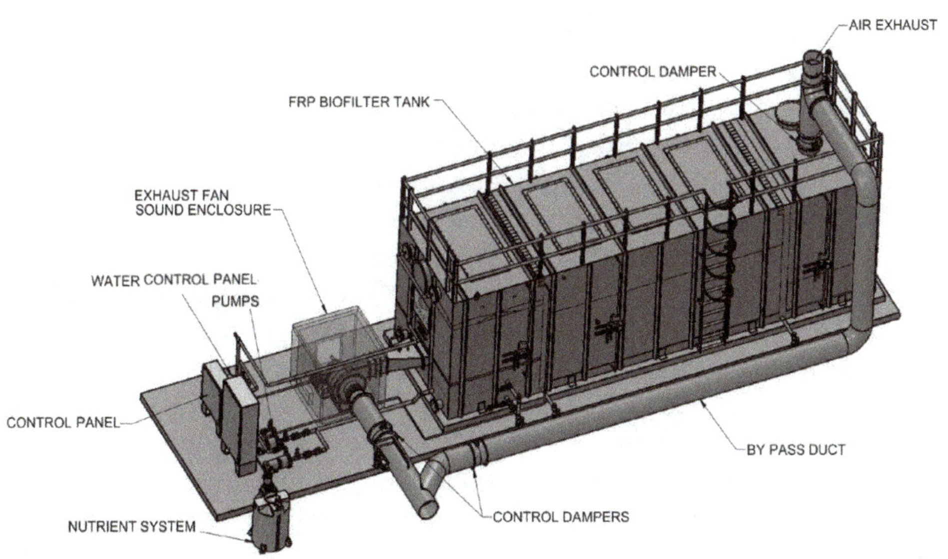

FIGURE 8.3 Typical modular biofilter system for 10 000 m^3/h (courtesy of Biorem Technologies).

long-term performance and replacement frequency) and reduce pressure drop (affecting the operating costs for the fan to move the air through the biofilter media). Sensitivity analysis of economics of odor control technologies have shown that the biofilter media type is a key parameter affecting their operating costs up to 50% of the total operating costs (Estrada et al., 2011).

Biofilter media should (a) be chemically and physically stable; (b) present appropriate physical characteristics such as high porosity, large void volume, large surface area, high water buffering capacity, high adsorptive properties, high pH buffering capacity, and low density; (c) retain microbes and nutrients strongly; (d) present low-pressure drop; and (e) exhibit good air or water distribution properties. Selecting biofilter media should consider local availability and economics. Several biological odor control supply companies also provide custom biofilter media, which is constantly being improved; however, no single type of media has been shown to provide optimal treatment characteristics under all operating conditions.

3.2.1 Organic Biofilter Media

Organic biofilter media may have advantages such as low cost, ready availability, and the natural presence of macro- and micro-nutrients and microorganisms. Compost, peat, bark, coconut fiber, wood chips, and soil have been applied successfully over the years, mostly in various mixtures.

Compost has been the most frequently used because it contains abundant nutrients, a wide microbial flora, and a large surface area, and can be made of multiple prime materials such as yard waste, sludge, wood residuals or agricultural, or source-separated waste. Peat is composed of stabilized carbohydrates, minerals, and humic acids that present interesting adsorptive properties compared with compost. Peat, however, is naturally acidic and hydrophobic, so additional pH buffers may be required; moisture control can also be more difficult. Peat also contains much less nutrients and microbial content than compost requiring inoculation and nutrient supply.

Bark can be relatively stable depending on the type of tree used or whether it has been stabilized in a composting process. Bark as the main material provides an improved open structure resulting in improved pressure drop characteristics. Coconut fibers are extremely resistant to biodegradation, which results in improved long-term performance and, in addition, has a high void volume; it has been successfully applied when mixed with peat or compost. Soil is composed of both inorganic minerals and organic matter and, contrary to several other natural materials, is more hydrophilic and, therefore, less sensitive to fluctuations in moisture content. However, its low permeability and low void volume require a relatively large footprint, especially to limit pressure drop. Other natural materials used in biofilter media are rice hulls, heather, and grape seeds.

Unfortunately, many of the organic media present a limited lifetime of 2 to 5 years because of biodegradation of the material, which compacts the media and generates increased pressure drop, or because of nutrient depletion. Organic media often compromises the long-term performance and/or the operating costs of a biofilter system.

3.2.2 Inert Biofilter Media

Inert media typically lasts significantly longer than organic media and minimizes the pressure drop and air channeling problems, but, in general, is more expensive. Its composition is not only relatively stable, but typically better defined and more homogeneous. Many engineered biofilter media use inert materials, which often reduces their footprint, the frequency of media replacements, the background odor concentration, and the risk of declining performance over time associated with organic media.

Several inert materials have been used successfully such as lava rock, clay balls, activated carbon, perlite, vermiculite, and sand, and synthetic materials such as polystyrene balls, plastic, ceramic or glass beads, wires, or rings. Many materials have been applied successfully over the years, often providing improved structural integrity in various mixtures with organic materials.

Biofilter media should maintain its structural integrity and resist filter bed compaction during long-term operation. Nevertheless, some inert biofilter media such as lava rock, clay balls, and activated carbon still may lose their structural integrity over time when biofilters are operated at relatively high hydrogen sulfide loads or too low irrigation rates. This type of operation would make the material extremely acidic (pH around 1), leading to chemical erosion, and often have resulted in increased pressure drop and reduced performance over time, requiring media replacement.

Plastic and polymeric materials may pose a disposal problem of solid-waste generation compared with biofilter media that uses natural inert or organic materials. Moreover, inert materials typically do not contain enough various microbial populations nor have strong biomass adhesion capabilities. In addition, inert materials do not contain many nutrients or, if nutrients are present, they are typically not readily available or in the right balance. The moisture-buffering capacities and pH are also limited and typically require biofilter media additives.

3.2.3 Biofilter Media Additives and Coatings

Different additives can be added to biofilter material to enhance the performance and/or the life of a biofilter media. Properly stabilized compost may be used as an excellent nutrient source and provide a diverse microbial population and an increased moisture buffering capacity. Nutrients are often added in soluble or powder form, but coated pellets may also be used to relinquish the nutrients in a slower release of time.

Enhanced pH buffering can be obtained with phosphate buffers or limestone, dolomite, and other carbonate-based rock, either in powder or fine granular form. Care must be taken using carbonated materials in the lower part of a biofilter because they can form calcium sulfate (gypsum). Calcium precipitates with sulfuric acid, the product of the degradation of reduced sulfur compounds such as hydrogen sulfide, and can foul and plug media and air-distribution piping.

Enhanced buffer capacities of a biofilter media toward hydrophobic compounds may be obtained with the addition of activated carbon; however, silicon oil has also been proposed. In addition, thin, adhesive polymer plastic coatings on inert biofilter material can increase general buffer capacities and could include nutrients or even microorganisms (Estrada et al., 2015).

3.3 General Biofilter Design Guidelines

3.3.1 Pressure Drop

The pressure drop of a biofilter system is an important design parameter because it impacts both the operational costs as well as overall removal performance. The operational costs are required for the energy input to the fans moving the air through the biofilter system. The removal performance is impacted by pressure drop because it dictates gas residence time distribution and preferential airflows in the biofilter media. The pressure drop is preferably less than 0.25 kPa (1.0 in. H_2O) and should be limited to 1.0 kPa (4.0 in. H_2O) with limited pressure drop increases over time, particularly when organic materials are used for the biofilter media.

The overall pressure drop is dictated by the gas load (superficial gas velocity), the media height, the type of biofilter media, and the pollutant load. The gas load is typically limited to 250 $m^3/(m^2{\cdot}h)$, also expressed as 7 cm/s or 14 cfm/sq ft, and the media height is typically limited to 2 m (6 ft). The average pollutant load is typically limited to 50 $g/(m^3{\cdot}h)$ to avoid biomass accumulation. The pressure drop depends on media material size distribution, shape, and moisture content, and will have lower and more consistent pressure drops at higher void volumes and larger particle sizes, especially when inert materials are used.

3.3.2 Gas Contact Time

The empty bed gas contact time is typically used as a design parameter for odor control systems. The empty bed contact time (EBCT) is calculated using the amount of treated airflow divided by the media volume in the odor treatment system. The EBCT is commonly expressed in seconds of contact time.

The required EBCT is dependent on foul air composition, inlet airflow loading rate, and the biofilter system, including biofilter media and required performance; it is based on empirical experience rather than theoretical calculations. Biofilters applied for foul air treatment at wastewater collection or treatment processes typically require between 45 to 60 seconds for 95 to 98% odor removal. The required EBCT might be significantly higher (90 to 120 seconds) when, for example, treating relatively concentrated odorous air containing odors with reduced biodegradability or water-solubility odorous compounds such as foul air from autothermal thermophilic aerobic digestion processes.

The EBCT strongly depends on the media type because media material and characteristics such as void volume can be quite different among biofilter media. The void volume directly influences the minimum required gas contact time with a larger void volume, resulting in a longer actual gas contact time (AGCT). The AGCT is an important design parameter and different than the often-used theoretical EBCT because it represents the real time in seconds that the air is in contact with the biofilm.

$$\text{Theoretical EBCT (s)} = \text{media volume (m}^3\text{) / air flow (m}^3\text{ s}^{-1}\text{)}$$

$$\text{Actual AGCT (s)} = \text{media volume (m}^3\text{)} \times \text{void fraction (\%) / airflow (m}^3\text{ s}^{-1}\text{)}$$

The AGCT will always be lower than the EBCT and its extent depends on the porosity of the media and the water and biomass attached to the media. The porosity of biofilter typically ranges from about 30% (soil or compost) to 60% (pre-engineered biofilter media), whereas the water and biomass attached to the media are typically less than 25% of the media bed volume for odor control applications.

3.3.3 Air Distribution

The airflow distribution through the biofilter system can vary greatly and, as such, its significance should not to be underestimated. An optimum air distribution through the media is important to obtain high odor removal efficiencies. When high removal efficiencies are required (>99%), all the air must be treated effectively. To treat the air effectively, all the air needs to exchange the odorous compounds to the biofilm on the biological media. An optimum air distribution preferably moves the air through the biofilter reactor as a plug flow, which is, in reality, never the case.

In addition, the inlet air plenum is of importance and should be considered an important design aspect. Two commonly used inlet air plenum configurations, may include simple means to prevent extensive preferential airflows, are discussed in the following sections.

3.3.3.1 Open Air Plenum

Open air plenums do not contain media and use a specialized support material to hold the biofilter media. The media is typically directly supported on a modular plastic grid system, which is made of, for example, aeration floors, raised flooring system, stormwater holding boxes, or just a structural floor supported from the biofilter wall.

3.3.3.2 Media-Supported Air Plenum

Media-supported air plenums contain some sort of open structural material that surrounds inlet air-distribution piping as well as supports the biofilter media. The material is typically a granular material such as river gravel rock or lava rock and should be relatively high acid resistant. The air-distribution piping is typically perforated at specific hole sizes to accommodate equal pressurization.

3.3.3.3 Prevention of Preferential Airflows

An optimum air distribution is critical, especially when high odor removal efficiencies are required. The biofilter media is typically a randomly placed material with irregular or differently shaped media particles. The foul air is always forced through the irregular-shaped media following the path of least resistance (e.g., along the boundary between the media and the biofilter wall).

Different methods can be used to minimize preferential airflow (also called "short-circuiting"), among them a horizontal flap of liner material securely attached to the biofilter wall so that air cannot pass and is forced back into the media. In addition, air-distribution plenums could be made smaller than the media bed (see also Figure 8.1), such that air along the wall is prevented. Open air plenums tend to generate relatively high-pressure zones near the wall opposite the air inlet, and airflow barriers inside the air plenum could be used to slow down the air velocity. Moreover, foul air distribution pipe headers can be installed in the plenum to optimize air distribution before it enters the biofilter media.

Moreover, larger-sized biofilter media materials are typically used in the bottom (foul air inlet side) of a biofilter and finer-sized media material toward the top (clean air outlet side). This would accommodate improved redistribution of the air as it travels up through the biofilter media, experiencing increasing pressure losses.

3.3.4 Moisture Control

Biofilter systems must be maintained in a moist condition to sustain biological activity. On the other hand, too much water may lead to anaerobic zones and less optimal air distribution. Both pre-humidification of the incoming air as well as irrigation of the biological filter media should be applied for biofilter systems.

3.3.4.1 Pre-Humidification

Pre-humidification systems are like wet scrubbers, where water is added to the top of packing material and the inlet air is introduced at the bottom. The air and water travel countercurrent while the air picks up moisture. In addition, an in-duct spray system using fine-mist nozzles may also be used immediately upstream of the biofilter system. The fine-mist droplets are light enough to not drop out, but stay floated in the air while they evaporate. When fine-mist nozzles are used, upfront strainers and/or water softener should be used to eliminate the risk of clogging up the nozzles because of calcium deposition or other fine particles.

Alternatively, steam injection is a very effective way to increase the relative humidity close to 100% without reducing the air temperature, although it is relatively costly unless excess steam is easy available. A media-supported air plenum, as discussed in Section 3.3.3.2, may be a source of moisture when porous inert materials are used such as clay balls, and can be considered a low-efficiency humidification system for the airstream.

3.3.4.2 Surface Irrigation

The inlet air is typically less than 100% in relative humidity, even when pre-humidification is applied, causing evaporation of water from the biological filter media and drying it out. Even when the inlet air is fully water saturated, the small amount of heat generated from the exothermic biological activity will still slowly dry out the media, requiring the irrigation of water.

Media irrigation is typically done with surface irrigation using sprinklers on an on/off time base. The sprinkler can be any type, but preferably one that ensures the full media surface is completely covered without limited spray overlap between the sprinklers. It is typically detrimental to biofilter system operation to try to add water only with surface irrigation because of preferential water streams draining through the media, creating inadequate or inhomogeneous moist conditions near the bottom of the media. The bottom of the media is typically the air inlet side, where water is needed most.

3.3.4.3 Subsurface Drip Irrigation

Subsurface irrigation using drip irrigation hoses is an extra method and can be an effective way to provide water near the bottom of the media, which is typically the initial zone with the highest biological activity where water is needed most. Irrigation hoses are laid horizontally across the media to provide a layering effect of irrigation hoses and media. Subsurface irrigation may be accomplished with soaker hoses, preferably with discrete, non-plugging orifices installed at small (10 to 15 cm [4 to 6 in.]) regular intervals.

4.0 BIOSCRUBBERS AND BIOTRICKLING FILTERS

4.1 Defining Bioscrubbers and Biotrickling Filters

Biological gas treatment techniques are traditionally classified as biofilters, bioscrubbers, and biotrickling filters. Biofilters are systems where both the microorganisms and the water phase are immobilized on the biofilter media. Bioscrubbers are traditionally defined as two reactor units with water recirculating between them. In the first reactor unit, contaminants are absorbed from the gas phase in the water phase through mass transfer. In the second unit, the contaminants are converted by microorganisms suspended in the water (suspended growth biomass). Biotrickling filters are typically defined as reactor units with a packing through which water is typically trickled. The absorption of contaminants from the gas phase to the liquid phase and degradation of the contaminants take place in the same unit. The microorganisms are mostly immobilized on the packing (fixed-film biomass).

The mobile liquid phase in bioscrubbers and biotrickling filters provides benefits to better control the internal process conditions such as pH, salt content, and nutrient concentrations. Bioscrubbers and biotrickling filters may, for certain applications, be more effective than biofilters, especially for the treatment of compounds difficult to degrade or high concentrations of compounds that generate acidic byproducts such as sulfuric acid.

The terms *biotrickling filter* and bioscrubber have been used interchangeably in the wastewater industry; as such, there is sometimes confusion about the two technologies. Biotrickling filters have sometimes been called "bioscrubbers", mostly to differentiate between chemical and biological gas scrubbing and not to confuse biotrickling filters for odor control with biotrickling filters used for wastewater treatment. In the wastewater industry, bioscrubbers are referred to as fixed biofilm bioscrubber and could also be referred to as a recirculating biotrickling filter. A biotrickling filter, on the other hand, could be described as a once-through type odor control system with intermittent irrigation. Figure 8.4 summarizes the most significant differences between the three traditional biotechnologies used for gas treatment and the biological odor control technologies referred to in the wastewater industry. In this chapter, fixed-film bioscrubbers are further referred to as bioscrubbers, and once-through operated biotrickling filters are further referred to as being biotrickling filters.

4.2 Bioscrubber and Biotrickling Filter Reactor Configurations

A bioscrubber or biotrickling filter system can have different reactor configurations depending on their application, size, location, and operational requirements. Regardless of their configuration, all properly designed bioscrubbers and biotrickling filters should have the following components:

- Air extraction system (air ducting with balancing dampers and fan),
- Inlet air-distribution system,
- Bioscrubber/biotrickling filter packing media,
- Water irrigation system, and
- Instrumentation for process control and monitoring.

Bioscrubbers and biotrickling filter reactors can be built taller than biofilters to accommodate for the increased height of the packing media, the structure of which is typically relatively open with a low pressure drop. The relatively tall bioscrubber and biotrickling filter reactors are typically round because this type of reactor may provide increased strength more economically than square reactors. In addition, round reactors better accommodate the round spray pattern of irrigation nozzles, providing water distribution more homogeneously on the packing media. The height of the packing media in bioscrubbers or biotrickling filters may be up to 4 or 6 m (12 to 18 ft) compared to biofilters, which are typically limited to 1 or 2 m (3 to 6 ft) of media. Bioscrubber and biotrickling filter systems all have a small dispersion stack on the top of the relatively high reactor for improved dilution of residual odors. Some basic bioscrubber and biotrickling filter configurations are discussed further in the following subsection.

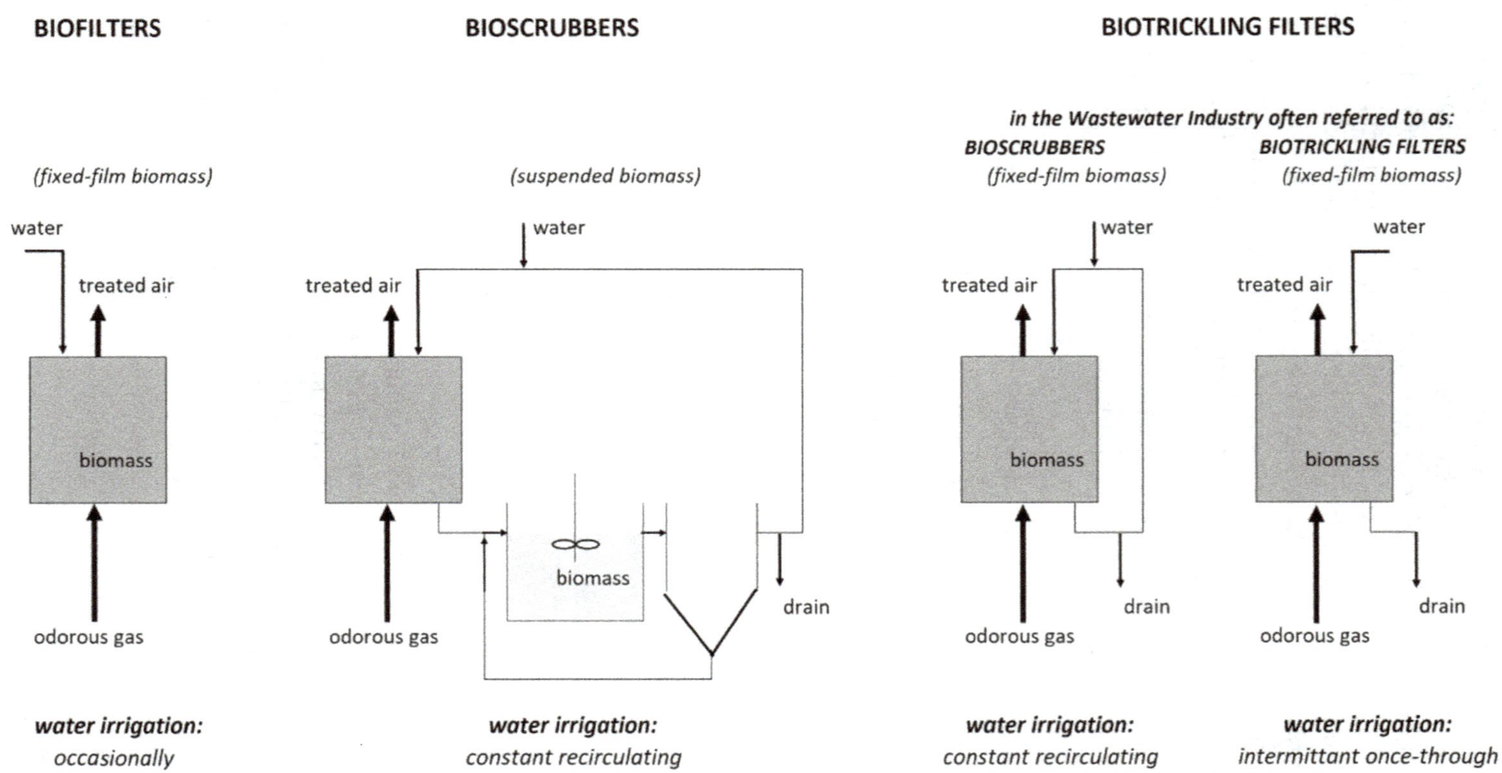

FIGURE 8.4 Schematic representation of traditionally classified biological vapor-phase treatment technologies.

4.2.1 Recirculating and Once-Through Operating Mode

The method by which water is added to a bioscrubber or biotrickling filter reactor is important because water not only prevents the biofilm in the bioreactor system from drying out, it also serves as a means to remove the degradation product (e.g., sulfuric acid from the packing media). Water can be recirculated over the media or can be added to the top of the media, and then, after passing through the media, be directly removed from the bottom of the system. The water can be supplied to the top of the media continuously or intermittently. As discussed in Section 4.1, biotrickling filters using continuous recirculation are often called "bioscrubbers".

In summary, a traditionally classified biotrickling filter reactor that has an operating mode with continuous recirculation water flow is called a "(fixed-film) bioscrubber". A traditionally classified biotrickling filter reactor that has an operating mode with water intermittently irrigated and flowing once through the packing media is called a "(once-through operated) biotrickling filter" as illustrated in Figure 8.4.

A bioscrubber configuration will need a pump to recirculate the process liquid. The liquid is recirculated from either a reservoir in the bottom of the reactor or a separate recirculation sump. As discussed above in Section 4.1, biotrickling filters are herein referred to as "once-through operated", with water irrigated on the top and trickling down to the bottom, after which the liquid is directly removed from the reactor. Biotrickling filters will need a temporary recirculation pump that is only used during the startup phase to build up biomass on the packing media. The recirculation pump of a biotrickling filter may be removed after the startup phase or may be kept as standby to restart the biotrickling filter system, if required.

The continuous recirculation of liquid in a bioscrubber may have advantages when the objective is maintaining a relatively constant pH in the recirculation liquid. Constant, relatively high recirculation rates with or without the addition of pH corrective chemicals represents an effective method to obtain a relatively constant pH in the liquid phase and throughout the packing media.

In the intermittent once-through operating mode of a biotrickling filter, water is intermittently sprayed on top of the media and directly removed from the bottom of the bioreactor system (e.g., Figure 8.5). Typically, pH-neutral water is intermittently irrigated on the top of the packing media and trickles through the packing media, removing the sulfuric acid form from the conversion of hydrogen sulfide.

The intermittent once-through operating mode may have some important benefits when removing a mixture of odorous compounds. First, and more importantly, a pH gradient is obtained within the packing media with the once-through operating mode, making it possible to convert odorous compounds that require microorganisms within a more pH-neutral environment. Many highly odorous compounds, such as volatile organic reduced sulfur and nitrogen compounds, need a near-pH-neutral (pH 5 to 8) environment to be able to be properly degraded.

Secondly, a thinner water film can be maintained on the biofilm layer. A thin water film is often preferred to minimize the resistance for mass transfer of more poorly water-soluble pollutants from the gas phase to the biofilm. The thinner water film on the packing media may also result in a larger void volume in the media, and thus actual gas contact time.

4.2.2 Single-Stage and Multi-Stage Treatment

Polluted airstreams almost always contain a mixture of many compounds. These many compounds typically have different chemical properties, as can be seen from, for example, their water solubility and their biodegradability. For air purification, the pollutants need to be transported from the gas phase to the biofilm, where they will be degraded by the microorganisms attached to a carrier (media) of a reactor. To degrade all compounds, an optimal mix of microorganisms is required. Microorganisms differ from each other in their capacities to obtain their energy, carbon, and nutrients.

Microorganisms also differ in their affinity and degradation capacity for compounds, their ability to form a good biofilm structure, their growth rate, and their nutrient requirements for growth. Unfortunately, the optimal environmental conditions for the microorganisms also differ. Therefore, because of the many different compounds in the airstream, a mix of microorganisms is required and different environmental conditions for the microorganisms are preferred in a biological air treatment system. Using different stages or layers optimized for the different microorganisms, a potentially broader range of compounds can be degraded in the bioreactor system more effectively.

Different stages can be a physical separation of different layers of media or can be versatile zones layered with the packing media. Water can be irrigated on top of each stage or can be irrigated only on the top stage. Different reactor stages can be positioned on top of each other (Figure 8.5) or can be placed next to each other in separate reactor shelves.

4.2.3 Cocurrent or Countercurrent

The air through the biological odor treatment system can flow cocurrent or countercurrent with the water flow direction. Cocurrent operation means that both the air- and water flow direction is downward. Countercurrent operation means that the airflow direction is upward with the water direction downward.

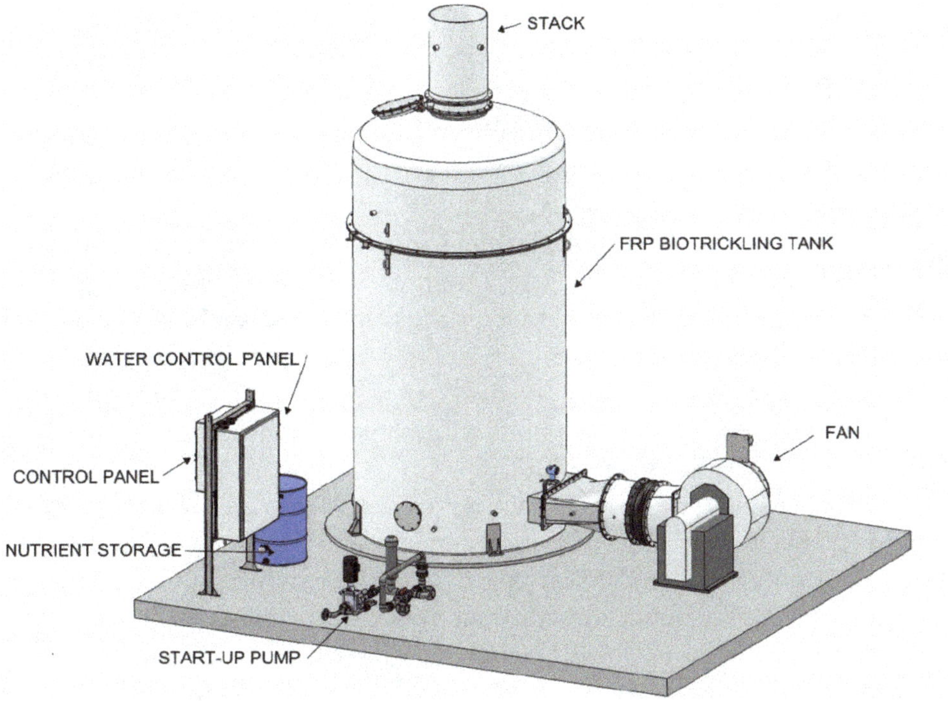

FIGURE 8.5 Typical biotrickling filter system treating 12 500 m^3/h (courtesy of BioAir Solutions).

The choice of direction will influence the removal of the individual compounds because water may transport compounds or degradation products through the system, influencing the mass-transfer dynamics and the pH of the liquid. The difference can be easily illustrated by the pH gradient in a biological odor treatment system that is generated by the sulfuric acid form as an end product from the degradation of reduced sulfur compounds.

Countercurrent is typically used to better maintain a higher pH above the initial media zone at the air inlet side or to better maintain a pH gradient throughout the packing media, improving the mass transfer of hydrogen sulfide and other odorous compounds. In addition, countercurrent operation avoids the "treated" clean air to be in contact with the relatively "dirty" process liquid when the goal is to obtain low outlet odor concentrations.

4.3 Media Types

The packing media in the bioscrubbers and biotrickling filters for odor control at wastewater application typically use special low-pressure, high-efficiency inert or plastic packing. Examples are the smaller-diameter versions of Jaeger Tri-packs, Intalox Snowflakes, Tellerettes, and more conventional Raschig and Pall rings and saddles, which are all randomly packed packing media. Bioscrubber types of odor treatment systems tend to use more open, larger-size material compared to biotrickling filters to accommodate the higher liquid irrigation rates.

Recently, structured pre-engineered packing materials have been used more often to improve performance. Structural packing made of inert or plastic combining a high-porosity, optimum air and water distribution and large specific surface are still under development. Structural packing may prevent deformation and may better limit preferential air- and water stream through the packing media. Higher porous materials such as open-pore polyurethane foam products or other densely wired plastic materials have been successfully used in many full-scale applications, exhibiting not only good hydrogen sulfide, but also high odor removal efficiencies (Gabriel & Deshusses, 2004; Kraakman, 2008).

The surface material of the packing media can be important because many microorganisms have a natural tendency to attach better to hydrophobic, nonpolar materials with a rough surface, which may be critical when treating foul air with low odorant concentrations. At temporary low concentrations, the biomass may switch into starvation mode and starts the removal of cells from a biofilm (a process called "dispersal"). Starvation-induced dispersal is an active and regulated process in biofilms that may be crucial for species in the biofilm community to survive.

Today, multiple types of materials are typically applied with increased density or decreased particle size toward the outlet of the system to improve air distribution and increase mass-transfer rates while minimizing the risk of pressure drop increases over time.

4.4 General Design Guidelines

4.4.1 *Pressure Drop*

The pressure drop of bioscrubber and biotrickling filter systems is a critical design parameter because it affects both the operational costs as well as odor removal performance. The operational costs are required for the energy input to the fans that move the air through the odor treatment system. The odor removal performance is affected by pressure drop as it partly dictates air and liquid distribution through the packing media.

The pressure drop is preferably limited to 0.5 to 1.0 kPa (2 to 4 in. H_2O) and is dictated by the gas load, the total media height, the pollutant load, and the type of packing media. The gas load for odor removal at WRRFs is typically limited to 1500 $m^3/(m^2{\cdot}h)$ (40 cm/s) or 80 cfm/sq ft), the media height is typically limited to 5 or 6 m (15 or 18 ft), and the average pollutant load is typically limited to 100 $g/(m^3{\cdot}h)$ to avoid biomass or elemental sulfur accumulation. The packing media will typically have lower pressure drop at larger void volumes, because the actual gas face velocity is lower.

A more constant pressure drop will be sustained when using acid-resistant inert materials because compaction is avoided. Compaction of certain packing media types such as polyurethane foam cubes or non-acid-resistant materials have frequently resulted in operational problems. Increased pressure drops reduce performance, especially when operated at relatively high pollutant loads where biomass and/or elemental sulfur accumulates over time, increasing the density of the packing media. This often dictates that design requirements be focused on long-term reliable performance rather than short-term maximum performance.

4.4.2 Gas Contact Time

The empty bed gas contact time (EBCT) is typically used as design parameter for odor control systems. As described in section 3.3.2, the EBCT is calculated using the amount of treated airflow divided by the volume of the media in the odor treatment system. The minimum required EBCT depends on multiple factors such as inlet odor loading, required removal efficiencies, reactor type and operation mode, packing media type, and configuration. For bioscrubbers and biotrickling filters treating odors at WRRFs and/or wastewater collection systems, at least 15 seconds is typically required to obtain around 95% of the total odor concentration removed, whereas less than 10 seconds may be enough for a 95 to 99% removal of only hydrogen sulfide. This assumes proper design and operation including measures taken and maintained to provide a sufficient level of reliability for its specific application (see also Section 5.4).

Longer gas contact times are typically required to treat inlet hydrogen sulfide concentrations higher than about 150 ppmv to eliminate incomplete conversion and elemental sulfur accumulation. Larger systems may also be required for obtaining low outlet odor concentrations because of reduced conversion rates or increased mass-transfer limitation at lower odor compound concentrations.

Multilayered biotrickling systems with more diverse pH conditions and increased microbial diversity typically improved removal of residual odors better than single-stage recirculation bioscrubber-type systems. In addition, biological media characteristics such as gas velocity, surface area, and airflow distribution characteristics as well as void volume of the media affect the minimum reactor size as well.

The void volume directly influences the minimum required gas contact time with a larger void volume, resulting in a longer actual contact time. The AGCT will be lower depending on the porosity of the media and the water and biomass attached to the media. The porosity of bioscrubbers and biotrickling filters ranges from about 50% (lava rock) to 95% (pre-engineered structured media), whereas the water and biomass attached to the media are typically less than 10% of the media bed volume for odor control applications.

Gas velocity through the system is also critical for the overall sizing of the system because gas velocity can increase mass transfer (potentially improving performance), as shown by Dorado et al. (2009) and Estrada et al. (2014). Increased gas velocities, on the other hand, may result in increased pressure drop, potentially decreasing cost-effectiveness of the odor control system.

4.4.3 Gas Distribution

The media type and media configuration influence the airflow characteristics through the biological gas treatment system; as such, its significance should not to be underestimated. An optimum air distribution is important to obtain high odor removal efficiencies. When high removal efficiencies are required (>99%), all the air must be treated effectively. To treat the air effectively, all the air needs to exchange the odorous compounds to the biofilm on the biological media. An optimum air distribution preferably moves the air through the reactor as a plug flow, which is, in reality, not possible to obtain.

The gas distribution within a biological media can be tested using tracer gasses and may be expressed as the number of mixtures using the number of mixer-in-series model (Kraakman, 2008; Mendoza et al., 2004). In addition, large preferential airstreams through the media or partial bypasses of the air along the reactor walls must be avoided and should always be tested during the commissioning stage using smoke emitters to confirm the inlet air distribution system and biofilter media are properly installed. A reactor with one layer of randomly packed media is likely to be subject to suboptimal air distribution. On the contrary, better air distribution can be achieved in multi-layers of pre-engineered structured media with open spaces at different heights in the reactor or separate reactors in series, which levels out pressure difference of the air in the media bed as the air travels through.

5.0 DESIGN AND OPERATIONAL CONSIDERATIONS

5.1 General Advantages and Limitations of Biological Odor Treatment Technologies

Biological odor treatment technologies have improved and become more cost-effective, particularly over the last two decades. They have been shown to be most applicable for odor treatment in wastewater treatment processes when inlet hydrogen sulfide concentration exceeds about 10 to 20 ppmv, especially for large airflow rates (Estrada et al., 2011; Kraakman et al., 2014).

Bioscrubber and biotrickling filter systems recently have been applied more often to overcome some of the biofilter design and operational limitations, sometimes in combination with conventional biofilters. Conventional biofilters have limited capability of sustainably treating elevated hydrogen sulfide concentrations because sulfuric acid accumulates in a biofilter media, reducing the overall odor removal efficiencies over time. In addition, biofilters are relatively sensitive to poor moisture control and require a relatively large footprint. The conventional biofilter limitations of acidification, drying out, the requirement of media replacement, and the relatively large footprint required are partly solved by the development of the now more advanced bioscrubber and biotrickling filter systems.

Biotrickling filters and bioscrubbers contain a liquid phase in which important process conditions like pH, salt content, nutrients, toxic intermediates, or end products of the microbial degradation are easier to monitor and control. Biotrickling and bioscrubbing technologies require less or no media change out because acid-resistant inert plastic packing materials are typically used, which saves operational costs associated with media renewal in biofilters.

Conversely, important disadvantages associated with biotrickling and bioscrubbing technologies compared to biofilters may be the relative complexity of construction and the startup of the biological process. The startup of biotrickling and bioscrubber reactors can be complicated because the inert media do not initially contain microorganisms. When the microorganisms are present, they could be washed out by the recirculating water, which is a problem that is sometimes encountered with full-scale operation of bioscrubbers with low inlet pollutant concentrations.

Biofilters and once-through operated biotrickling filters (sometimes called "trickling biofilters") make it possible to create different zones of environmental conditions, which can be important when treating complex mixtures of odorants for removing certain organic odor contaminants or obtaining low outlet concentrations.

Bioscrubbers and biotrickling filters are typically applied to treat odors from the liquid stream (headworks, primaries) because these foul airstreams contain relatively high concentrations of hydrogen sulfide and other readily water-soluble compounds. Biofilters are typically applied more to treat odors from the solids processing stream (biosolids dewatering, sludge holding tanks) to remove relatively high amounts of organic odors with limited amounts of hydrogen sulfide.

Limitations of biological odor treatment technologies include the potential long startup times; limitations to treat large, fluctuating inlet concentrations; and the potential larger footprint requirements compared to chemical scrubber and activated carbon systems. Table 8.1 summarizes the advantages and disadvantages of biofilters, bioscrubbers, and biotrickling filters. Figures 8.6 and 8.7, which are found later in this chapter, show examples of a full-scale biofilter and a full-scale biotrickling system, respectively.

TABLE 8.1 Advantages and disadvantages of vapor-phase biological odor treatment systems.

Technology	Advantages	Disadvantages
BIOFILTERS (water phase and microorganisms immobile)	• Typically more economical for the treatment of odor constituents that are relatively poorly water soluble. • Relatively easy to construct.	• Limited process control. • Not applicable for the treatment of contaminants that are degraded to acid end products. • Require regular media change out. • Relatively large footprint. • Require pre-humidification system.
BIOSCRUBBERS (water phase mobile and microorganisms mobile and/or immobile)	• Enhanced process control possible because of the presence of a mobile liquid phase. • Applicable for odor constituents that are degraded to acid end products. • A relatively small foot print. • Typically lower operational costs. • No pre-humidification required.	• More complex to build and operate than biofilter. • Not applicable for odor constituents that are poorly water soluble (limited total odor removal).
BIOTRICKLING FILTERS (water phase mobile and microorganisms immobile)	• Enhanced process control possible because of mobile liquid phase. • Applicable for odor constituents that are relatively medium poorly water soluble or degraded to acid end products. • Relatively small footprint required. • Typically lower operational costs. • No pre-humidification required.	• More complex to build and operate.

5.2 Best Practices and General Considerations

5.2.1 Startup

The startup time to form optimal performing active biofilms for optimal performance typically ranges between a few weeks to a few months, with temperature being one of the most determining factors. Biotrickling and bioscrubbing technologies, typically using inert materials for gas-liquid contact and biofilm support, require inoculation of microbial cultures. Fresh activated sludge from wastewater treatment processes or existing gas treatment bioreactors is typically used. For startup microbes, the tolerance to suboptimal or changing inlet conditions and the competitiveness in the complex biological ecosystem is important to obtain a resilient biological odor treatment system.

5.2.2 Water Supply Sources

Water is the most important input parameter for a biological odor treatment system. Microorganisms require a moist environment to thrive, and also the supply of water may be critical to remove the degradation products such as sulfuric acid. The water supply source that is typically used at WRRFs is recycled water such as secondary plant effluent water.

Secondary plant effluent water has proven to provide a more reliable operation of a biological odor treatment system than potable water, mainly because it contains all of the required nutrients. It also contains some enhanced pH-buffering capacity and may provide sources to enrich the microbial diversity. Recycled water from membrane bioreactors (MBRs) can be used as well, although this may require additional dosing of some of the nutrients. Secondary effluent or recycled water may preferably be collected before chlorination to eliminate any negative impacts during upsets of the chlorine dosing system.

Because water is critical for a biological odor treatment system, an automatic backup water supply should be installed. Secondary water is typically used as the main water supply with potable water as the backup water supply. Potable water would require the additional dosing of nutrient solution, especially when inert media materials are used.

Odor control systems located at pump stations or along wastewater collection systems cannot easily be supplied with secondary plant effluent water and, therefore, rely on potable water with nutrient dosing. Sometimes, groundwater or surface (river) water may be used as the main water supply with potable water as backup water, which may reduce the amount of nutrients required to maintain optimal performance.

5.2.3 Drain Water Considerations

Drain water from biological odor control systems is typically clear, contains some minor amounts of chemical oxygen demand (COD), and is very low in pH. Because of the potentially extreme low pH (around pH 1 or 2), drain pipes need to be fully protected again corrosion, while the drain water needs further processing or dilution before release back into the environment. Bioscrubbers and biotrickling filters typically generate more drain water because they are typically applied to treat foul airstreams containing relatively high concentrations of hydrogen sulfide generating more acidic drain water. Drain water may be generated in biofilters as a result of over-irrigation or water condensation from the incoming humidified air.

Drain water is typically drained back into the wastewater liquid stream, which provides dilution and pH buffering. The alkaline reduction of the wastewater might need to be considered to prevent potential negative effects for downstream wastewater processes.

5.2.4 Instrumentation and Controls

Instrumentation and controls should be focused on monitoring and maintaining continuity of the water and air supply and, sometimes, the nutrient supply. Instruments should always involve water flow detection, airflow detection, and nutrient dosing detection, and may include the monitoring of pressure drop, pH, and relative humidity and the removal efficiency of hydrogen sulfide or other odor constituents. The moisture content in biofilters may be monitored by measuring the filter medium's weight with load cells, which have been applied in modular biofilter system or moisture probes along the organic biofilter media bed height.

The quality of certain instruments requires extra care because the airstreams are typically saturated with water, creating corrosive conditions, which may require preconditioning of the air for instruments such as hydrogen sulfide monitors.

5.2.5 Operations and Maintenance Considerations

Operations and maintenance (O&M) requirements are considered relatively low for biological odor control systems, which does not mean it is not important. A biological odor control system is often designed for remote operations, and operations typically involve a daily check for alarms, performance, and water leakages. Weekly monitoring involves several critical parameters like moisture content (biofilters), water irrigation rate (bioscrubbers and biotrickling filters), pH of the drain water, nutrient dosing (level nutrient tank), pressure drop, and odor removal.

5.3 Performance Parameters

Key performance parameters for biological odor control technologies are the total odor removal, specific odorant removal capacity such as hydrogen sulfide, ammonia, reduced sulfur or reduced nitrogen compounds, the total water usage, pressure drop, media life, and O&M requirements.

Technologies are typically evaluated for each application in terms of their energy consumption to achieve the target odor removal efficiencies during operation. The energy requirements typically entail a significant share of the annual operating costs and are responsible for a part of the environmental impacts, such as greenhouse gas emissions, in odor treatment technologies. Biological odor control technologies such as biofilters, bioscrubbers, and biotrickling filters exhibit moderate energy consumptions. Process-integrated biotechnological odor control technologies, discussed in Section 6.0 of this chapter, generally require less energy, resulting in a lower environmental impact (IWA Publishing, 2015).

Economic criteria have traditionally driven process selection for full-scale applications. However, today, standards have shifted toward more holistic approaches including environmental aspects and process sustainability. Tools have been developed and recently applied to odor control technologies at WRRFs to quantify their environmental and social impacts (Alfonsína et al., 2014; IWA Publishing, 2015) and provide a wider picture of the economic/environmental/social performance of each specific odor control technology.

5.4 Reliability

5.4.1. Defining Robustness and Resilience

Robustness reflects the capacity of a treatment system to maintain functionality with certain changes such as inlet fluctuations and operational failures. *Resilience* is the rate at which a treatment system returns to its original state after being disturbed. Biological systems may be impacted by sudden changes in parameters such as inlet concentrations, pH, water content, temperature, and nutrients. The level of robustness and resilience of biological systems has been demonstrated with well-designed pilot- and full-scale experiments in the laboratory. It has helped designers to design the right installation for each application and operators to define effective standard operating procedures for various situations that may be encountered.

Reliability (R), the combination of robustness and resilience, may be quantified as the sum of multiple risks for change or upsets.

$$R = \Sigma\ (p \times E)$$

Where p = probability of occurrence of a change or an upset, and
E = the negative effect of a change or an upset.

The probability of occurrence of a change or an upset (p) such as an equipment failure may be expressed as the expected number of occurrences per year (number/year) or as the percentage of operating time during which an occurrence (e.g., a critical pump failure or a water supply failure) is likely to occur. The negative effect of the change or upset (E) may be expressed as the loss of the removal efficiency (%) or the impact on the people living near the installation (e.g., the number of occurrences during which the odor concentration in the emitted airstream exceeded the odor threshold in the neighborhood).

5.4.2. Examples of Upset Conditions

5.4.2.1 Water Supply Failure

An example of a water supply failure was the failure of a valve supplying secondary effluent water to a large-scale biotrickling filter system treating foul air from the headworks at Hyperion Water Reclamation Plant in Los Angeles. The water contained approximately 40 ppm hydrogen sulfide, and water irrigation was interrupted for 18 hours. The hydrogen sulfide removal dropped from 99.5 to 85% over the 18 hours; meanwhile, performance was fully restored in 6 hours after the repair of the water supply.

5.4.2.2 Fan Failure

Interruption of the foul air supply for 2.5 hours to the same system mentioned in section 5.4.2.1 resulted in an initial hydrogen sulfide removal of 65% after the air supply was restored; recovery to 95% removal took about 1 hour. Fan failures offer the unique opportunity to examine the resilience of a biological odor treatment system during periods when inlet pollutants are absent. There is experience that has (semi) quantified the recovery time of biotrickling filter systems treating odors at a WRRF after long periods without input of odorous air into the biotrickling filter system.

Based on this, it can be concluded that, first, the bacteria in the biofilm of the biotrickling filter system do not directly die off during periods without the input of odors, even after several weeks. Second, the recovery is relatively fast (hours to days depending on the duration) and does not require a new startup procedure (typically many weeks) when the biofilm is kept wet (e.g., by keeping water irrigation operational).

5.4.2.3 Power Outage

An example of this type of upset condition is the absence of both foul air and water supplied to a biotrickling filter treating 47 600 m^3/h (28 000 cfm) at the Arlington WRRF in Florida from different sources for 12 hours, followed by a second power outage a few days later of 15 hours. At the inlet concentration of about 300 ppm, the hydrogen sulfide removal (which was 99.5% before the power failure) was 75% at the restart of the system in both cases. Full recovery after the first power outage was just over 2 days, whereas the second time the full recovery was shorter and just under 2 days.

5.4.2.4 Nutrient Dosing Failure

The failure of a nutrient dosing pump at a small biotrickling filter treating 50 ppm of hydrogen sulfide system at a pump station resulted in a hydrogen sulfide removal decline not until after 2 days and reduced, over a week's time, from over 99% to about 90% removal. Recovery occurred within hours after restoring the nutrient supply.

5.4.2.5 High Chlorine in Supply Water

The failure of hypochlorite dosing at a facility effluent disinfection station resulted in elevated chlorine concentration for over a week, with chlorine concentrations in water above 10 mg/L for several days. The high chlorine containing facility effluent water was supplied to a full-scale biotrickling filter system in Florida treating 250 ppm of hydrogen sulfide and temporarily reduced the hydrogen sulfide removal from over 99% to about 95%. Further analysis showed that the chlorine concentrations in water higher than 4 mg/L seemed to start to impact the performance of this biological odor control system.

5.4.2.6 Hypochlorite in Inlet Air

During the startup of a biotrickling odor treatment system at a headworks, replacing a chemical scrubber resulted in hypochlorite vapors sucked into the biological system during the chemical scrubber fans duty/standby autorotation. This resulted in an extremely slow startup and sudden increased outlet hydrogen sulfide concentrations during the startup of the biological odor treatment system. Hypochlorite vapors are detrimental to any biological odor treatment system, and any risks for its presence should be avoided at all times.

5.4.2.7 Toxics in Inlet Air

A newly installed small biotrickling filter system at a wastewater collection network could not be started up properly for several months, even after trying inoculation with fresh activated sludge on multiple occasions. The cause appeared to be sewer rehabilitation works that involved the use of rodent-killing and hypochlorite-containing cleaning agents. After rehabilitation work was finished, startup was completed with no delays.

5.4.2.8 Low Air Inlet Concentrations

A full-scale bioscrubber system during the startup phase at a headworks WRRF in Sydney, Australia, experienced extreme low inlet concentrations for several days on multiple occasions during several raining events. Biomass sloughing of the media resulting from biofilm dispersion was observed to cause an extreme slow recovery of bioscrubber performance over periods up to several weeks. Periods of extreme low inlet concentrations contain the risk of the biomass switching into starvation mode. Particularly in bioscrubbers, this may lead to dispersal, the removal of cells from a biofilm, which may be crucial for species in the biofilm community to survive. This biofilm dispersal process requires energy, and starvation-induced dispersal is an active and regulated process in biofilms.

Several mechanisms contribute to dispersal, including decreases in bacterial adhesiveness and degradation of the biofilm matrix resulting in biomass sloughing off the media. Relatively open media material with limited water holding and biomass holding capacity are more vulnerable to degradation of the biofilm matrix and may require long recovery periods up to several days or a few weeks to fully recover after periods of extreme low inlet concentrations.

5.4.2.9 Sudden Air Inlet Peak Concentrations

Once or twice per day, sudden extreme high inlet hydrogen sulfide concentrations (>600% of average) were experienced in a small biotrickling filter system at a pump station. The removal efficiency dropped to about 60 to 70% during the peak, frequently resulting in odor complaints. The problems ceased after the pump control of one of the upstream wastewater pumps was changed by installing variable frequency drive (VFD) controls. Fluctuation of wastewater flows resulting from the fluctuating discharges of wastewater into the collection system by the community results in gradual fluctuations of odors such as hydrogen sulfide. Extreme sudden fluctuations in odors such as hydrogen sulfide are not the result of these typical diurnal discharges, but rather the result of infrequent pump operations at the pump station or the stop-start of process units at the WRRF.

An average-to-peak concentration ratio of one to three can typically be handled without loss of performance by a biological odor treatment system when the increase of the concentration from the average to peak is over a period of 15 to 30 minutes or more. The frequency of inlet peaks is also important because the alternation of inlet concentrations (also called "feast–famine conditions") is known to increase the resilience of a biological system. More extreme average-to-peak ratios may require a larger system or a combination with other technologies.

5.4.2.10 High Inlet Concentration for Extended Period

An example of this type of upset condition was the failure of a chemical dosing system in the sewage network resulting in extreme high inlet hydrogen sulfide into a large-scale bioscrubber system (treating 500 ppm of hydrogen sulfide) for a period longer than a year. This higher-than-design inlet load resulted in increased pressure drop over the bioscrubber system because of the accumulation of the elemental sulfur-reducing foul air extraction rate as well as the hydrogen sulfide and odor removal performance.

The inlet average mass load is the quantity of contaminant that enters the reactor over a long period and is generally normalized to the reactor volume. To maintain long-term stable operation and prevent the accumulation of biomass or degradation intermediates, a maximum mass load should be maintained with long-term operational performance in mind, regardless of the performance at that moment. Excess biomass can cause an increase in pressure drop, air channeling, and the creation of anaerobic zones, which ultimately leads to a reduction in performance. Treatment systems for certain inorganic compounds like hydrogen sulfide can be designed for relatively high mass loads because the biomass growth yield from hydrogen sulfide degradation is lower than the growth yield from organic compounds. Conversely, biomass control strategies can be incorporated to the design and operation such as periodic packing media cleaning.

5.4.2.11 Extreme Weather Conditions

Biological odor treatments are successfully installed in systems that experience cold winter conditions like Canada and northern Europe as well as hot summers like Texas, Florida, and the Middle East. The microbial community typically adapts to the prevailing temperature, but stops being active when the temperature gets below the freezing point. Completely frozen biofilters have been reported to stop or reduce foul air extraction, but the biological recovery is fast (hours) after being unfrozen. Temperatures above about 40 °C (100 °F) significantly reduce the microbial diversity in the biological odor treatment system, but systems are still capable of treating odors from wastewaters. In extreme temperatures such as in the Middle East, a biological system can function well, but becomes more sensitive to suboptimal moisture control because evaporation rates are much higher.

5.4.3 Opportunities to Improve Reliability

To achieve and maintain a highly resilient engineered system involving biological processes, the biology itself does not seem to need a stable or specific persistent microbial community (Cabrol et al., 2012). Instead, a rich, dynamic diversity of microbial interactions seems more important to achieve enhanced process stability. The resilience capacity is typically correlated to a flexible microbial community able to adapt by both metabolic and microbial community composition changes, where even minor species have significant importance. Short-term stress microbial responses are regulated in terms of enzymatic activity, while responses over longer periods involve shifts in microbial community composition.

The design of engineered biological systems should nevertheless always incorporate features to enhance resilience toward critical performance indicators. General measures that can be undertaken to improve the overall reliability can be targeted towards sustaining biological availability and increasing buffering capacity.

5.4.3.1 Sustaining Biological Activity

Some compounds can be temporarily stored in the biofilm for use later. For example, hydrogen sulfide is degraded via elemental sulfur, which at high peak loads is used by microorganisms as temporary energy storage. It can be observed that the biofilm of a biofilter or biotrickling filter system turn completely creamy white when temporarily exposed to extreme inlet hydrogen sulfide loadings. This can be explained by the formation and accumulation of elemental sulfur sometimes excreted by the microorganisms outside their cells. After lowering the inlet hydrogen sulfide concentrations to average loadings again, this creamy white color slowly disappears over time (hours or days). Examples of possible interventions to maintain the biological activity include

- the addition of substrates to the water supply to keep the bacteria active (alive or prevent them from becoming dormant) by dosing thiosulfate or elemental sulfur slurry during wet weather events to prevent hydrogen sulfide breakthrough during a fast incline of hydrogen sulfide concentrations after wet weather events;
- the addition of substrates other than hydrogen sulfide replacements for certain (mixotrophic) hydrogen sulfide oxidizing bacteria to stay

active during periods of hydrogen sulfide absence by, for example, the addition of organic materials. This can be obtained by providing facility effluent water, which always contains some organic material; and

- the use of media types that enable holding relatively large amounts of biomass inside the system (which allows for abundant use of extracellular microbial polymeric substance present in a biofilm enhancing the robustness of the system against changes).

5.4.3.2 Increasing the Buffering Capacity

The liquid and the media are typically not designed to accommodate a large buffer capacity, but they offer opportunities to increase resilience against large inlet fluctuations depending the origin of the odorous compound. Examples include

- maintaining a high-water content inside the system that enhances the absorption capacity of the reactor system for hydrophilic pollutants or using water with high alkalinity to absorb pH changes inside the system caused by peak inlet concentrations and water supply upsets;
- using packing material that improves absorption capacity of the reactor system (e.g., adding adsorbents or polymers based on copolymers of polyurethane, vinyl acetate, and ethylene); and
- adding substrates to the water supply to increase the absorption capacity of the reactor system (e.g., liquid solvents to increase the liquid sorption capacity).

Other examples of enhancing reliability of biological odor control system may include the following:

- Eliminate single point of failures—Through system design, the reliability can be enhanced by, for example, making provisions for an automatic backup water supply or installing automatic duty/standby rotating equipment for critical items such as valves, pumps, or fans.
- Enhance the microbial diversity—Enhanced microbial diversity can be obtained through spatial stratification along the biological filter height by providing different environmental conditions (e.g., pH, odorant concentrations, media type, and composition) within the odor control system.
- Select the optimal biological odor treatment system—This depends on the application and the required performance, including reliability,

determine the type of biological system. For example, bioscrubbers are generally more susceptible to upsets than biotrickling filters and biotrickling filters are generally more susceptible to upsets than biofilters. Bioscrubber entail more equipment with moving parts such as recirculation pumps, whereas biofilters contain organic materials that typically provide more additional buffer capacity than the synthetic medias of bioscrubbers and biotrickling filters.

- Increase buffer capacity—As previously mentioned, some additional robustness can be built in by providing more buffer capacity (e.g., by the addition of additives, as discussed in Section 3.2.3, or by increasing water holding capacity).
- Integrate automation—Include instrumentation for alarming and automated controls such as temporarily slowing down the fan speed during peak emissions because of irregular operation of process equipment at the source (i.e., pump stop and starts).
- Incorporate physical-chemical polishing units—Combine biological technologies with other physical-chemical technologies such as adsorption (activated carbon filtration), oxidation (chemical treatment using hypo chlorite or ozone), or a stack (enhancing dispersion of any residual odors, as shown in Figure 8.6).
- Bring together multi-stage treatment steps—An odorous airstream can be treated in a hybrid system (e.g., a combination of a bioscrubber with a biofilter) or the use of a multi-stage biotrickling filters for spatial stratification along the biological filter height to enhance microbial diversity (Figure 8.7).

Table 8.2 presents a list of potential upset conditions and some opportunities that can be considered to enhance reliability when designing or operating biological odor control technologies.

6.0 PROCESS INTEGRATED BIOTECHNOLOGY

6.1 Introduction

Odor emissions at WRRFs are typically solved by implementing odor treatment technologies. This involves covering and extracting foul air from odorous process units and treating the gas using specialized odor treatment systems before atmospheric discharge. This general approach can entail important costs and typically requires operator efforts. The costs associated

FIGURE 8.6 In-vessel biofilter system with dispersion stack treating 50 000 m^3/h (courtesy of Metro Denver and Jacobs Engineering).

FIGURE 8.7 Multi-stage biotrickling filter system treating 125 000 m^3/h (courtesy of Melbourne Water and Jacobs Engineering).

TABLE 8.2 Potential upset conditions and opportunities to improve reliability of biological odor treatment systems.

Potential Upset	Likely Consequence	Resolution
Water Interruption	• Reduced microbial activity resulting from too dry or acidic conditions within hours.	• Install a water storage tank. • Install a water supply from a second source as automatic backup. • Install instrumentation for water flow and pressure alarming. • Keep critical spare parts such as valves and supply pumps.
Power Outage	• Reduced microbial activity at the power restore because of too reduced enzyme activity for hours/few days.	• Connect power supply to a back-up power supply (e.g., diesel generator). • Run-signal to supervisory control and data acquisition (SCADA) to generate alarm during local power outage.
Nutrient Dosing Failure	• Microbial activity slowly reducing after several days.	• Install a second dosing pump in duty/duty arrangement at 50% each. • Install level switch in nutrient dosing tank with low-level alarm. • Install a flow switch after each dosing pumps. • Eliminate nutrient dosing by selecting a water source rich in nutrients (e.g., secondary facility effluent water).
High Chlorine in Supply Water	• Instant reduction of the microbial activity, potentially killing most of the microbes in the system biofilm.	• Take facility effluent water before chlorine disinfection step. • Install a water storage tank with dechlorination (e.g., thiosulfate dosing). • Implement chlorine alarming instrumentation.

(*continued*)

TABLE 8.2 (*Continued*).

Potential Upset	Likely Consequence	Resolution
High Inlet Air Concentration (sudden peaks)	• A breakthrough of inlet concentrations within minutes after the peak event initiation.	• Design foul air extraction from multiple source to dampen out fluctuating emissions from the specific processes. • Change pump control of sewage pumps station by installing VFD controls. • Temporarily reduce airflow to allow for better inside the treatment system. • Install robust inlet and outlet H_2S detection instrumentation with alarming.
High Inlet Air Concentration (long period)	• Slow increase of pressure drop over the system because of accumulation of elemental sulfur or biomass reducing foul air extraction rate as well as the overall odor removal performance.	• Install inlet air monitoring equipment. • Design and install for additional treatment capacity.
Low Inlet Air Concentration	• Biomass sloughing of a bioscrubber media because of biofilm dispersion with slow recovery of performance over periods of up to several weeks.	• Install a media that has improved biomass-holding capacities. • Use a water source rich in nutrients and carbon source. • Install biofilter using organic media.
Fan Failure	• Reduced microbial activity at the fan restore resulting from too-reduced enzyme activity for hours/few days at a failure period of half a day or more.	• Install second fan in duty/standby arrangement with autorotation. • Install instrumentation such as flow or pressure transmitters with alarming. • Keep critical spare parts of fan.

with treating odors using these traditional methods, both on a capital and operating basis, add significantly to the total costs of a WRRF. Moreover, this end-of-pipe treatment approach addresses odor problems once odorants have been produced and released from the wastewater.

A more desirable approach could be the prevention of odorant formation and/or release from the wastewater. Limited options are available for the prevention of odorant release at WRRFs beyond proper design and good operating practices.

This section discusses widely applicable existing and emerging odor control biotechnologies that use readily available WRRF byproducts with a minimum facility upgrading requirement. The relatively low investment and operating costs may contribute to the sustainability and economic efficiency of odor control at wastewater treatment facilities. The three facility-integrated approaches that will be discussed in the following section are activated sludge recycling, activated sludge diffusion and oxidized ammonium recycling for odor control.

Activated sludge recycling (ASR) is a strategy where activated sludge from secondary clarifiers or aerobic activated sludge from aerated biological reactors is recycled to the inlet of the WRRF headworks to instantly adsorb and degrade odorous compounds. Activated sludge diffusion (ASD) for odor control re-uses odorous air emissions from unit processes as the process feed air to a traditional activated sludge process. Rather than using ambient air for the oxygen supply in the aeration basins, odorous air is used. Oxidized ammonium recycling (OAR) is based on the recycling of streams with high nitrate or nitrite concentration at a WRRF to the inlet works of a WRRF or upstream in the sewer system creating anoxic conditions.

6.2 Activated Sludge Recycling

6.2.1 Introduction

Activated sludge recycling (ASR) is a strategy for odor control consisting of the recycling of activated sludge from secondary clarifiers or aerobic activated sludge bioreactors to the inlet of the WRRF headworks (Figure 8.8). ASR can reduce the release of odorous compounds at the inlet works, pretreatment, pumping stations, and primary settlers, which are typically reported as the main responsible process units for malodorous emissions at WRRFs.

The implementation of ASR would require relatively low investment costs, accounting for the pipeline to transfer the sludge to the headworks and the pumping equipment needed. The extra operating costs would derive from the power needed for pumping and the maintenance of this equipment.

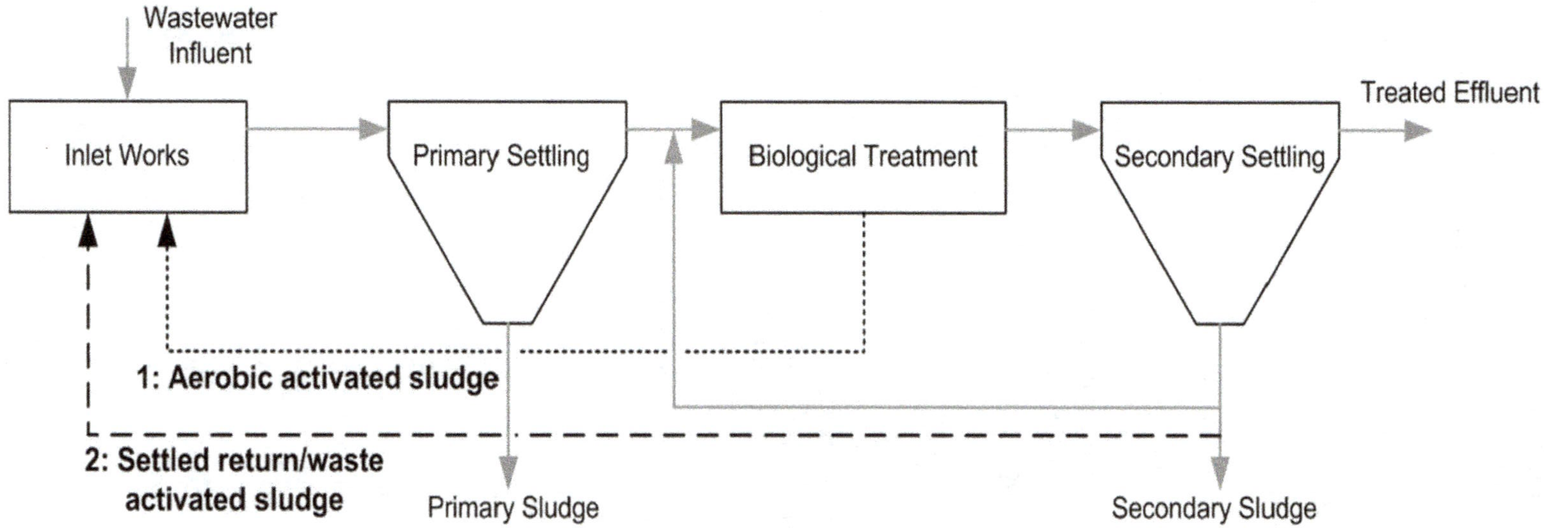

FIGURE 8.8 Typical WRRF flow diagram including two different options for activated sludge recycling (ASR) operation with (1) direct ASR from the aerobic activated sludge reactor (dotted line) and (2) ASR from the secondary clarifier (dashed line).

The ASR strategy promotes the biological uptake of odorous compounds before they volatilize from the liquid phase. Adsorption to the activated sludge flocs followed by oxidation of potential malodorous compounds is assumed to be the mechanism preventing their release from the subsequent wastewater treatment units. The recycled activated sludge contains concentrations of oxygen and/or nitrate that will be used as electron acceptors for the oxidation of the odorants or the malodorous compound precursors. Biological odorant oxidation can thus take place by aerobic oxidation or anoxic oxidation coupled to denitrification.

Activated sludge typically exhibits a high biological diversity, holding the potential to adsorb and biologically oxidize most biogenic compounds responsible for odor nuisance, mainly reduced volatile organic or inorganic compounds such as hydrogen sulfide, mercaptans, amines, indoles, and fatty acids. In fact, the diffusion of malodorous emissions into aeration basins, known as activated sludge diffusion, has been used as a method for odor control, and activated sludge is commonly used as inoculum for standard biological odor treatment systems such as biofilters, bioscrubbers, and biotrickling filters.

The effect of iron salts, often added during wastewater treatment for phosphorus precipitation, present in the activated sludge streams can be beneficial for this odor prevention process by promoting the precipitation of dissolved sulfide as ferrous sulfide.

6.2.2 Critical Design and Operating Parameters

Design guidelines for this odor technology are still under development and full-scale pilot testing for a few weeks is, therefore, recommended. The sulfide concentration in the raw wastewater is a key parameter to determine if an ASR strategy for reducing odor problems in WRRFs is envisaged. As a general rule, it seems reasonable to assume that a 0.5 activated sludge recycling ratio (activated sludge flow to influent wastewater flow) reduces hydrogen sulfide gas concentrations ranging from 60 to 75% when dealing with average sulfide concentrations in the wastewater of up to 5 mg/L. WRRFs with lower sulfide concentrations may obtain removal efficiencies as high as 99% or benefit from operating at lower activated sludge recycling ratios. Conversely, higher sulfide concentrations (e.g., because of industrial effluents or warmer climates) would require higher recycling ratios, which may lead to operational problems if the design hydraulic capacity of the WRRF is exceeded.

Facility operators might have the concern that recycling strategies increase the abundance in sulfur-consuming and filamentous bacteria, having a negative influence on the biological wastewater treatment process or activated sludge floc sedimentation properties. However, no negative experiences on this issue have been reported at full scale. Full-scale operations have also confirmed no increase in biosolids production, oxygen requirements in the aerobic tanks, or nitrate levels in the facility effluent.

A potential issue that might be addressed is the influence of ASR strategies on the performance of enhanced biological phosphorus removal processes because ASR may partially deplete the readily biodegradable volatile fatty acids present in the influent wastewater. In addition, grit settling properties caused by the mixture of the raw wastewater and return activated sludge might affect the grit removal process. However, when the amount of RAS to be mixed with the raw wastewater is limited, the impact on the grit removal process can be prevented while still providing significant odor control (Schmidt et al., 2014). The characteristics of the primary sludge will change and may require greater care in the handling/thickening of the primary sludge.

Estrada et al. (2015) reviewed and discussed multiple full-scale studies and implementation. They concluded that ASR has been implemented successfully and should be considered when developing facility-wide odor management strategies. Implementation of ASR can eliminate the covering of process units, foul air extraction ductwork, fans, and end-of-pipe odor treatment systems.

6.3 Activated Sludge Diffusion

6.3.1 Introduction

Activated sludge diffusion (ASD) for odor control refers to the reuse of odorous air emissions from other unit processes as the process feed air (oxygen) to a traditional activated sludge process. The adsorption of odorous compounds to the activated sludge flocs combined with biological activity of the activated sludge provides odor control. The odorous compounds are co-degraded with the contaminants dissolved in the inlet wastewater. Compared to vapor-phase biofilter systems, where a variety of conditions can exist within the biofilm, it is easier to control the process where complete mixture can be assumed. Electron acceptors (e.g., oxygen), pH, and metabolite buildup can be readily controlled in the liquid phase. In addition, mass transfer can be optimized by maintaining minimum concentrations in the liquid. ASD reactors may also be called "suspended growth bioreactors" and "airlift bioreactors" when baffles are used to cause a distinct pattern of liquid recirculation (Kennes & Veiga, 2001).

Activated sludge diffusion has been practiced at many wastewater treatment sites since the 1950s as an odor control measure, mostly with positive results. However, not all design limitations are well defined yet and odor removal performance data are still limited (Bowker & Webster, 2005; Kiesewetter, et al., 2012; Vertz, Van Durme, & McKnight, 2006). Figure 8.9 presents a diagram of the ASD process.

In ASD systems, the malodorous air collected from covered or contained process units is directly sparged into the aeration tank as the air needed to satisfy the biological oxygen demand of the wastewater treatment process. Odorous compounds diffuse into the mixed liquor together with oxygen and are subsequently adsorbed and/or converted by the vast activated sludge community.

6.3.2 Critical Design and Operating Parameters

Activated sludge diffusion requires little in terms of extra facilities, and might be able to use existing aeration equipment, such as blowers and diffusers. Minor to moderate modifications are then required, such as air pre-filters, covering odorous process units and extending ducting from the odorous process unit(s) to the blower inlets. The type of diffuser selected is an important factor, with fine-bubble diffusers providing significantly better performance than coarse-bubble diffusers. Fine-bubble diffusers are generally made of ceramic or membrane material on a plate, disc, or tube generating tiny gas bubbles ranging in diameter from 1 to 4 mm.

The use of ASD poses limitations in the amount of foul air that can be treated because of a preset bioreactor air demand. The odorous airflow may not be a good match for the process air demand in the biological aeration tank. Specific design controls may have to be implemented to accommodate both minimum oxygen demand and minimum required foul air extraction and must be dealt with on a case-by-case basis.

There is a potential for corrosion in ductwork, process air blowers, process air piping, and diffusers. However, corrosion is not a proven concern, but corrosion-resistant materials must be specified to eliminate the risk of corrosion. The blowers and ductwork are typically operated in a manner that leads to condensation, and the combination with hydrogen sulfide-laden air may lead rapidly to the corrosion of inlet air filters and steel inlet ducts.

There are reports of tarry coatings on process blowers and there is a potential for particulate contamination of diffusers. Moisture traps (coarse grease screens) or inline mist eliminators should also be considered. Such systems would have to be checked and maintained once placed in service.

The degree of odor control performance is difficult to determine upfront. Typically, hydrogen sulfide is removed with an efficiency of more than 99% and the odor concentration is reduced by about 95%. However, the concentration of dimethylsulfide and dimethyldisulfide emitted from the aeration basin may slightly increase when introducing odorous air into the aeration

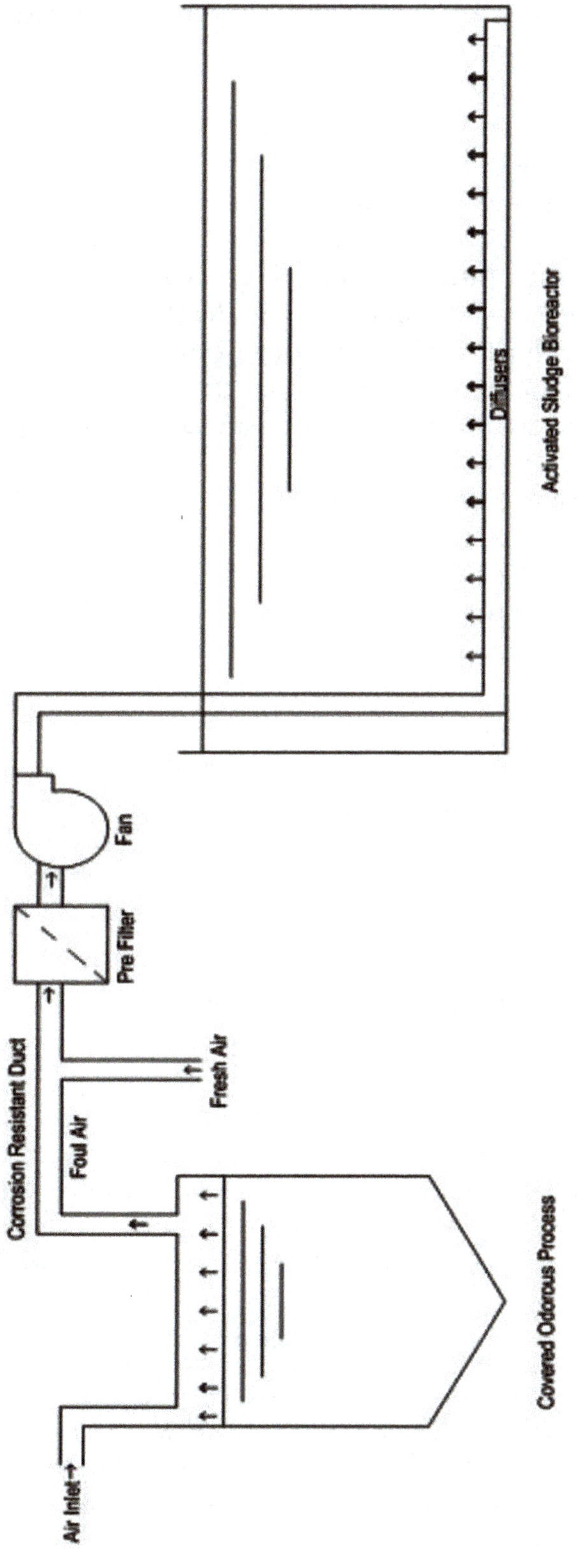

FIGURE 8.9 Typical flow diagram of an ASD operation.

basin (compared to using ambient air as oxygen supply). So, while significant net odor removal can be achieved, there is likely the potential to add to the background odors coming from the aeration basins.

The potential for increased filamentous activated sludge growth appears to be unfounded under normal operating conditions. However, a change in the microorganism population dynamics of the activated sludge will most likely be observed when it is exposed to significant hydrogen sulfide loads for an extended period. At extreme high hydrogen sulfide loadings (>5 mg hydrogen sulfide/(g MLSS·d)) an increase in the sludge volume index (SVI) can be expected, with the risk that bulking of the sludge may occur under these unusual conditions. A proper upfront design of ASD is necessary as well as an implementation that includes startup and testing procedures monitoring of SVIs and floc settling before and after ASD is implemented.

ASD has been applied at multiple large-scale WRRFs without any negative impact on the performance of activated sludge systems for COD and nutrient removal. ASD offers the possibility of highly efficient odor removal that has been demonstrated to be highly robust toward fluctuations commonly found under routine operation. Good design is required to prevent unwanted side effects, and implementation of ASD requires relatively low capital expenditure and helps a facility save on its energy consumption.

6.4 Oxidized Ammonia Recycling

6.4.1 Introduction

Oxidized ammonium recycling (OAR), also sometimes referred to as "sweet water recycling", is based on the recycling of streams such as nitrified centrate, nitrified wastewater, or dissolved air flotation filtrate. These streams contain relatively high nitrate or nitrite concentrations and are injected at the inlet works of a WRRF or upstream in the sewer system. OAR reduces septic (anaerobic) conditions, eliminating the production of odorous compounds. In addition, OAR would promote available microorganisms to oxidize dissolved sulfides and any readily biodegradable odorants, preventing their further release as malodorous emissions. Figure 8.10 shows a diagram with the elements included in an OAR system for odor control.

6.4.2 Critical Design and Operating Parameters

The OAR strategy is commonly implemented in existing denitrification-nitrification WRRFs to reduce nitrogen levels discharged to receiving waterbodies to meet discharge limits or facility expansion/upgrading to include anaerobic digestion (Estrada et al., 2015). Effluents with high ammonia concentrations are nitrified and recycled to the inlet works, where they

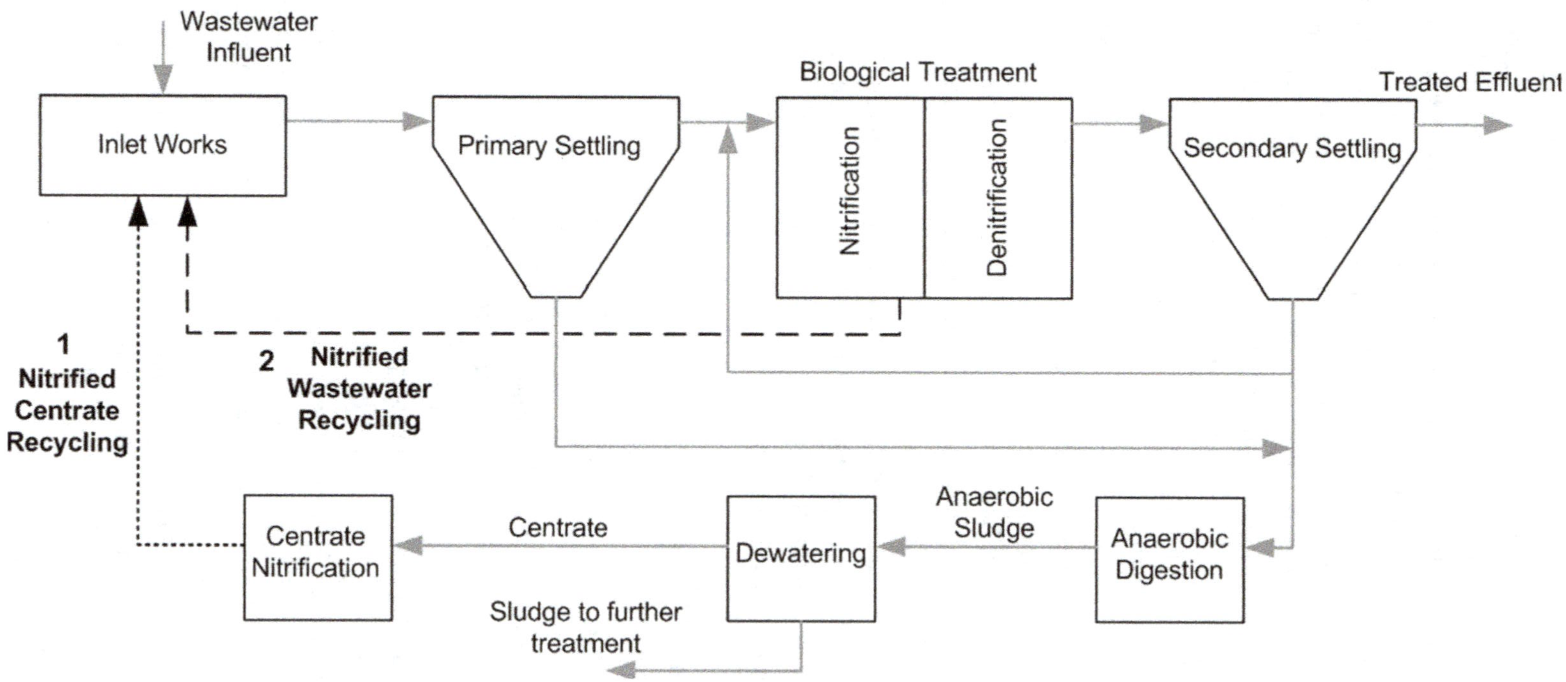

FIGURE 8.10 WRRF flow diagram including the sludge line and two different options for an OAR operation. 1) dotted line: OAR from centrate nitrification units, 2) dashed line: OAR from the nitrification stage in the biological reactor.

undergo denitrification. The addition of nitrate to the wastewater influent promotes anoxic conditions, where nitrate is used as an electron acceptor by microorganisms to oxidize dissolved sulfides and any readily biodegradable odorants. This is similar to dosing nitrate-containing chemicals to a sewer collection system to eliminate odor and corrosion problems.

The quest for energy and economic efficiency in WRRFs has made anaerobic digestion a widespread technology for sludge management to generate part of the electricity consumed during wastewater treatment. The dewatering of the anaerobically digested sludge, however, generates ammonia-rich effluents (typically 500 to 1000 mg/L), commonly referred to as "centrates" or "reject waters". Traditionally, this ammonia-rich effluent, representing up to 20% of the total ammonia load to the WRRF, is recycled to the biological treatment systems to be removed by conventional nitrification-denitrification process. This would increase the overall wastewater treatment costs and challenge compliance with the nitrogen discharge limits. Thus, innovative treatment technologies for these centrate streams have been developed, some of which include biological oxidation of ammonia to nitrate and its further recycling to the WRRF inlet works.

In conclusion, OAR is a proven process that could eliminate the requirement for covering of process units, foul air extraction ductwork, fans, and end-of-pipe odor treatment. It should be considered when developing facility-wide odor management strategies.

7.0 REFERENCES

Alfonsína, C., Lebrero, R., Estrada, J. M, Muñoz, R., Kraakman, N. J. R., Feijoo, G., & Moreira, M. T. (2014). Selection of odour removal technologies in wastewater treatment plants: A guideline based on Life Cycle Assessment. *Journal of Environmental Management, 149*, 77–84.

Barton, J. W., Davison, B. H., Klasson, K. T. & Gable, C. C. (1999). Estimation of mass-transfer and kinetics in operating trickle-bed bioreactors for removal of VOCs. *Environmental Progress, 18*(2), 87–92.

Bowker, R., & Webster, N. (2005). Odor and air emissions reductions using existing bioreactors. *Proceedings of the Annual Water Environment Federation Technical Exhibition and Conference*. Alexandria, VA: Water Environment Federation.

Cabrol, L., Malhautier, L., Poly, F., Lepeuple, A. S., & Fanlo, J. L. (2012). Bacterial dynamics in steady-state biofilters: Beyond functional stability. *FEMS Microbiology Ecology, 79*(1), 260–271.

Dorado, A. D., Rodríguez, G., Ribera, G., Bonsfills, A., Gabriel, D., Lafuente, J., & Gamisans, X. (2009). Evaluation of mass-transfer coefficients in biotrickling filters: Experimental determination and comparison to correlations. *Chemical Engineering & Technology, 32*(12), 1941–1950.

Estrada, J. M., Bernal., O. I., Flickinger, M. C., Manoz, R., & Deshusses, M. A. (2015). Biocatalytic coatings for air pollution control: A proof of concept study on VOC biodegradation. *Biotechnology Bioengineering, 112*(2), 253–271.

Estrada, J. M., Dudek, A., Muñoz, R., & Quijano, G. (2014). Fundamental study on gas–liquid mass-transfer in a biotrickling flter packed with polyurethane foam. *Journal of Chemical Technology Biotechnology, 89*(9), 1419–1424.

Estrada, J. M., Kraakman, N. J. R., Muñoz, R., & Lebrero, R. (2011). A comparison analysis of odour treatment technologies in wastewater treatment plants. *Environmental Science & Technology, 45*(3), 1100–1106.

Gabriel, D., & Deshusses, M. A. (2004). Technical and economic analysis of the conversion of a full-scale scrubber to a biotrickling filter for odor control. *Water Science & Technology, 50*(4), 309–318.

IWA Publishing. (2015). *Sewage treatment plants: Economic evaluation of innovative technologies for energy efficiency*. London, UK: IWA Publishing.

Kennes, C., & Veiga, M. C. (2001). *Bioreactors for waste gas treatment.* Springer Science & Business Media, B.V.

Kennes, C., & Veiga, M. C. (2013). *Air Pollution Prevention and Control.* Bioreactors and Bioenergy. Wiley.

Kiesewetter, J., Kraakman, N. J. R., Cesca, J., Trainor, S., & Witherspoon, J. (2012). Expanding the use of activated sludge at biological wastewater treatment plants for odor control. *Proceedings of Water Environment Federation specialty conference*, Louisville, Kentucky, United States, April 2012. Alexandria, VA: Water Environment Federation.

Kim, S., & Deshusses, M. A. (2007). Determination of mass-transfer coefficients for packing materials used in biofilters and biotrickling filters for air pollution control. 1. Experimental results. *Chemical Engineering Science, 63*(4), 841–855.

Kraakman, N. J. R. (2008). Multi-layered bioreactors: Design considerations, implementation issues and operating experiences. *Proceedings of the WEF Odors and Air Pollutants Conference*, Phoenix, Arizona, United States, April 2008. Alexandria, VA: Water Environment Federation.

Kraakman, N. J. R., Estrada, J. M., Lebrero, R., Cesca, J., & Muñoz, R. (2014). Evaluating odour control technologies using reliability and sustainability criteria. *Water Science & Technology, 69*(7), 1426–33. doi: 10.2166/wst.2014.036

Kraakman, N. J. R., Roche Rios, J., & van Loosdrecht, M. C. M. (2011). Review of mass-transfer aspects for biological gas treatment. *Applied Microbiology and Biotechnology, 91*(4), 873–886.

Mendoza Perez, J. A., Prado-Rubianes, Ó. J., Veiga, M. C., & Kennes, C. (2004). Hydrodynamic behaviour and comparison of technologies for the removal of excess biomass in gas-phase biofilters, *Water Research, 38*(2), 404–13.

Myung Cha, J., Suk Cha, W., Lee, J. H. (1999). Removal of organo-sulphur odour compounds by Thiobacillus novellus SRM, sulphur-oxidizing microorganisms. *Process Biochemistry, 34*(6), 659–665.

Schmidt, D., Karkeskint, J., Porter, J., Parker, D., & Hoskins, K. (2014). Mixing RAS with sewage to reduce odor emissions while avoiding adverse impacts to processes. *Proceedings of the Water Environment Federation Odors and Air Pollutants Conference*, Miami, Florida, United States. Alexandria, VA: Water Environment Federation.

Shareefdeen, Z., & Singh, A. (2005) *Biotechnology for odor and air pollution control.* Heidelberg, Germany: Springer-Verlag.

Vertz, J., Van Durme, G. P., & McKnight, M. D. (2006). Activated sludge diffusion provides cost effective biological odor control at two Texas wastewater treatment facilities. *Proceedings of the Annual Water Environment Federation Technical Exhibition and Conference*. Alexandria, VA: Water Environment Federation.

9

Chemical or Oxidation Odor Control

1.0	INTRODUCTION	279
2.0	WET CHEMICAL GAS ADSORPTION SYSTEMS	280
	2.1 Mass Transfer Concepts and Chemistry of Wet Chemical Absorption Systems	280
	2.2 Wet Chemical Scrubber Configurations	283
3.0	PACKED TOWER SCRUBBER DESIGN PARAMETERS AND SIZING	287
4.0	DETERMINING REQUIRED PACKING DEPTH	288
5.0	MAKEUP WATER REQUIREMENTS	292
6.0	ESTIMATING CHEMICAL USE	296
7.0	MIST ELIMINATION	301
8.0	MIST SCRUBBER SYSTEMS	303
9.0	IONIZATION SYSTEMS	304
10.0	CATALYTIC OXIDATION	305
11.0	REFERENCES	306

1.0 INTRODUCTION

This chapter will discuss the theoretical and practical aspects of removing odorous compounds from an airstream fouled by contact with water resource recovery facility (WRRF) processes or conveyance systems by absorption and chemical reaction in a wet chemical scrubber. The concepts of gas-gas oxidation by ions and catalytic oxidation will also be introduced. Sufficient theoretical background will be provided to illustrate the academic aspects of developing the size and capacity of packed tower odor control systems. However, because there is copious data available from installed and working systems, empirical data and practical experience will be emphasized with regard to system development. It is believed that this is a pragmatic approach to understanding, designing, and operating real-world applications of the technology.

Wet chemical odor scrubbing systems have been in use at WRRFs for decades. Their basic configurations have included traditional packed tower systems and atomized mist systems. Both rely on providing sufficient gas-liquid interface area and time within the scrubber to allow odorous compounds to move through the foul air as the carrier gas phase to the interface with liquid and absorb into the scrubbing bulk liquid.

Once in the liquid, those compounds are altered into ionic forms or different chemical species that cannot again enter the gas phase. Most scrubbers used in WRRFs are designed to remove hydrogen sulfide (H_2S) (because of its prevalence and very low odor detection and recognition thresholds) as well as other reduced sulfur and organic sulfide compounds. Because these compounds tend to form acids in solution, a solution with basic chemistry is required. When ammonia (NH_3) or amines are the target odor an acidic scrubber solution chemistry is used. Because ammonia has an odor recognition threshold several orders of magnitude higher than that of hydrogen sulfide, it is less frequently the target of odor control.

2.0 WET CHEMICAL GAS ADSORPTION SYSTEMS

Wet chemical gas absorption systems at WRRFs fall into two general categories: packed towers and atomized mist systems. Both systems will be discussed in this chapter; however, it should be noted that while packed towers have stood the test of time and are still widely used, atomized mist systems have fallen out of favor because they have proven to less reliably remove odors at the rate of packed towers. In both systems, it is necessary to provide a large area of contact between the gas and liquid phases to achieve adequate mass transfer between the two during the time the gas passes through the scrubber. While the basic method of providing large surface areas of contact differed between the two scrubber types, the chemistry of their scrubbant liquids was essentially the same.

2.1 Mass Transfer Concepts and Chemistry of Wet Chemical Absorption Systems

Wet chemical scrubbers rely on chemical addition to facilitate the quick transfer of the gases targeted for control into the liquid phase and to render them into forms that will not reenter the gas phase. To understand the importance of the chemistry necessary to facilitate absorption, it is important to understand the basics of gas phase to liquid phase mass transfer which, at a fundamental level, is fairly simple.

First the gases targeted for transfer must be brought from the interior of the bulk gas volume to the interface between the gas and liquid phases. To do so, those gases must diffuse from the interior of the bulk gas phase to the interface with the liquid before there is any hope of transferring the target gas between the two phases. Obviously, if the distance between the interior of the gas phase to the gas-liquid interface is relatively great, it will take a relatively long time for the gas in the middle of the gas phase to diffuse to the gas-liquid interface.

To minimize the distance and time necessary for this target gas to reach the surface of the area of contact between the two phases, we can use the turbulent mixing inherent in the gas and liquid interaction and the high surface area interface within the scrubber. Because the volume of the gas phase does not change, maximizing its surface area minimizes the third dimension of volume, thus reducing the time the target gas requires to reach to the gas-liquid interface. Once near the interface, the flow is essentially laminar and the gas must pass the remaining distance by molecular diffusion to be absorbed into the bulk liquid (Wark & Warner, 1981). Once the target gas has reached the gas-liquid interface, it is important to facilitate its transfer between the two phases. This is where the chemistry of the liquid phase comes into play.

The driving force for absorption from the gas to the liquid phase is dependent on the concentration of each gas in those two phases. Gas will diffuse more easily from areas of high concentration to areas of low concentration. The greater the difference in concentration between the gas and liquid phases, the higher the driving force and the more quickly the target gas will pass from the gas to the liquid phase.

Assuming that the incoming gas-phase concentration is only controllable by actions outside of the scrubber system, then the only thing that we can hope to control in the scrubber is the dissolved gas concentration in the liquid phase. By altering liquid-phase parameters, predominantly pH, we can control which ionic forms of odor-causing compounds are present in solution and the solubility of gases into the liquid phase. If a gas such as hydrogen sulfide (H_2S) were to dissolve into the liquid phase at neutral pH, about half of it would dissociate into the HS^- ion (and H^+ ion), as shown in Figure 9.1 below. In this case, the buildup of dissolved hydrogen sulfide gas in the liquid phase would begin to inhibit the absorption of additional hydrogen sulfide from the gas phase.

If we can increase the pH to above 9, however, virtually all the hydrogen sulfide will dissociate into ionic forms, and the driving force will remain strong. This pH shift is typically accomplished by adding sodium hydroxide

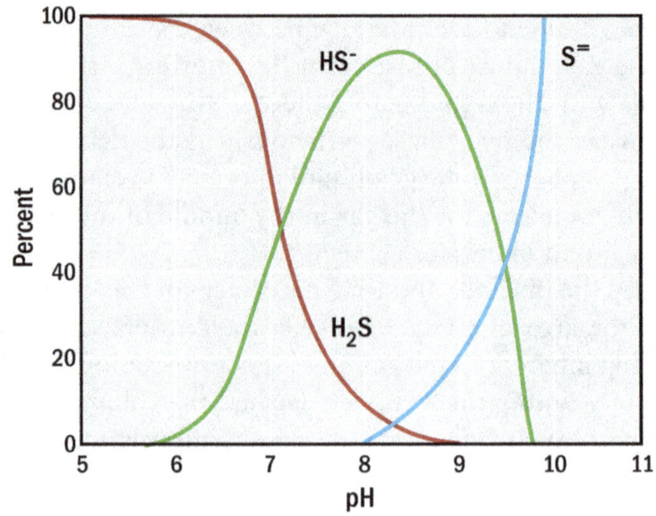

FIGURE 9.1 Sulfide equilibrium diagram (adapted from U.S. EPA, 1985).

(NaOH). Sodium hypochlorite (NaOCl) is also added to the solution to oxidize the dissolved sulfides. The following equations illustrate the reaction:

$$H_2S + 2NaOH \leftrightarrow Na_2S + 2H_2O \tag{9.1}$$

$$Na_2S + 4NaOCl \leftrightarrow Na_2SO_4 + 4NaCl \tag{9.2}$$

Hydrogen peroxide (H_2O_2) has also been used as an oxidant in scrubbers targeting sulfides for removal. However, because of difficulties monitoring its concentration in solution, its use is uncommon. Equation 9.3 illustrates hydrogen peroxide's reaction with hydrogen sulfide:

$$H_2S + H_2O_2 \leftrightarrow 2H_2O \tag{9.3}$$

Chemical addition to maintain sodium hydroxide and sodium hypochlorite at appropriate concentrations is accomplished using chemical metering and injection pumps. Sodium hydroxide injection is done to maintain a pH of 9 to 10.5, and sodium hypochlorite injection is done to maintain an oxidation–reduction potential (ORP) of 600 to 800 mV and convert the sulfur into nonvolatile sodium sulfate (Na_2SO_4) and sodium chloride (NaCl). The setting of pH and ORP levels can vary between scrubber installations as a result of the individual needs and odorous compounds present. Because the compounds present and their relative concentrations can vary over time, adjustments may be appropriate to maintain odor control and the best economic scrubber performance.

When ammonia (NH_3) or amines are the targeted gases for removal, a pH of 1.5 to 6.5 is typically used. The addition of sulfuric acid (H_2SO_4) can be used to maintain pH. The chemical equation for ammonia removal is as follows:

$$H_3 + H_2SO_4 \leftrightarrow NH_42O_4 \tag{9.4}$$

2.2 Wet Chemical Scrubber Configurations

Packed tower wet chemical scrubbers are typically arranged in a countercurrent flow configuration, as shown in Figure 9.2.

This configuration introduces the foul air near the bottom of the scrubber. It then flows up through an inert packing material. Scrubbing solution is applied by spray nozzles or a trough and weir irrigation system to the top of the packing. The solution then flows downward through the packing over which it disperses, creating a large surface area maximizing the interface area between the gas and liquid phases. The gas exits the packing, traveling upward through a mist eliminator that captures solution droplets carried in the gas stream and prevents them from exiting the scrubber. After passing through the mist eliminator, the scrubbed air then exits either to the atmosphere or to another stage of treatment.

After flowing though the packing, the scrubber solution drains to either an integral sump, as shown in Figure 9.2, or to a sump system external to the scrubber. This latter case is sometimes used when multiple scrubbers are operated in parallel. The scrubber solution is recirculated from the sump to the top of the packing. Chemical probes are used to monitor the scrubber solution chemistry. In Figure 9.2, pH and ORP probes are used with controllers to trigger and regulate the injection of NaOH to adjust pH and NaOCl to adjust ORP levels when they fall below set-point values. Water is added to the sump to flush salts (e.g., NaCl, Na_2SO_4) accumulation from the solution and replace water lost to evaporation.

The basic concept of a wet chemical scrubber is quite simple when one considers that the amount of pollutant flowing into the scrubber in the gas and liquid phases must equal the amount of pollutant leaving the scrubber in the liquid and gas phases. Figure 9.3 presents a schematic of the liquid and gas phases entering and leaving the scrubber, and Equation 5 describes the material balance considering each phase.

$$G_{m,i}y_i + L_{m,i}x_i \leftrightarrow G_{m,o}y_o + L_{m,o}x_o \tag{9.5}$$

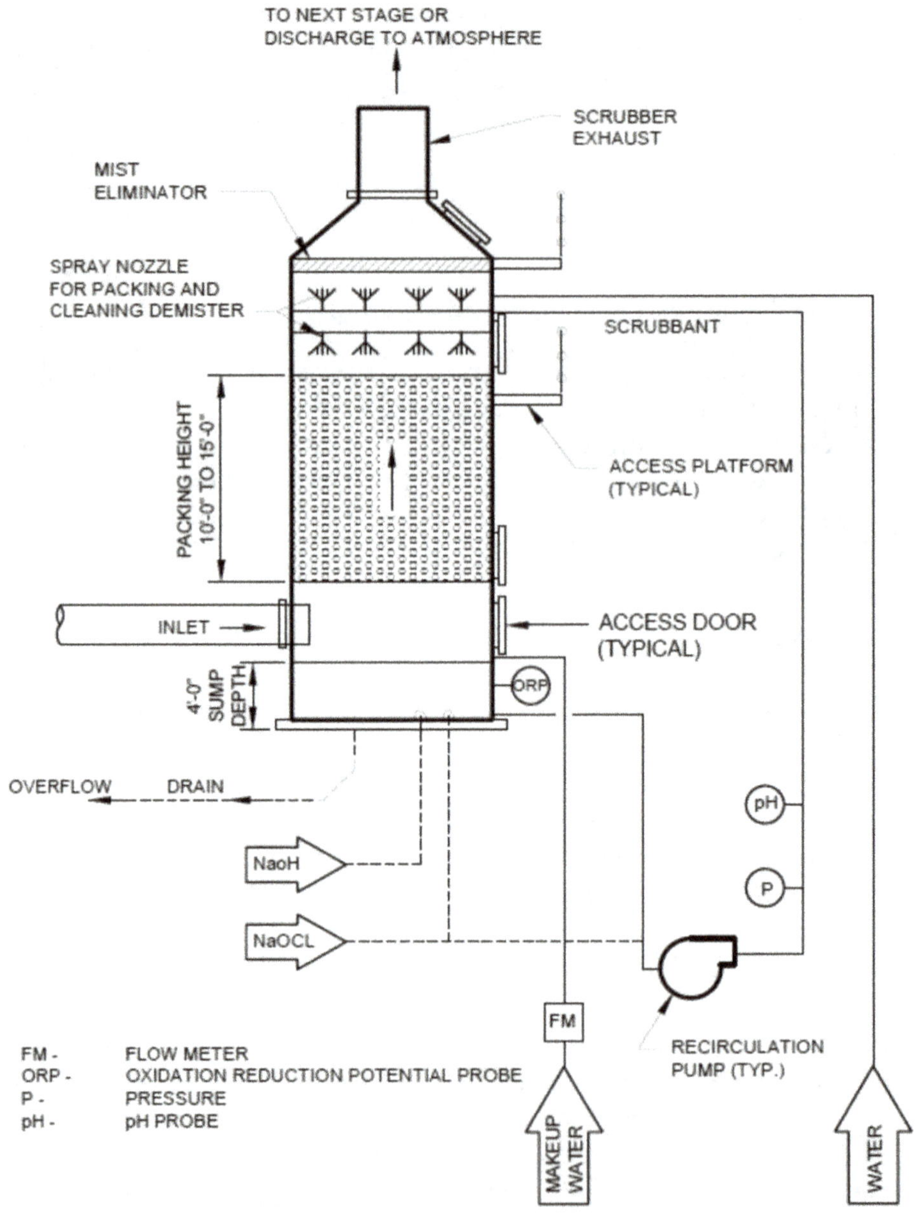

FIGURE 9.2 Typical counter-current packed tower scrubber for wastewater odor control.

Where subscripts i and o indicate material into and out of the scrubber, respectively;

$G_{m,i}$ = total moles of gas into the scrubber;
$G_{m,o}$ = total moles of gas out of the scrubber;
$L_{m,i}$ = total moles of liquid into the scrubber;

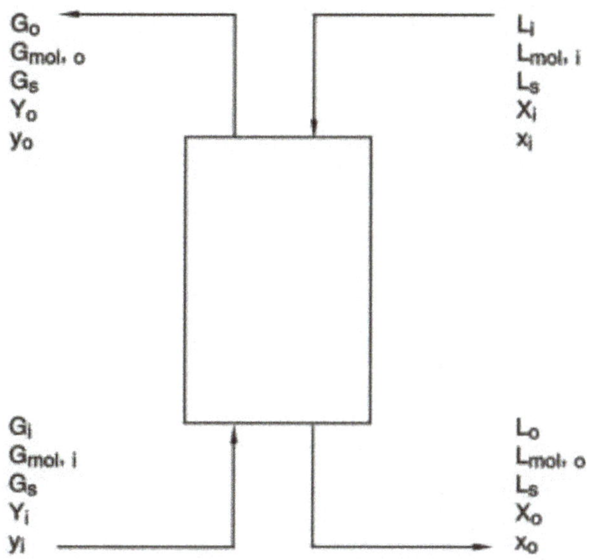

FIGURE 9.3 Wet chemical scrubber mass balance schematic.

$L_{m,o}$ = total moles of liquid out of the scrubber;
y_i = mole fraction of target gas coming in with the carrier gas;
y_o = mole fraction of target gas exiting with the carrier gas;
x_i = mole fraction of target gas coming in with the scrubber liquid; and
x_o = mole fraction of target gas exiting with the scrubber liquid.

Other scrubber configurations are available. Crossflow scrubbers pass foul air horizontally through packing while the scrubbing solution flows from top to bottom similar to the countercurrent flow scrubber. Crossflow scrubbers are typically not as efficient at removing pollutants (odors) as countercurrent flow scrubbers because the high-strength odors only encounter fresh scrubbing solution at the top of the packing. High-strength odors entering lower in the packing encounter partially spent solution that has less ability to absorb and react the target gasses. Additionally, some of the incoming gases can pass around the top of the packing and receive less-efficient pollutant removal.

Concurrent flow scrubbers introduce both foul air and scrubbing liquid at the top of the scrubber and both flow down to the bottom. These scrubbers are not as efficient as the countercurrent variety because the average driving force of absorption is lower.

Low-profile multi-stage scrubbers (Figure 9.4) have alternating countercurrent and concurrent stages arranged in series. These are typically used when obstructions prevent installation of a taller, single-stage countercurrent scrubber or when the aesthetic impact of a tall scrubber is not desired.

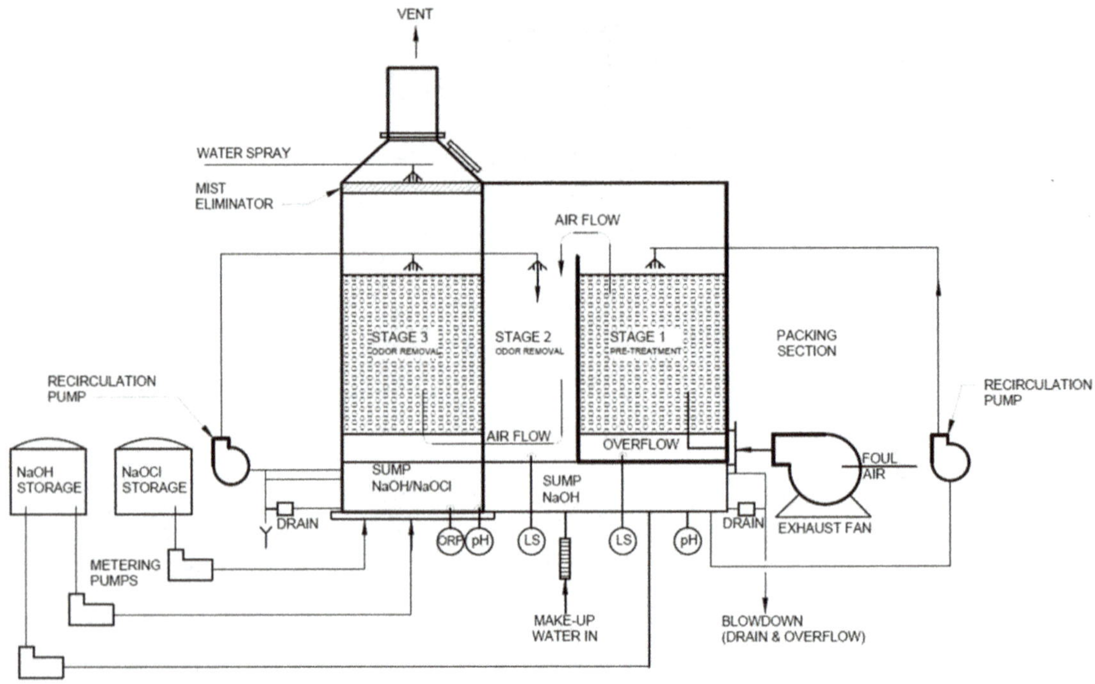

FIGURE 9.4 Low-profile three-stage scrubber for hydrogen sulfide and odor removal.

These scrubbers can be configured to provide differing scrubbing solutions in each stage to target multiple specific gases in a single scrubber package.

In the configuration shown in Figure 9.4, the first stage can be used as a roughing stage to remove the bulk of the incoming hydrogen sulfide by being built with its own dedicated sump, separate from the other stages, into which only sodium hydroxide is added. This dedicated stage 1 sump can receive its makeup water as overflow from the stage 2 and 3 sump, into which both sodium hydroxide and sodium hypochlorite are injected. The residual sodium hypochlorite from the downstream stages will oxidize some of the sulfide absorbed in the first stage. This can provide a more economical application compared to using a sump and chemistry common to all stages.

Because unreacted sulfide may remain in the stage 1 scrubbing solution, discharging overflow from the first stage sump to a WRRF process or sewer of near-neutral pH can result in hydrogen sulfide volatilizing out of solution, causing odor emissions and subsequent corrosion problems. If NH_3 is targeted for removal, H_2SO_4 can be added to the first stage while NaOH and NaOCl are added to stages 2 and 3 for sulfide removal. In this case, it is important to separate the sumps of the stages using different chemistries and provide robust mist elimination to minimize carryover between stages that would cause the acid and base to react with one another and increase chemical use and costs.

3.0 PACKED TOWER SCRUBBER DESIGN PARAMETERS AND SIZING

The design of a packed tower scrubber is primarily based on the target gas removal efficiency desired within a scrubber volume of acceptable height and cross section with reasonable liquid flowrates and air pressure differential across the scrubber. Typical wastewater odor control packed tower scrubbers targeting hydrogen sulfide and other sulfides are designed with gas flowrates of 1.5 to 2.5 m/s (300 to 500 fpm) face velocity (volumetric flow divided by the scrubber cross-sectional area). Higher face velocities will require greater packing depth to achieve the same collection efficiency, so there will be a tradeoff between diameter and height of the scrubber. Higher face velocities will result in greater air pressure drops across the scrubber resulting in higher energy costs.

Liquid flowrates are typically about 12 $m^3/(m^2{\cdot}h)$ (5 gpm/sq ft) of scrubber cross section, but can be higher or lower depending on the packing design and incoming gas concentrations. Higher liquid flowrates will result in less packing height required. Determining the required packing height will be discussed later in this section, but, for now, we can use airflow rate and face velocity to begin sizing the scrubber.

As an example, we will consider a packed scrubber to treat 4.7 m^3/s (10 000 cfm) and initially choose a gas face velocity of 2.0 m/s (400 fpm). The cross section of the scrubber is then calculated by dividing the volumetric airflow rate by the selected gas velocity, as follows:

$$\begin{aligned}\text{Scrubber cross section} &= \text{Volumetric airflow/face velocity}\\ &= (4.7\ \text{m}^3/\text{s})/(2.0\ \text{m/s})\\ &= 2.3\ \text{m}^2\\ \\ &= (10\ 000\ \text{ft}^3/\text{min})/(400\ \text{ft/min})\\ &= 25\ \text{ft}^2\end{aligned}$$

Assuming a circular cross section scrubber design and the fact that the circular area is equal to $\pi d^2/4$, the scrubber diameter can then be calculated by taking the square root of area multiplied by $4/\pi$, as follows:

Scrubber diameter = Square root (Scrubber cross section × $4/\pi$) or, in our case,

$$\begin{aligned}&= \sqrt{2.3\ \text{m}^2 \times \frac{4}{\pi}}\\ &= 1.7\ \text{m}\end{aligned}$$

$$= \sqrt{25 \text{ ft}^2 \times \frac{4}{\pi}}$$
$$= 5.6 \text{ ft}$$

Most packed tower scrubbers are fabricated from fiberglass reinforced plastic (FRP). The plastic resin-wetted fiberglass filaments are wound around mandrels that are typically sized in increments of 0.3-m (1-ft) diameter in North America, so, in our case, we would have to select a diameter of 1.5 or 1.8 m (5 or 6 ft). Outside North America, local FRP fabricators should be consulted for diameter increments locally available. The selection of either diameter will likely depend on how much clearance is available or necessary at the location of the scrubber installation. Assuming that sufficient clearance is available to allow the installation of the 1.8-m (6-ft) diameter option, we can back calculate the face velocity as follows:

$$\text{Face velocity} = \frac{4.7 \text{ m}^2/\text{s}}{\frac{\pi(1.8 \text{ m})^2}{4}}$$
$$= 1.8 \text{ m/s}$$

$$\text{Face velocity} = \frac{10\ 000 \text{ ft}^3/\text{min}}{\frac{\pi(6 \text{ ft})^2}{4}}$$
$$= 354 \text{ ft/min}$$

Had we selected the 1.5-m (5-ft) diameter option, we would have found that the face velocity would have slightly exceeded the maximum of our typical range. This does not mean that selecting the smaller diameter is out of the question. However, the higher velocities will result in more energy to move the gas through the scrubber and overall scrubber height will increase.

4.0 DETERMINING REQUIRED PACKING DEPTH

Conventional scrubber theory found in chemical engineering texts is useful to understanding, on a conceptual level, the pollutant mass transfer from the gas to the liquid phase described in Equation 9.5 to a point. However, the theory has some significant shortcomings when applied to a typical wastewater odor scrubber targeting hydrogen sulfide and other reduced sulfur compounds. The conventional scrubber theory can be described using the graph in Figure 9.5.

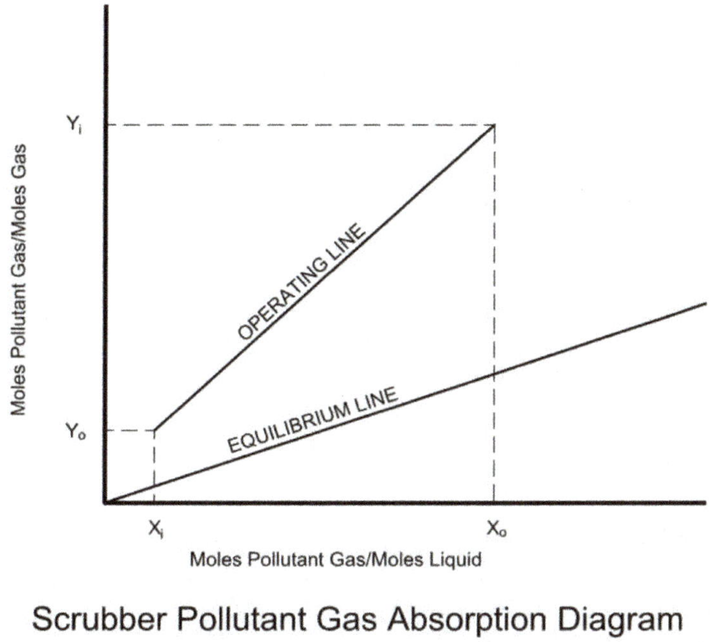

FIGURE 9.5 Scrubber pollutant gas absorption diagram.

In Figure 9.5, the equilibrium line shows the liquid-phase concentration of pollutant gas in equilibrium with the gas-phase concentration throughout the scrubber. The operating line shows the design gas-phase concentrations at the inlet (y_i) and exhaust (y_o), and is assumed to be linear in between. It is necessary for the gas and liquid concentrations to not be in equilibrium so that a driving force will exist and absorption will occur. The distance between the operating and equilibrium lines provides an indication of the magnitude of the driving force.

At this point in the discussion of scrubber theory, it is important to point out that in a scrubber with active chemical reactions occurring, not just simple absorption, defining the liquid-phase concentrations of the gases targeted for removal is difficult at best. Sulfide removal scrubbers using sodium hydroxide and sodium hypochlorite or ammonia/amine scrubbers using H_2SO_4 are just such devices. Therefore, it is better to rely on experimental or empirical data to properly size a scrubber where chemicals are being added and in which reactions are taking place. Having said that, we will continue with just a little more theory and then more practical application.

The amount of packing required to facilitate the absorption as the airstream passes through the scrubber is described as follows:

$$Z = H_{tu} \cdot N_{tu} \tag{9.6}$$

Where Z = packing height,

H_{tu} = height of a mass transfer unit, and
N_{tu} = number of mass transfer units.

The number of mass transfer units (N_{tu}) required may be calculated using the following equation (U.S. EPA, 2002):

$$N_{tu} = \frac{\ln\left[\left(\frac{y_i - mx_i}{y_o - mx_i}\right)\left(1 - \frac{1}{AF}\right) + \frac{1}{AF}\right]}{1 - \frac{1}{AF}} \quad (9.7)$$

Where

y_i = mole fraction of pollutant gas in incoming gas,
y_o = mole fraction of pollutant gas in exiting gas,
x_i = mole fraction of pollutant gas in incoming liquid,
m = equilibrium line slope,

$$AF = \frac{L_{\text{mol}, i}}{mG_{\text{mol}, i}}$$

L = incoming liquid molar flowrate, and
G = incoming gas molar flowrate.

The reader is referred to the referenced U.S. Environmental Protection Agency (U.S. EPA) document for the expanded explanation of general scrubber theory. Although it is deemed important to understand the complexities of the theory from an academic sense, significant empirical data exist in the published literature to estimate necessary packing depths required to achieve desired target gas removal efficiencies. Additionally, packed tower equipment suppliers possess sufficient proprietary information to assist in the appropriate sizing of packed tower scrubbers. These suppliers should be consulted to verify packing depth dimensions for specific applications.

In the case of a wastewater hydrogen sulfide odor scrubber using sodium hydroxide and sodium hypochlorite, we can reasonably assume $x_i \approx 0$ (i.e., a negligible amount of pollutant enters the absorber in the liquid stream) and $1/AF \approx 0$ (i.e., the slope of the equilibrium line is very small and/or the L_{mol}/G_{mol} ratio is very large). In this case, Equation 9.7 simplifies to the following:

$$N_{tu} = \ln\left(\frac{y_i}{y_o}\right) \quad (9.8)$$

Packing manufacturers can provide dimensions of H_{tu} (height of mass transfer unit), which they determine experimentally for various liquid and gas flowrates and liquid chemistry. It is extremely important to communicate and coordinate any scrubber design with the packing manufacturer.

Consider the following example data to begin determining the packing height in a countercurrent packed tower scrubber:

Inlet H_2S concentration: 100 ppmv
Temperature: 17.2 °C (63 °F)
Relative humidity: 40%
H_2S exhaust concentration desired: 0.1 ppmv (removal efficiency desired: 99.9%)
Airflow rate: 425 m^3/min (15 000 cfm)

Then, to first determine number of mass transfer units,

$$N_{tu} = \ln\left(\frac{100}{100-99.9}\right) = \ln\left(\frac{100}{0.10}\right) = 6.9$$

At this point, a packing material must be found that will have an acceptable transfer unit height (so your scrubber does not get too tall) and an acceptable airflow resistance (that will have to be overcome by your system fan) so energy costs will be reasonable. Packing manufacturers can provide this data, typically in the form of curves or tables showing transfer unit height for gases targeted for removal and various scrubber airflow rates, liquid flowrates, and chemistry concentrations/set points.

Here, we will consider the data from one manufacturer presented in Table 9.1.

For the sake of illustration, we can use the height of the transfer unit for the 2″ packing size in the table. From this data, it is easy to calculate the height of the packing needed at the given gas (G) and liquid (L) flowrates using Equation 9.6:

$$Z = 0.493 \text{ m} \times 6.9 = 3.4 \text{ m}$$
$$Z = 19.4 \text{ in.} \times 6.9 = 133.9 \text{ in. or about } 11.2 \text{ ft}$$

TABLE 9.1 Typical mass-transfer performance data for hydrogen sulfide removal.[a]

Absorption Chemistry	G Kg/m^2-hr (lb/hr-ft^2)	L Kg/m^2-hr (lb/hr-ft^2)	Temp. °C (°F)	H_{tu} m (in.) for Various Packing Sizes		
				1″	2″	3.5″
H_2S – NaOH	6001 (1229)	6499 (1331)	20.6 (68)	0.33 (13.0)	0.493 (19.4)	55.9 (22.0)

[a]data from Jaeger Tri-Pacs® Product Bulletin 600, Raschig USA Inc.®

If that packing height would cause the overall scrubber height to be too tall, one could opt to use the 2.5-cm (1-in.) packing with the H_{tu} = 0.33 m (13.0 in.), so the packing height would only be about 2.3 m (90 in. or 7.5 ft). However, it should be noted that the data presented are for gas and liquid flowrates at 1.4 m/s (273 fpm) and 6.6 $m^3/(m^2{\cdot}h)$ (2.7 gpm/sq ft), respectively, which would be considered low for most wastewater packed tower odor control scrubbers.

As previously mentioned, typical odor control packed tower scrubbers will operate with a gas flowrate between 1.5 to 2.5 m/s (300 and 500 fpm) face velocity and liquid flowrates of about 12 $m^3/(m^2{\cdot}h)$ (5 gpm/sq ft). The packing manufacturer should be able to assist with the calculation of transfer unit heights for a variety of liquid and gas flowrates, so a specific packing can be selected that will fit the needs of the individual application.

It cannot be overemphasized that the selection of packing and design of the packing height should be done in consultation with the packing manufacturer who should be willing to assist. The manufacturer's data and assistance will also help you avoid a condition known as "flooding" that occurs when the airflow is too high and prevents the scrubbing liquid from flowing evenly down through the packing, significantly compromising scrubber efficiency and wasting energy. If there is any question about the ability to properly design a packed tower scrubber system using the data available from the manufacturer and corroborating the findings with information on similar systems at other treatment plants or in the literature, pilot testing a small scrubber on the odors' source should be considered.

5.0 MAKEUP WATER REQUIREMENTS

Water addition must flow into the scrubber for two main reasons. The first reason is that water must be continuously drained from the scrubber to remove byproducts of the reactions necessary to remove odorous compounds from the air. These are mainly the salts NaCl and Na_2SO_4, shown in Equations 9.1 and 9.2. The second reason is that moisture will be evaporated from the scrubber liquid and carried out with the cleaned airstream.

To calculate the water necessary to flush the reaction byproducts from the sump, it is necessary to determine the rate of salt formation. Then, using the rule of thumb that the salts in the scrubber liquid should be kept at no more than 5% by weight, this flushing makeup rate can be determined. The rate of salt formation is proportional to the rate of absorption of H_2S in the scrubber as calculated below.

$$\text{H}_2\text{S volumetric flow rate} = 425\frac{\text{m}^3 \textit{Foul Air}}{\text{min}} \times \left(\frac{100-0.1}{10^6}\right)\frac{\textit{Parts}\ \text{H}_2\text{S}}{\textit{Parts Foul Air}} = 0.042\frac{\text{m}^3\ \text{H}_2\text{S}}{\text{min}}$$

$$H_2S \text{ molar flow rate} = 0.042\ \frac{m^3\ H_2S}{\text{min}} \times \frac{1\ \text{mol}}{0.0224 m^3 \cdot \dfrac{(273+17.2)K}{(273+0)K}}$$

$$= 0.042\ \frac{m^3}{\text{min}}\ H_2S \times \frac{1\ \text{mol}}{0.0238 m^3} = 1.764\ \frac{\text{mol}\, H_2S}{\text{min}}$$

$$H_2S \text{ molar flow rate} = 1.764 \frac{\text{mol}\, H_2S}{\text{min}} \times 60 \frac{\text{min}}{\text{hr}} = 105.8 \frac{\text{lb mol}\, H_2S}{\text{hr}}$$

$$H_2S \text{ volumetric flow rate} = 15000 \frac{ft^3 \textit{Foul Air}}{\text{min}} \times \left(\frac{100-0.1}{10^6}\right) \frac{\textit{Parts}\ H_2S}{\textit{Parts Foul Air}} = 1.5 \frac{ft^3}{\text{min}}\ H_2S$$

$$H_2S \text{ molar flow rate} = 1.5 \frac{ft^3}{\text{min}} H_2S \times \frac{1\ \text{lbmol}}{359 ft^3 \cdot \dfrac{(460+63)R}{(460+32)R}} = 1.5 \frac{ft^3}{\text{min}} H_2S \times \frac{1\ \text{lbmol}}{382 ft^3}$$

$$= 0.004 \frac{\text{lbmol}\ H_2S}{\text{min}}$$

$$H_2S \text{ molar flow rate} = 0.004 \frac{\textit{lbmol}\, H_2S}{\textit{min}} \times 60 \frac{\textit{min}}{\textit{hr}} = 0.24 \frac{\textit{lbmol}\, H_2S}{\textit{hr}}$$

Using the H_2S absorption rate, NaCl and Na_2SO_4 salt formation based on the stoichiometric rates shown in Equations 9.1 and 9.2 can then be calculated.

NaCl formation rate:

$$= \frac{4\ \text{mol NaCl} \times 58.44\ \dfrac{\text{g NaCl}}{\text{mol NaCl}}}{1\ \text{mol}\ H_2S \times 34.09\ \dfrac{\text{g}\ H_2S}{\text{mol}\ H_2S}} \times 105.8\ \frac{\text{mol}\ H_2S}{\text{hr}} \times 34.09\ \frac{\text{g}\ H_2S}{\text{mol}\ H_2S}$$

$$\text{NaCl formation rate} = 24.731 \frac{\text{g NaCl}}{\text{hr}} = 24.7 \frac{\text{kg NaCl}}{\text{hr}}$$

$$= \frac{4\ \text{lbmol NaCl} \times 58.44\ \dfrac{\text{lbm NaCl}}{\text{lbmol NaCl}}}{1\ \text{lbmol}\ H_2S \times 34.09\ \dfrac{\text{lbm}\ H_2S}{\text{lbmol}\ H_2S}} \times 0.24\ \frac{\text{lbmol}\ H_2S}{\text{hr}} \times 34.09\ \frac{\text{lb}\ H_2S}{\text{lbmol}\ H_2S}$$

$$\text{NaCl formation rate} = 56.1 \frac{\text{lbm NaCl}}{\text{hr}}$$

Na_2SO_4 formation rate:

$$= \frac{1 \text{ mol } Na_2SO_4 \times 142.05 \dfrac{Na_2SO_4}{\text{mol } Na_2SO_4}}{1 \text{ mol } H_2S \times 34.09 \dfrac{\text{g } H_2S}{\text{mol } H_2S}} \times 105.8 \frac{\text{mol } H_2S}{\text{hr}} \times 34.09 \frac{\text{g } H_2S}{\text{mol } H_2S}$$

$$Na_2SO_4 \text{ formation rate} = 15028 \frac{\text{g } Na_2SO_4}{\text{hr}} = 15.0 \frac{\text{kg } Na_2SO_4}{\text{hr}}$$

$$= \frac{1 \text{ lbmol } Na_2SO_4 \times 142.05 \dfrac{\text{lbm } Na_2SO_4}{\text{lbmol } Na_2SO_4}}{1 \text{ lbmol } H_2S \times 34.09 \dfrac{\text{lbm } H_2S}{\text{lbmol } H_2S}} \times 0.24 \frac{\text{lbmol } H_2S}{\text{hr}} \times 34.09 \frac{\text{lb } H_2S}{\text{lbmol } H_2S}$$

$$Na_2SO_4 \text{ formation rate} = 34.1 \frac{\text{lbm } Na_2SO_4}{\text{hr}}$$

Na_2SO_4 formation rate:

$$= \left(24.7 \frac{\text{g NaCl}}{\text{h}} + 15.0 \frac{\text{g } Na_2SO_4}{\text{h}}\right) \times \frac{1 \text{ kg } H_2O}{0.05 \text{ kg } H_2O} \times \frac{1 \text{ L } H_2O}{1 \text{ kg } H_2O} \times \frac{1 \text{ hr}}{60 \text{ min}}$$

$$= 13.2 \frac{\text{L } H_2O}{\text{min}}$$

$$= \left(56.1 \frac{\text{lbm NaCl}}{\text{hr}} + 34.1 \frac{\text{lbm } Na_2SO_4}{\text{hr}}\right) \times \frac{1 \text{ lbm } H_2O}{0.05 \text{ lbm salt}} \times \frac{1 \text{ gal } H_2O}{8.34 \text{ lbm } H_2O} \times \frac{1 \text{ hr}}{60 \text{ min}}$$

$$\text{Flushing water req.} = 3.6 \frac{\text{gal } H_2O}{\text{min}}$$

Water from the scrubber that is lost to evaporation must be included in the makeup water calculations. To do this, the use of a psychrometric chart is helpful and simple to use. The example chart shown in Figure 9.6 is a simplified version for illustration of the methodology and should not be used for actual calculations. Psychometric charts are available online and can be selected for the appropriate air density/altitude of the actual application.

For the sake of example, it will be assumed that our 425 m^3/min (15 000 cfm) scrubber will have an incoming foul airstream at 17.2 °C (63 °F) and 40% relative humidity (RH). The assumption is that the relatively dry airstream will become saturated with moisture as it passes through the wetted packing. The process of

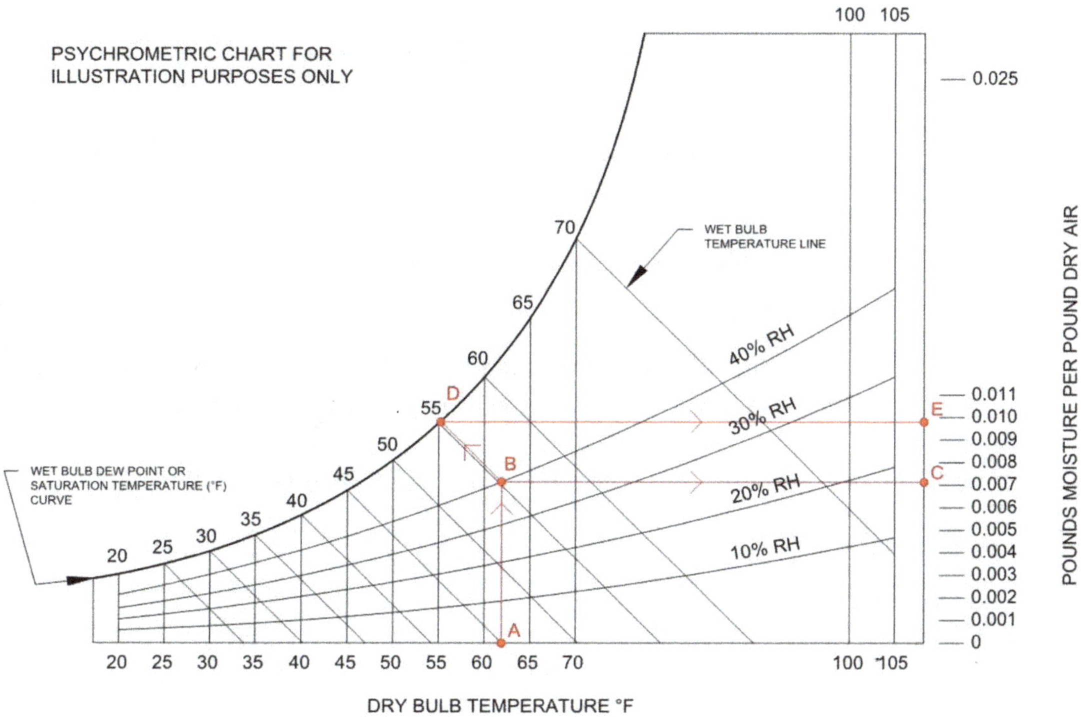

FIGURE 9.6. Simplified psychrometric chart for determining scrubbing liquid evaporation.

determining the water lost to evaporation begins by drawing a vertical line from the dry bulb temperature at point A to the 40% RH curve at point B. From point B, a horizontal line is drawn to point C, where it intersects the scale showing on the right-hand side of the chart where it reads about 0.007 lb of moisture per pound of dry air (0.0075 kg of moisture per kilogram of dry air).

Also from point B, a line is drawn parallel to the wet bulb temperature lines to point D on the saturation temperature curve. From point D, a horizontal line is drawn to where it intersects the vertical scale at about 0.010 lb of moisture per pound of dry air (0.010 kg of moisture per kilogram of dry air). It is this difference in moisture content that we can use to determine the moisture lost from the scrubber to evaporation, as follows:

Evaporative moisture loss = airflow rate × change in moisture content

$$= 425\ \frac{\text{m}^3}{\text{min}} \times 1.20138\ \frac{\text{kg dry air}}{\text{m}^3} \times (0.010 - 0.007)\ \frac{\text{kg moisture}}{\text{kg dry air}}$$

$$= 511\ \frac{\text{kg dry air}}{\text{m}^3} \times (0.003)\ \frac{\text{kg moisture}}{\text{kg dry air}} \times \frac{1\ \text{L H}_2\text{O}}{\text{kg moisture}}$$

$$= 1.5\ \frac{\text{L H}_2\text{O}}{\text{min}}$$

$$= 15000\ \frac{ft^3}{min} \times 0.075\ \frac{lbm\ dry\ air}{ft^3} \times (0.010 - 0.007)\ \frac{lbm\ moisture}{lbm\ dry\ air}$$

$$= 1125\ \frac{lbm\ dry\ air}{min} \times (0.003)\ \frac{lbm\ moisture}{lbm\ dry\ air} \times \frac{1\ gal\ H_2O}{8.34\ lbm\ moisture}$$

$$= 0.4\ \frac{gal\ H_2O}{min}$$

Total makeup water requirements would be the amount of flushing water required and the water lost to evaporation, as follows:

Water addition required = flushing water + evaporative moisture loss

$$= 13.2\ \frac{L\ H_2O}{min} + 1.5\ \frac{L\ H_2O}{min}$$

$$= 14.7\ \frac{L\ H_2O}{min}$$

$$= 3.6\ \frac{gal\ H_2O}{min} + 0.4\ \frac{gal\ H_2O}{min}$$

$$= 4.0\ \frac{gal\ H_2O}{min}$$

6.0 ESTIMATING CHEMICAL USE

Chemical use in a wet chemical scrubber must take into account that required to remove foul air from the airstream and the amount wasted because of the flushing water added to the scrubber to prevent the buildup of reaction byproducts, as previously shown. Building on the examples earlier in this chapter, we will consider the 15 000 cfm (425 m^3/min) scrubber with the 6-ft (1.8-m) diameter and a 4-ft (1.2-m) deep integral sump, as shown in Figure 9.2. Using this information, we can determine the rate of flushing water as a fraction of the sump volume and conservatively assume that some of the chemical will be immediately wasted because of the flushing water addition. Then, this will be incorporated into the chemical addition required for scrubber operation. Calculating the sump volume, we find the following:

$$V_{\text{sump}} = \frac{\pi d^2}{4} \times D_{\text{sump}}$$

$$= \frac{\pi(1.8\text{m})^2}{4} \times 1.2\ \text{m} \times 1000\ \frac{\text{L}}{\text{m}^3}$$

$$= 3{,}120\ \text{L}$$

$$= \frac{\pi(6\ \text{ft})^2}{4} \times 4\ \text{ft} \times 7.48\ \frac{\text{gal}}{\text{ft}^3}$$

$$= 846\ \text{gal}$$

Given that our flushing water requirement was calculated as 4.0 gpm (14.7 L/min = 0.25 L/s) then the turnover or waste rate of the scrubbing liquid as a fraction of the sump volume is then calculated as follows:

$$\text{Scrubber liquid waste rate} = \frac{\text{flushing water rate}}{\text{sump volume}}$$

$$= \frac{0.25\ \frac{\text{L}}{\text{sec}}}{3120\ \text{L}} \times 60\ \frac{\text{sec}}{1\ \text{min}} \times 60\ \frac{\text{min}}{1\ \text{hr}}$$

$$= 0.29\ /\ \text{hour}$$

$$= \frac{4.0\ \frac{\text{gal}}{\text{min}}}{846\ \text{gal}} \times 60\ \frac{\text{min}}{\text{h}}$$

$$= 0.29\ /\ \text{hour}$$

To estimate the amount of chemical required to remove hydrogen sulfide from an airstream, it is, of course, first necessary to determine the amount of hydrogen sulfide in the airstream. Hydrogen sulfide monitoring typically provides the volumetric concentration of that species in parts per million parts by volume of the total gas. The molar flowrate of hydrogen sulfide can be calculated as follows:

$$H_2S\ \text{molar flow rate} = Q_{\text{Foul Air}} \times C \times \frac{1}{\text{MV}} \times \frac{(460+32)\text{R}}{(460+T_{\text{act}})\text{R}} \times \frac{60\ \text{min}}{\text{h}} \quad (9.9\text{a})$$

$$H_2S \text{ molar flow rate} = Q_{\text{Foul Air}} \times C \times \frac{1}{MV} \times \frac{(273)K}{(273 + T_{\text{act}})K} \times \frac{60 \text{ min}}{\text{h}} \quad (9.9b)$$

Where

$Q_{\text{Foul Air}}$ = Foul airflow rate in cubic feet per minute,
C = Volumetric concentration of H_2S in parts per million by volume (ppmv),
MV = Molecular volume (359 ft^3/ lbmoles at 32 °F), and 0.0224 m^3/moles at 0 °C,
T_{act} = Temperature at the actual foul air conditions when C was measured in °F,
R = Rankine, and
K = Kelvin.

As an example, we can assume the following conditions:

Q = 15 000 cfm, 425 m^3/min
C = 100 ppmv = 100 parts $H_2S/10^6$ parts foul air
T_{act} = 63 °F, 17.2 °C

Then,

$$H_2S \text{ flow rate}\left(\frac{\text{mol}}{\text{hr}}\right) = \left\{425\frac{m^3}{\text{min}} \times \frac{100 \text{ parts } H_2S}{10^6 \text{ parts air}}\right\} \times \left\{\frac{1}{0.0224\,\frac{m^3\ H_2S}{\text{mole } H_2S}}\right\} \times \frac{(273)K}{(273+17.2)K} \times \frac{60 \text{ min}}{\text{hr}}$$

$$H_2S \text{ flow rate} = 107\frac{\text{mol}}{\text{hr}}$$

$$H_2S \text{ flow rate}\left(\frac{\text{lbmol}}{\text{hr}}\right) = \left\{15\,000\frac{ft^3 \text{air}}{\text{min}} \times \frac{100 \text{ parts } H_2S}{10^6 \text{ parts air}}\right\} \times \left\{\frac{1}{359\,\frac{ft^3\ H_2S}{\text{lb mole } H_2S}}\right\} \times \frac{(460+32)R}{(460+63)R} \times \frac{60 \text{ min}}{\text{hr}}$$

$$H_2S \text{ flow rate} = 0.24\frac{\text{lbmol}}{\text{hr}}$$

With the molar flowrate of hydrogen sulfide coming in with the foul airstream calculated, we can now calculate the amount of chemicals necessary for its removal. Referring back to Equations 9.1 and 9.2, we can see that 2 moles of sodium hydroxide (NaOH) and 4 moles of sodium hypochlorite (NaOCl) will be required to remove each mole of incoming hydrogen sulfide.

This assumes we want 100% removal, which is not a bad assumption even if the actual target capture efficiency is less than 100%. This assumption provides a seemingly conservative estimate that will still be too low, as will be shown, because there will be other reactions besides those with hydrogen sulfide that will consume NaOH and NaOCl, and some of those chemicals will also be lost with the scrubbant liquid that must be flushed out to prevent the buildup of reaction products.

Stoichiometric NaOH requirement

$$\frac{2\text{ mol NaOH}}{\text{mol }H_2S} \times \frac{40\text{ g NaOH}}{\text{mol NaOH}} \times \frac{107\text{ mol }H_2S}{\text{hr}} = 8560\frac{\text{g NaOH}}{\text{hr}}$$

$$= \frac{2\text{ lbmol NaOH}}{\text{lbmol }H_2S} \times \frac{40\text{ lbm NaOH}}{\text{lbmol NaOH}} \times \frac{0.24\text{ lbmol }H_2S}{\text{hr}} = 19.2\frac{\text{lbm NaOH}}{\text{hr}}$$

Assuming 25% caustic (NaOH) solution at 10.6 lbm/gal, 1.27 kg/L, the addition rate would be as follows:

$$= 8.56\frac{\text{kg NaOH}}{\text{hr}} \times \frac{1}{\dfrac{0.25\text{ kg NaOH}}{\text{kg 25\% NaOH solution}} \times \dfrac{1.27\text{ kg 25\% NaOH solution}}{\text{L 25\% NaOH solution}}}$$

$$= 27\frac{\text{L 25\% NaOH solution}}{\text{hr}}$$

$$= 19.2\frac{\text{lbm NaOH}}{\text{hr}} \times \frac{1}{\dfrac{0.25\text{ lbm NaOH}}{\text{lbm 25\% NaOH solution}} \times \dfrac{10.6\text{ lbm 25\% NaOH solution}}{\text{gal 25\% NaOH solution}}}$$

$$= 7.2\frac{\text{gal 25\% NaOH solution}}{\text{hr}}$$

Including the amount of chemical wasted as a result of sump flushing water added, calculated above as 0.29/hr or 29%/hr, we increase the stoichiometric requirement as follows:

$$= 27\frac{\text{L 25\% NaOH solution}}{\text{hr}} \times 1.28$$

$$= 34.6\frac{\text{L 25\% NaOH solution}}{\text{hr}}$$

$$= 7.2 \frac{\text{gal 25\% NaOH solution}}{\text{hr}} \times 1.28$$

$$= 9.2 \frac{\text{gal 25\% NaOH solution}}{\text{hr}}$$

Stoichiometric NaOCl requirement

$$= \frac{\text{4 gm NaOCl}}{\text{mol } H_2S} \times \frac{\text{74.4 gm NaOCl}}{\text{mol NaOCl}} \times \frac{\text{107 mol } H_2S}{\text{hr}}$$

$$= 31.8 \frac{\text{kg NaOCl}}{\text{hr}}$$

$$= \frac{\text{4 lbmol NaOCl}}{\text{lbmol } H_2S} \times \frac{\text{74.4 gm NaOCl}}{\text{lbmol NaOCl}} \times \frac{\text{0.24 lbmol } H_2S}{\text{hr}}$$

$$= 71.4 \frac{\text{lbm NaOCl}}{\text{hr}}$$

Assuming 12.5% sodium hypochlorite solution at 10.0 lbm/gal, 1.2 kg/L, the addition rate would be

$$= 31.84 \frac{\text{kg NaOCl}}{\text{hr}} \times \frac{1}{\frac{\text{0.125 kg NaOCl}}{\text{kg 12.5\% NaOCl solution}} \times \frac{\text{1.2 kg 12.5\% NaOCl solution}}{\text{L 12.5\% NaOCl solution}}}$$

$$= 212.3 \frac{\text{L 12.5\% NaOH solution}}{\text{hr}}$$

$$= 71.4 \frac{\text{lbm NaOCl}}{\text{hr}} \times \frac{1}{\frac{\text{0.125 lbm NaOCl}}{\text{lbm 12.5\% NaOCl solution}} \times \frac{\text{10.0 lbm 12.5\% NaOCl solution}}{\text{gal 12.5\% NaOCl solution}}}$$

$$= 57.1 \frac{\text{gal 12.5\% NaOH solution}}{\text{hr}}$$

Including the amount of chemical wasted as a result of sump flushing water, added as calculated above, we increase the stoichiometric requirement as follows:

$$= 212.3 \frac{\text{L } 12.5\% \text{ NaOCl solution}}{\text{hr}} \times 1.28$$

$$= 271.7 \frac{\text{L } 12.5\% \text{ NaOCl solution}}{\text{hr}}$$

$$= 57.1 \frac{\text{gal } 12.5\% \text{ NaOCl solution}}{\text{hr}} \times 1.28$$

$$= 73.1 \frac{\text{gal } 12.5\% \text{ NaOCl solution}}{\text{hr}}$$

As mentioned above, the stoichiometric chemical requirements will be lower than what is actually required to do other reactions with chemical species that also enter the scrubber with the foul air. Carbon dioxide (CO_2), for instance, can result in high demands on sodium hydroxide because it forms carbonic acid (H_2CO_3) in solution. For each mole of carbon dioxide, one mole of sodium hydroxide is consumed in the process. When foul air from biological processes is to be treated or there is the possibility of combustion source exhaust entering the scrubber, the carbon dioxide present should be considered in addition to hydrogen sulfide.

7.0 MIST ELIMINATION

Mist formed within a wet chemical scrubber must be removed to prevent the loss of the scrubbant liquid and its escape to the atmosphere or downstream ductwork, where it can cause damage to equipment or the environment. Mist eliminators (Figure 9.7) for wet chemical scrubbers come in three basic types: chevron blades, mesh pads, and simple random dump packing. The selection of a specific type of mist eliminator should be made on a case-by-case basis in consultation with the scrubber system manufacturer.

Chevron blade mist eliminators mainly rely on inertial impaction of mist droplets on the angled or curved blades. Capture efficiencies of essentially 100% of droplets in the 10- to 40-µm range can be achieved at air pressure drops of 0.10 to 0.9 kPa (0.4 to 3.5 in. water column) (Koch-Glitsch, 2015) depending on the airstream velocity. The liquid accumulates on the mist eliminator surfaces and flows by gravity to the bottom of the mist

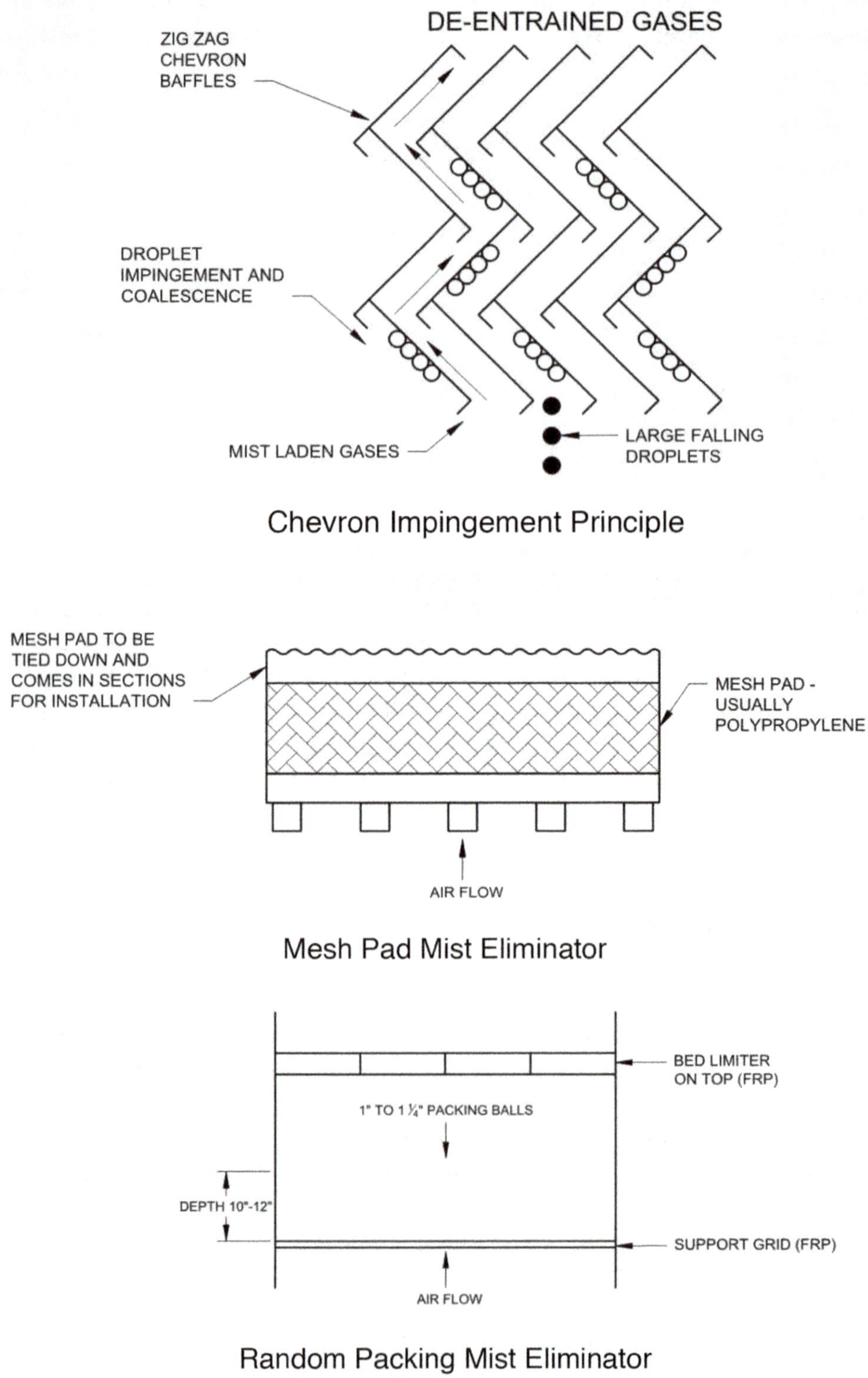

FIGURE 9.7 Mist eliminators commonly used in wet chemical scrubber.

eliminator and then drips back into the scrubber. The blades typically have baffles integrally formed to them to prevent the moisture coalesced on them from flowing up and past the mist eliminator. Chevron blade mist eliminators typically operate with relatively low air pressure drops and are less susceptible to fouling, but with lower overall control efficiency than mesh pad mist eliminators.

Mesh pad mist eliminators rely on inertial impaction and Brownian diffusion of droplets and vapor to the surface of the pad. Capture efficiencies of essentially 100% of droplets in the 3- to 10-µm range can be achieved at air-pressure drops of less than 0.2 kPa (1 in. water column) (Koch-Glitsch, 2015) depending on the airstream velocity. These mist eliminators can be of single-diameter stainless steel wire or plastic strands (commonly polypropylene) arranged in what is termed a "knitted pad" or a more structured "filament pad". Structured pads can be arranged in successively coarser to finer and more densely arranged filaments to achieve ever greater mist removal efficiency depending on the needs of the application. With greater mist removal efficiency comes greater air pressure drop across the pads and a greater probability of contaminant buildup and more frequent maintenance to clean the pads. Cleaning can typically be accomplished with a high-pressure water spray.

Random packing mist eliminators can be made up of the same type of packing used in the scrubber for target gas capture, but are commonly a small-diameter spherical packing design (Lannon, 2016). The current authors have observed that this method of mist elimination can be effective, but there is a dearth of hard data reported in the literature regarding the control efficiencies obtainable.

8.0 MIST SCRUBBER SYSTEMS

Mist scrubber systems (Figure 9.8), as the name implies, rely on a highly atomized spray of solution introduced at the top of a chamber where it interacts with the foul airstream and absorbs the targeted odorous compounds.

The liquid scrubbant is used as a once-through treatment and is collected and discharged from the bottom of the scrubber and discarded. Because the scrubbant is not recirculated, there is no accumulation of reaction byproducts.

The mist formed in this type of scrubber must be very fine, on the order of 10 to 20 µm. The many small droplets provide a large interface area between the gas and liquid for a given scrubber volume. To atomize the scrubbing liquid into such fine droplets, compressed air is necessary with fine titanium nozzles that required filtered water softened to remove

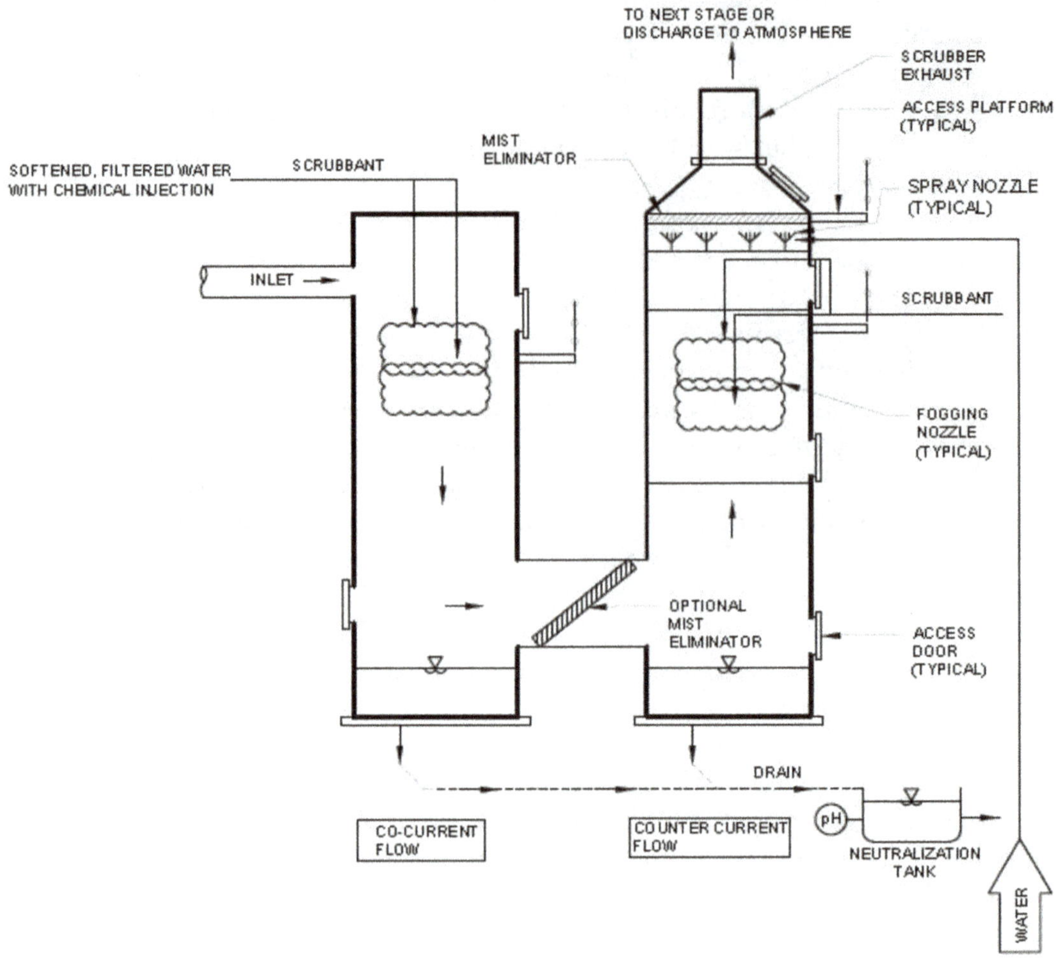

FIGURE 9.8 Typical mist scrubber system schematic diagram.

minerals that can cause clogging. If the water is not appropriately treated, nozzle clogging will result, adding to required maintenance. The complexity of forming such a fine mist and the associated maintenance involved has resulted in few applications of this technology at WRRFs, and many of those that were installed have been replaced with other odor control devices.

9.0 IONIZATION SYSTEMS

Ionization systems used in gas-phase odor control systems are a relative newcomer to the field of wastewater gas-phase odor control. The basic method of these systems is the production of reactive oxygen species (ROS), such as hydroxyl (HO^-) or ozone (O^3), that will react with odorous compounds forming non-odorous compounds such as water and carbon dioxide or a relatively high odor threshold compound such as sulfur dioxide. The three main methods of forming the ROS species are non-thermal plasma (NTP), thermal plasma, and photoionization (Wolstenholme, 2016). The ROS is formed by passing air in close proximity to high voltage electrodes in the case of NTP or through ultraviolet light, typically between 200 and 300 nm in wavelength in the case of photoionization. Thermal plasma has no known applications in wastewater odor control systems, which is likely attributable to the need to produce temperatures in the range of several thousands of degrees Fahrenheit.

After ROS formation, the airstream may be directed through activated carbon that acts as a catalyst to increase the reaction efficiency. Promising results from the application of these technologies have been reported by reputable odor control practitioners (Webster et al., 2014), but most papers in the literature have been produced by employees of the equipment manufacturers. Few long-term follow-up reports have been made by neutral parties. It would be prudent to perform a long-term pilot test of a variety of these technologies before their use on large-scale installations.

10.0 CATALYTIC OXIDATION

Catalytic oxidation in and of itself has been little used in the treatment of wastewater odors with the exception of catalytic carbons and that associated with thermal oxidation, both of which are covered elsewhere in this manual of practice. One proprietary method that has been used is the LO-CAT® process treating high levels of hydrogen sulfide in digester gas. Strictly speaking, this process is for reducing sulfur oxide (SO_X) emissions from digester gas combustion and not for odor control.

The process was originally used to remove hydrogen sulfide from natural gas streams and has subsequently been used in other industries. The process uses chelated iron as the catalyst, allowing hydrogen sulfide to react with oxygen introduced to the digester gas stream (Merichem Company, 2018) in the following overall reaction:

$$H_2S + \frac{1}{2}O_2 \leftrightarrow H_2O + S \tag{9.10}$$

The process requires the introduction of sufficient air as a source of oxygen and replacement of iron and chelates to make up for losses with the sulfur cake produced. Caustic is also required to maintain the catalytic solution in the alkaline pH range (Merichem Company, 2018). The manufacturer of the process claims that it can treat digester gas containing up to 500 ppmv H_2S and discharge a gas at or below 40 ppmv, which would indicate approximately 92% hydrogen sulfide removal efficiency.

11.0 REFERENCES

Koch-Glitsch. (2015). Mist elimination liquid-liquid coalescing (Koch-Glitsch Bulletin MELLC-02, Rev. 3-2015). https://www.koch-glitsch.com/getattachment/feac490a-4b5b-48db-af28-c832117ddb7a/attachment.aspx

Lannon, M. (2016). Demystifying mist elimination. *Proceedings of the Water Environment Federation Odors and Air Pollutants Conference*, Milwaukee, Wisconsin, United States. Alexandria, VA: Water Environment Federation.

Merichem Company. (2018). LO-CAT process for waste water treatment plants. Merichem Company website. http://www.merichem.com/images/casestudies/Waste_Water_Treatment.pdf

U.S. Environmental Protection Agency. (1985). *Design manual: Odor and corrosion control in sanitary sewerage systems and treatment plants* (EPA/625/1-85/018). Washington, D.C.: U.S. Environmental Protection Agency.

U.S. Environmental Protection Agency. (2002). *SO_2 and acid gas controls* (EPA/452/B-02-001). Washington, D.C.: U.S. Environmental Protection Agency.

Wark, K., & Warner, C. F. (1981). *Air pollution: Its origin and control* (2nd ed.). New York, NY: Harper and Row, Publishers, Inc.

Webster, N., Oppenheim, S., Schneider, J., Marts, M., & Newbanks, P. (2014). Odor control planning for the Edwards WWTF—Keeping air fresh in the Colorado Rockies. *Proceedings of the Water Environment Federation Odors and Air Pollutants Conference*, Miami, Florida, United States. Alexandria, VA: Water Environment Federation.

Wolstenholme, P. (2016). The latest odor control issues and solutions for wastewater systems. *Proceedings of the Water Environment Federation Odors and Air Pollutants Conference*, Milwaukee, Wisconsin, United States. Alexandria, VA: Water Environment Federation.

10

Physical and Thermal Odor Control

1.0 INTRODUCTION 308

2.0 DRY MEDIA ADSORPTION 308

- 2.1 Physical Adsorption 309
- 2.2 Chemical Adsorption 309
- 2.3 Reactive Adsorption 309
- 2.4 Media Types 310
  - *2.4.1 Virgin Activated Carbon* 310
    - *2.4.1.1 Coal-Based Carbon* 311
    - *2.4.1.2 Coconut Shell Carbon* 311
  - *2.4.2 Impregnated Medias* 311
  - *2.4.3 High-Capacity Carbon* 312
  - *2.4.4 Water-Regenerable Carbon* 312
  - *2.4.5 Other Media* 312
  - *2.4.6 Advantages/Disadvantages* 313
- 2.5 Adsorption Mechanisms and Performances 313
  - *2.5.1 Volatile Sulfur Compounds* 314
  - *2.5.2 Volatile Nitrogen Compounds* 314
  - *2.5.3 Volatile Organic Compounds* 314
  - *2.5.4 Key Factors Affecting Media Performance* 314
    - *2.5.4.1 Media Characteristics* 314
    - *2.5.4.2 Adsorption Properties* 315
    - *2.5.4.3 Adsorption Competition in Multi-Compound Applications* 315
    - *2.5.4.4 Odorant Conversion and Transformation* 315
    - *2.5.4.5 Moisture/Humidity* 315
- 2.6 Design Considerations 316
  - *2.6.1 Configurations* 316
    - *2.6.1.1 Vertical* 316
    - *2.6.1.2 Horizontal* 316
    - *2.6.1.3 Radial* 316
    - *2.6.1.4 Single Bed* 317
    - *2.6.1.5 Multi-Bed* 317
    - *2.6.1.6 Other Configurations* 317
  - *2.6.2 Media Types* 317
  - *2.6.3 Bed Depth* 317
  - *2.6.4 Bed Velocity* 318
  - *2.6.5 Polishing Downstream of Primary Treatment* 318
  - *2.6.6 Bed Smoldering* 319
  - *2.6.7 Passive Versus Active Systems* 320

2.6.8 *Best Practices* 320

2.7 Operation and Maintenance Considerations 321

2.7.1 *Media Sampling* 321

2.7.2 *Monitoring* 321

2.7.3 *Media Change-Out Approaches* 321

2.7.3.1 *Manual* 321

2.7.3.2 *Outsource* 322

2.7.3.3 *Dense-Phase Conveyance and Vacuum* 322

3.0 THERMAL OXIDATION 322

3.1 Applicability 323

3.2 Technologies Available 323

3.2.1 *Direct-Fired Thermal Oxidizer* 323

3.2.2 *Recuperative Thermal Oxidizer* 323

3.2.3 *Regenerative Thermal Oxidizer* 323

3.2.3.1 *Catalytic Operation of Regenerative Thermal Oxidizers* 324

3.2.4 *Other Technologies* 324

3.2.4.1 *Flares* 324

3.2.4.2 *Incinerators* 325

3.2.4.3 *Catalytic Incinerators* 325

3.2.4.4 *Microturbines* 325

3.3 Design Considerations 325

3.3.1 *Minimum Destruction Efficiency* 325

3.3.2 *Combustion Chamber Temperature* 326

3.3.3 *Detention Time* 326

3.3.4 *Composition of Inlet Air* 326

3.3.5 *Pretreatment (Specific to Regenerative Thermal Oxidizers)* 326

3.3.6 *Media Type (Specific to Regenerative Thermal Oxidizers)* 326

3.3.7 *Materials of Construction* 327

3.3.8 *Air Permit Considerations* 327

4.0 REFERENCES 327

1.0 INTRODUCTION

Physical adsorption, because of its excellent odorant removal capabilities combined with its simplicity of operation, relatively low initial cost, and exceptional potential to target a wide range of odorants, is a widely used vapor-phase odor control approach in many sectors, including water resource recovery facility (WRRF) applications (SCORe, 2014). Similarly, thermal oxidation, while a more energy-intensive process than physical adsorption, can be a favorable approach for the destruction of odorants and volatile organic compounds (VOCs).

2.0 DRY MEDIA ADSORPTION

Dry media adsorption has a long history of application to abate wastewater odors. The adsorption of odorants onto dry media occurs when molecules (i.e., the adsorbate) in the gas stream contact and adhere to a solid surface (i.e., the adsorbent). This process differs from absorption, in which a fluid (the absorbate) permeates or is dissolved by a liquid or solid (the absorbent). Adsorption is a surface-based process, whereas absorption involves the whole volume of the material. The large surface area per unit weight of activated carbon is a key property that makes it ideal for contaminant control in water and air.

The two main mechanisms for adsorption include physical adsorption and chemical adsorption. Reactive adsorption is a combination of the two aforementioned mechanisms.

2.1 Physical Adsorption

Physical adsorption is generally considered the dominant mechanism for gas-phase adsorption (Le-Minh et al., 2018). Transport of the pollutant compounds through the dry media bed brings the adsorbate into direct contact with the adsorbent. The adsorbate molecules then diffuse into the adsorbent macro and micro pores and are attached to the surface by the potential energy from van der Waals nuclear forces of attraction. The size of the pore structure as well as the molecular size and polarity of the adsorbate all contribute to the extent of adsorption. Another contribution can be capillary condensation, in which the adsorbate vapor condenses within the pore structures at a vapor pressure below saturation due to surface tension effects (Le-Minh et al., 2018). Figure 10.1 shows the macro, meso, and micro pore structure of activated carbon.

Polarity can involve three distinct attractive interactions between adsorbates and adsorption sites: orientation effects between polar molecules, dispersion effects between nonpolar molecules and nonpolar surfaces, and induction effects between polar and nonpolar molecules. All three polarity interactions can occur on the adsorbent surface of activated carbon because it is nonpolar in nature and includes multiple polar functional groups (Le-Minh et al., 2018).

2.2 Chemical Adsorption

Chemical adsorption (a.k.a. chemisorption) includes chemical bonding between the adsorbate and the adsorbent. Contrary to physical adsorption, the energy required to desorb compounds that have been chemically adsorbed is significantly greater. Chemisorption also occurs during media activation or during chemical impregnation, which significantly changes the surface functional groups and surface chemistry of the media.

2.3 Reactive Adsorption

Reactive adsorption is a combination of physical and chemical adsorption in which adsorbate molecules are physically adsorbed onto the adsorbent

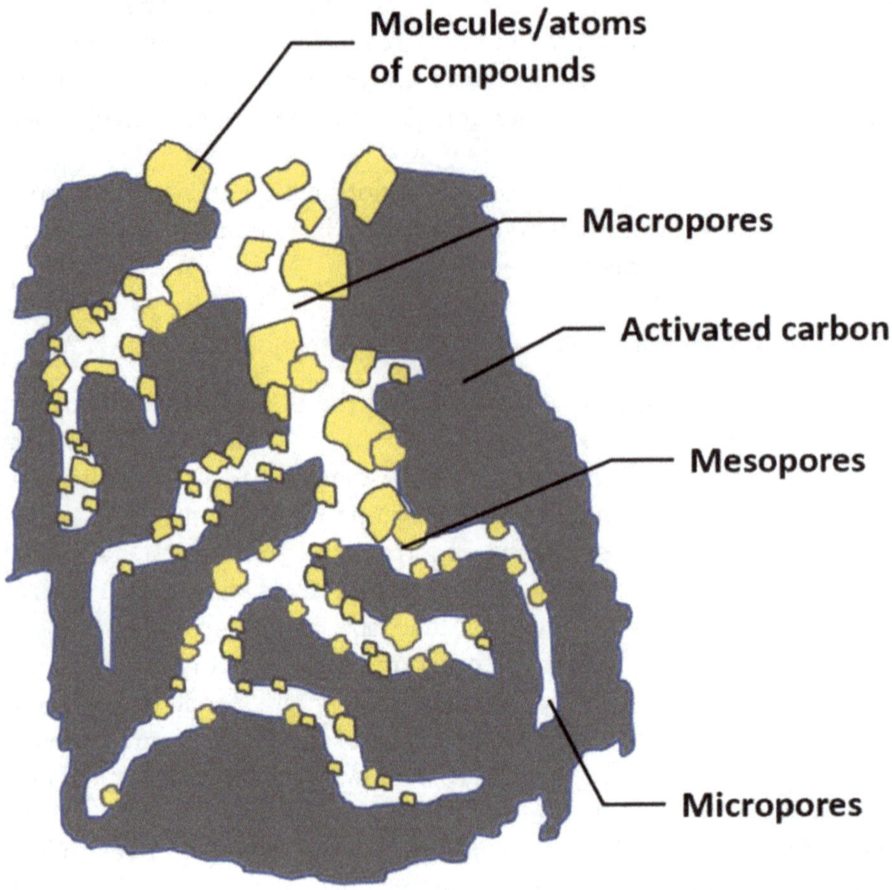

FIGURE 10.1 Macro, meso, and micro pore structure of carbon.

surface and then chemically transformed in the presence of catalysts. Oxidation of compounds can occur because of either catalytic reactive properties on the media surface or the action of impregnated oxidants. The presence of humidity improves oxidation and subsequent compound removal (Le-Minh et al., 2018).

2.4 Media Types

There are a number of common adsorption mediums used for the removal of typical odorants found at WRRFs.

2.4.1 Virgin Activated Carbon

Virgin carbon is often defined as any activated carbon that has not been chemically treated to alter its properties and works primarily through physical adsorption. Contaminants that are organic, have high molecular weights, and are neutral or nonpolar in their chemical nature are more readily adsorbed onto virgin activated carbon. However, the capture performance is reduced for small molecules and the relatively small reduced sulfur compounds. Virgin carbon is typically used in a polishing stage or for treatment of low-level odors. Common forms of virgin carbon used for odor control are derived from coal or coconut shell.

2.4.1.1 Coal-Based Carbon

Coal-based carbon is commonly produced from anthracite or bituminous grades. Coal-based carbon can be considered a general purpose carbon media used to control a wide spectrum of odorous compounds. It can be formed into a pellet shape or be granular (e.g., 8 x 30 mesh).

2.4.1.2 Coconut Shell Carbon

Coconut shell carbon is similar to coal-based carbon except it exhibits a greater percentage of micro pores and is a harder product. It is typically available in granular form.

2.4.2 Impregnated Medias

Chemically impregnated medias have been impregnated typically with a base or acid to enhance their chemisorption characteristics. For caustic impregnated activated carbons, catalytic reactions of alkaline hydroxides and acid gases such as hydrogen sulfide (H_2S) occur to form products such as sulfites, sulfides, and elemental sulfur. Acid-impregnated carbons are used for basic gases such as ammonia.

Oxidants such as potassium permanganate can also be added to enhance the chemical oxidation potential of the carbon. As an example, the following equations show the two reactions by which hydrogen sulfide can react with potassium permanganate (Nicolai & Smith, 2005), producing elemental sulfur in the first reaction and sulfate in the second. The pH conditions on the surface of the carbon determine which reaction occurs.

$$3H_2S + 2KMnO_4 \rightarrow 3S + 2H_2O + 2KOH + 2MnO_2 \text{ (Acidic pH)} \tag{10.1}$$

$$3H_2S + 8KMnO_4 \rightarrow 3K_2SO_4 + 2H_2O + 2KOH + 8MnO_2 \text{ (Basic pH)} \tag{10.2}$$

Other oxidation catalysts such as metallic oxides or noble metals can also be impregnated on the carbon surface to encourage chemisorption, enhancing overall adsorption capacity. Formation of metal sulfide occurs when reduced sulfur compounds react with the impregnated salt or oxide of a suitable metal. The resulting sulfide salts are fixed on the carbon.

2.4.3 High-Capacity Carbon

High-capacity carbon provides a greater capacity for capturing air contaminants when compared to other activated carbons. The advantages that it offers include a longer service life per volume of carbon, reduced frequency of carbon change out, and thus reduced disposal costs.

This product is an alternative for situations where water-regenerable activated carbon may be considered. It would likely be selected in situations where hydrogen sulfide or air loads were moderate and a water recharge system was not an option or to be avoided.

2.4.4 Water-Regenerable Carbon

Water-regenerable activated carbons, which are proprietary products, are carbons that can be washed to remove captured contaminants and restore adsorptive capacity. The process is not perfect, with the carbon achieving 75 to 85% of its previous capacity after washing.

Water regenerable carbon is generally manufactured from coal. In the last stages of the activation process, a nitrogen-containing gas (e.g., ammonia or urea) is introduced to form nitrogen functional groups on the surface of the carbon. These functional groups have catalytic properties and help oxidize hydrogen sulfide to sulfuric acid. The activation process does not go to completion in order to maximize the formation of the functional groups. This means that water-regenerable carbons have an underdeveloped pore structure with a lower surface area, and thus a lower capacity for VOCs compared to standard activated carbon.

2.4.5 Other Media

There are some other media types used for odor control applications. These include permanganate impregnated zeolite or alumina and iron sponge. Typically, potassium permanganate ($KMnO_4$) or sodium permanganate is used as a strong oxidant that reacts with most odorous compounds. A common approach for odor control applications uses a blended media containing permanganate-impregnated media mixed with carbon media to extend the life or treat airstreams with odors that do not adsorb well on carbon alone.

Iron sponge is a hydrated iron oxide applied to a substrate material such as woodchips. Used in similar adsorbers, iron sponge can remove significant levels of hydrogen sulfide from foul airstreams, but can require long residence times and, therefore, larger equipment (Apgar, 2016).

TABLE 10.1 Advantages and disadvantages of different activated carbon media.

Type	Advantages	Disadvantages
Virgin activated carbon	• Cheapest carbon by quantity.	• Lowest capacity of all for H_2S and methyl mercaptan. • Effective for larger nonpolar molecules.
Chemically impregnated carbon	• Improved H_2S capture over virgin carbon. • Can tailor carbon composition (blend) for widest range of capture.	• Concern about smoldering/ spontaneous combustion. • Less capacity for VOCs (because of additives occupying space). • Concerns if regenerated on-site because of chemical handling.
Metallic oxide	• Oxidizes to elemental sulfur (resulting in higher capacity, longer service).	• Not in common use (less supply and support). • Sulfur transformation to dimethyl disulfide (DMDS).
Water regenerable	• Regenerates on site, which avoids change out and saves disposal and new carbon costs. • Can be cost-effective in long term. • Can oxidize to sulfur.	• More costly (on a quantity basis). • Regeneration is not complete, deteriorates over time. • Sulfur transformation to DMDS. • Is primarily H_2S-specific. • Regenerate solution can exhibit low pH, requiring special handling.
High capacity carbon	• Very high H_2S capacity (approx. double that of chem. Impreg. carbon). • Oxidizes H_2S to sulfur (safer to handle and dispose of), not to sulfuric acid.	• More costly (on a quantity basis). • Sulfur transformation to DMDS. • Is primarily H_2S-specific.

2.4.6 *Advantages/Disadvantages*

Table 10.1 summarizes the advantages and disadvantages of different activated carbon media.

2.5 Adsorption Mechanisms and Performances

Different mechanisms for adsorption occur based on the type of compound, media, and operating conditions. Odor control applications involve a variety of conditions and different types of odorous compounds that make it difficult to predict the long-term performance.

2.5.1 Volatile Sulfur Compounds

Compounds in this group include hydrogen sulfide, mercaptans, dimethyl sulfide (DMS), and carbonyl sulfide. Removal of hydrogen sulfide occurs by chemical adsorption caused by the sulfate ion and/or elemental sulfur (SCORe, 2014). Oxidation reactions with adsorbed oxygen or catalysts occur to form sulfur dioxide or elemental sulfur. In the presence of water, further oxidation results in sulfuric acid, which is strongly adsorbed in the carbon pores (SCORe, 2014).

2.5.2 Volatile Nitrogen Compounds

This group includes ammonia, tri-methyl amine (TMA), and heterocyclic compounds such as indole, skatole, and pyridines. Virgin carbon has very low affinity for ammonia because of the low molecular weight and physical adsorption mechanism. Adsorption is enhanced using metal oxide impregnated carbon, where the mechanism is reactive adsorption. Water vapor will enhance the adsorption of ammonia and other amines because the ammonium ions react with acidic groups on the activated carbon surface (SCORe, 2014). Activated carbon will adsorb indole, skatole, and pyridines well; however, the adsorption mechanisms are not well understood.

2.5.3 Volatile Organic Compounds

Physical adsorption is the dominant mechanism for VOCs. Generally, higher molecular weight compounds and those with higher boiling points adsorb at higher capacity. Dry conditions are considered to be optimal for the physical adsorption of VOCs because there are more adsorption sites without competition by water vapor (SCORe, 2014).

2.5.4 Key Factors Affecting Media Performance

Physical characteristics including particle size, porosity, and surface area as well as chemical properties such as pH and the presence of other compounds are key factors that affect the performance in odor control applications.

2.5.4.1 Media Characteristics

Activated carbon media is typically produced as a granular or pelletized product. The granular media tends to settle more and create more fines than pelletized media. The particle shape affects the pressure drop and, therefore, power costs. For example, a 4-mm pellet media exhibits approximately half the pressure drop than a 4 x 8 mesh granular media. Pelletized media has greater mechanical strength because of the binders used in production.

2.5.4.2 Adsorption Properties

The adsorption capacity is influenced by the total surface area available, although other factors impact adsorption properties. Most activated carbons have typical total surface areas ranging from 600 to 1600 m^2/g (SCORe, 2014). Standard testing enables comparison and quality control for activated carbon media. Iodine number (America Society for Testing and Materials [ASTM] D4607) is a typical specification used to represent the adsorption properties, and butane activity (ASTM D5742) is a test method used to estimate the micropore volume of activated carbon.

2.5.4.3 Adsorption Competition in Multi-Compound Applications

Foul air contains various odorous compounds and, in most cases, non-ideal conditions for optimal adsorption. While some water vapor in the air is necessary for adsorption of odorous compounds, there is competition for adsorption sites. With multiple components being adsorbed, there is limited surface area, and the more weakly adsorbed odorants are replaced by others as the air flows through the media bed. This is typically the case in odor control applications where early breakthrough occurs for some compounds.

2.5.4.4 Odorant Conversion and Transformation

Odorous compounds adsorb onto the surface of activated carbon by different mechanisms. Surface functional groups formed during the activation process affect the chemical properties of the carbon and adsorption properties. The most prevalent are functional groups containing oxygen. Hydrogen sulfide adsorbed on the surface, along with the presence of water vapor and oxygen, is oxidized to sulfuric acid.

Other metal impurities such as oxides of iron, cobalt, molybdenum, and nickel may promote direct conversion of hydrogen sulfide into sulfate (SCORe, 2014). Methyl mercaptan (MM) oxidizes catalytically in the presence of water and converts to dimethyl disulfide (DMDS) (Bashkova, Bagreev, & Bandoosz 2002), which is strongly adsorbed on the carbon (SCORe, 2014). It is believed that the DMDS possibly replaces water from the carbon surface as well as weakly adsorbed compounds (SCORe, 2014).

2.5.4.5 Moisture/Humidity

The presence of water vapor is necessary for the adsorption mechanisms of many odorous compounds to take place. The performance of activated carbon is severely affected when condensation occurs on the surface of the carbon because the macropores fill and prevent further access. The use of a heater to increase the air temperature to lower the relative humidity will avoid condensation and extend the life of the carbon. It is reported that the optimal humidity for hydrogen sulfide adsorption is 10 to 60% (Le-Minh et al., 2018), whereas less than 50% humidity is desirable for VOCs.

2.6 Design Considerations

Carbon has had a long history and is one of the first methods used in controlling odors. Because of its long history, there are many lessons learned that have helped progress the technology and ensure its effectiveness.

2.6.1 Configurations

Multiple configurations exist for dry media adsorption systems. The selection of which configuration to implement is generally based on capacity, operating costs, footprint requirements, and designer preference. Furthermore, each configuration has its own unique design considerations to consider.

2.6.1.1 Vertical

Vertical vessel configurations are most typical, with the media bed(s) oriented horizontally. Both single bed and dual bed are available for this configuration. Although dual bed allows for an overall reduction in required footprint, its media change-out effort is typically more arduous compared to single bed.

2.6.1.2 Horizontal

Horizontal airflow/vertical bed configuration vessels provide some unique advantages. Two-, three-, and even four-bed configurations are available. One advantage of this configuration is its relative ease of media change out, with an access platform placed on top of the vessel with guardrails, allowing for media replacement via vactor truck or crane and super sacks without having to enter the vessel. An extension wand can be used for vacuuming out spent carbon where bed diameter is large. One disadvantage to this configuration is the risk of carbon media settling and subsequent bed density gradient because of the deeper vertical bed, resulting in short-circuiting potential. Also, while less likely, there is potential for void spaces to occur between the media and wall, resulting in short-circuiting (Apgar, 2014).

2.6.1.3 Radial

Freestanding, vertical, single-bed radial flow tower configurations provide outside-to-inside airflow pattern and generally exhibit smaller footprint requirements and lower pressure drop versus other configuration types (Wolstenholme et al., 2008). Some disadvantages of this type of configuration include the potential for shortened media life (bed depth is typically less than 1 m [3 ft]), rapid breakthrough as bed velocity increases across the media bed, and media density gradient and subsequent short-circuiting.

2.6.1.4 Single Bed

Single-bed units generally require a greater footprint compared to multi-bed units, but can require less effort during media change out. Where sufficient real estate is available, or where treatment capacity is small, single beds are typically preferred.

2.6.1.5 Multi-Bed

Multi-bed units are generally preferred for large-capacity systems where available real estate is limited.

2.6.1.6 Other Configurations

Other configurations exist including V-bank, elevated dual parallel bed with discharge hoppers, and custom configurations.

2.6.2 Media Types

Selecting the appropriate media is critical. Table 10.2 summarizes a comparison of media types (Graham, 2006).

2.6.3 Bed Depth

Bed depth in most adsorption systems is approximately 1 m (3 ft) as dictated by the mass transfer breakthrough curve for most odorous compounds. In dry media adsorption systems, adsorbates are adsorbed in an unsteady-state process until the bed is saturated (ARC, 2011). The adsorption bed can be divided into the following three zones:

- Zone 1—Saturated zone in which carbon pores are completely filled.
- Zone 2—Adsorption zone in which adsorption is actually occurring.
- Zone 3—Final zone in which carbon has little to no adsorbed compounds.

The saturated zone, or "mass transfer front," moves through the bed until it nears the outlet in which the pollutant concentration reaches breakthrough. This is when the carbon filter has reached the end of its useful life. The

TABLE 10.2 Comparison of odor control dry adsorption medias (adapted from Graham, 2006).

Item	Virgin Carbon	Impregnated Carbon	Water-Regenerable Carbon	High-Capacity Carbon	$KMnO_4$ Impregnated
Apparent density, g/c[c]	0.50	0.55	0.56	0.40	0.8–0.96
H_2S capacity, wt. %[a]	4–7	20–25	14–18	65–75	10–15
Impregnate	None	NaOH	None	None	$KMnO_4$
Fate of H_2S	Sulfur/ H_2SO_4	Sulfur/ H_2SO_4	H_2SO_4	Sulfur	Sulfur
Butane number[b]	23.5	NA	15.6	26.0	N/A
Ignition temperature	~380 °C (~720 °F)	~150 °C (~300 °F)	~420 °C (~790 °F)	~540 °C (~1000 °F)	—
pH of spent carbon	Acidic	Acidic	Acidic[c]	Neutral	—

[a]ASTM standard test method D-6646
[b]ASTM D-5742
[c]Unless thoroughly water-washed before disposal

shape of the breakthrough curve depends on the odorant type, bed velocity, concentration of the odorant, and bed depth (Schnelle & Brown, 2001). Figure 10.2 illustrates the mass transfer breakthrough curve.

Although bed depth can be greater, increasing bed depth increases the pressure drop across the media, which then increases the size of the extraction fan required as well as operating costs.

2.6.4 Bed Velocity

Bed velocity generally ranges between 0.2 and 0.3 m/s (40 and 60 fpm). However, depending on the media used, higher velocities can be tolerated, but with an increased risk of short-circuiting, reduced performance, and higher operating cost.

2.6.5 Polishing Downstream of Primary Treatment

It is not unusual to place adsorption downstream of primary treatment for polishing when strict odor limits apply. When placing carbon downstream of primary treatment, it is important to consider the effect of moisture

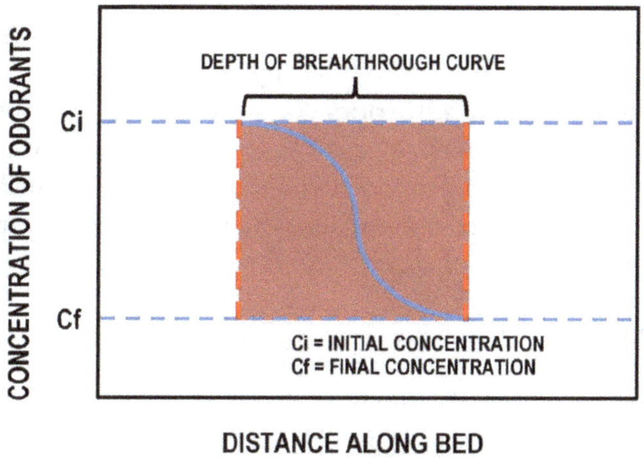

FIGURE 10.2 Dry media mass transfer breakthrough curve.

carryover. As mentioned previously, excessive moisture can be detrimental to performance. Mist eliminators and pre-heaters can be installed upstream of carbon. In addition, placement of the extraction fan between the primary treatment system and the polishing stage can impart energy to the airstream, increasing temperature, and reduce relative humidity.

2.6.6 Bed Smoldering

Smoldering or carbon bed fires in odor control systems at WRRFs have been observed to occur during a dormant state where heat of reaction occurs without heat dissipation (i.e., convective cooling). The ignition temperature of virgin or catalytic activated carbons is typically between 380 and 425 °C (716 and 797 °F), whereas the ignition temperature of chemically impregnated activated carbons is significantly lower (200 to 225 °C [392 to 437 °F]). Therefore, in hot climates like the Middle East, chemically impregnated carbons may not be appropriate.

When chemically impregnated carbon is used in a duty/standby arrangement, it is recommended that both vessels operate simultaneously or that one unit be operated in a duty mode and a moderate amount of air be fed to the standby unit to prevent smoldering. This would allow media to become spent in the duty unit first, thereby allowing media change out to occur in the one unit while the other duty unit remains operational, ensuring continuous odor treatment is maintained. Simultaneous operation reduces the energy requirements that would result from lower bed pressure losses through the carbon system, but it can represent challenges during media change out because both units would likely become spent at the same time, resulting in short-term zero treatment while both units receive media change out.

2.6.7 Passive Versus Active Systems

In most cases, odor control is provided by active systems in which air is delivered to the dry media adsorption vessel via a fan. However, in some cases, such as a sealed wet well, a passive system may be provided. One disadvantage of passive systems is that, depending on the dry media type, bed smoldering can pose greater risk because the unit can remain in a dormant state for extended periods of time.

2.6.8 Best Practices

Many applications include treatment of moist or unclean/dirty air. Prefiltration with grease filters and mist eliminators are commonly used to prevent potential buildup on the media, which can reduce performance over time (Pope, 2012).

Because of condensation effects, moisture accumulation may occur within media or interior vessel walls and collect in the bottom of the vessel. All carbon systems should be equipped with a drain at the bottom of the vessel. Because condensate is likely to be acidic, proper selection of materials of construction is necessary. Side-bottom or full-drain nozzles should be provided (Pope, 2012).

Many applications include permit requirements or project-specific requirements dictating that odor removal remain operational even under maintenance conditions. Because carbon requires replacement once exhausted, these applications require that redundancy be included in the design so that there is uninterrupted performance during media replacement.

Sample taps are commonly designed into the vessel walls to allow sampling media at different locations within the media bed for laboratory analysis of remaining media life and troubleshooting performance/flow issues.

Stacks at carbon vessels should be sized for a flowrate of no less 7.6 m/s (1500 fpm). Using lower velocities can allow rain to enter the stack and decreases the effectiveness of plume dispersion. Where possible, rain caps should not be used because they prevent upward plume trajectory and reduce dispersion (Pope, 2012). For applications where the vessel is operated intermittently, alternate duct, stack, and damper arrangements should first be considered before installing a rain cap. Alternatively, a V-shaped cap would allow upward plume trajectory while limiting rain impacts.

Some vessel orientations, including radial flow and vertical beds, exhibit a propensity for short-circuiting to occur when media settles over time. Where this risk is present, appropriate baffle plates should be installed.

Materials of construction for dry media adsorption vessels generally consist of fiberglass reinforced plastic (FRP) with vinyl ester resin for corrosion protection, but vessels constructed from Type 316 stainless steel are also available. Vessels are equipped with Type 316 stainless steel media grounding rods to prevent static electricity from accumulating.

2.7 Operation and Maintenance Considerations

2.7.1 Media Sampling

Dry media adsorption media life and replacement frequency are dictated more on empirical data and experience because vendors generally only provide hydrogen sulfide adsorption capacities of media. Because WRRF odors also contain other odorous gases and VOCs, the estimate of carbon life based solely on hydrogen sulfide adsorption capacity will be unreliable (SCORe, 2014). Media can be periodically sampled at multiple points along the media bed and samples delivered to a reputable laboratory for analyses for mass of hydrogen sulfide, percent moisture, and remaining hydrogen sulfide capacity.

Alternatively, remaining media life can be determined using a carbon saturation indicator that acts similar to a litmus test. When the media is spent, the media in the indicator change color from purple to brown. Finally, air samples can be collected and measured for odor concentrations using sample ports at different locations within the media bed for quantifying remaining media life.

2.7.2 Monitoring

Where consistent adsorbent performance is critical, vapor-phase odorant monitoring on the vessel outlet and, alternatively, on the vessel inlet should be provided. Increasing outlet concentrations may indicate that operating conditions are outside the designed range or that the media is reaching the breakthrough point. Multiple continuous odorant monitoring systems are available with varying accuracies and low detection limits. Other operating parameters such as relative humidity, inlet gas temperature, static pressure drop, and volumetric flowrate should also be monitored frequently because these factors are known to influence the adsorption capacity (SCORe, 2014).

2.7.3 Media Change-Out Approaches

Spent dry adsorption media must be replaced periodically to ensure consistent odorant removal performance. Several media change-out approaches exist.

2.7.3.1 Manual

Manual change out of spent media is most common, especially for small- to medium-sized systems. Depending on the volume of media to be replaced, manual media replacement generally consists of removing spent media via vactor truck and filling with new media using a crane and super sacks. Advantages to this approach include avoiding potential contracting and funding issues related to outsourcing. Spent carbon must be tested before disposal to determine landfill requirements.

2.7.3.2 Outsource

Outsourcing requires no staff involvement for carbon replacement. This is a full service where the service provider empties the vessels, places new carbon, and disposes of the spent carbon.

2.7.3.3 Dense-Phase Conveyance and Vacuum

A dense-phase carbon conveyance system is rare and generally only provided for very large systems in which carbon replacement is considered a significant effort. This system consists of a carbon fill station for filling a storage hopper with carbon from supersacks, a carbon blower, extensive conveyance piping, and fill nozzles at carbon vessels. A spent carbon vacuum system would consist of extensive vacuum piping, a spent carbon hopper, and vacuum blowers. The benefit of this system is that it requires less labor and time than conventional manual replacement. However, this approach is typically cost prohibitive.

3.0 THERMAL OXIDATION

Thermal oxidation is an energy-intensive process that destroys VOCs and odors. VOCs are broken down and recombined with oxygen to produce water vapor and carbon dioxide. Odorous compounds are oxidized into non-odorous or less odorous (e.g., hydrogen sulfide converted to sulfur dioxide) compounds. Thermal oxidation requires air to be heated to 704 to 980 °C (1300 to 1800 °F) for 0.5 to 1.5 seconds. The degree of destruction is a function of temperature, residence time, and turbulence.

The simplest oxidizer is a direct-fired burner that raises the air temperature to combustion levels. Because of the high cost of heating the process exhaust stream to the required oxidation temperature, most thermal oxidizers incorporate some type of primary heat recovery. Primary heat recovery transfers energy from the hot clean airstream exiting the oxidizer into the polluted inlet airstream. This process reduces the amount of energy required to achieve oxidation temperatures.

3.1 Applicability

Whereas other technologies may be applicable only to certain odors or VOCs, thermal systems are applicable to all. Thermal systems are clearly superior to other technologies in treating VOCs efficiently. Thermal systems can be designed with destruction efficiencies greater than 99% and should be considered when the chemical composition of the inlet airstream is very complex.

The application of thermal treatment to control odors at WRRFs is generally limited by cost. If energy is not recovered, capital costs are low, but fuel costs are high. If energy is recovered, capital costs are high, but fuel costs are reduced significantly.

3.2 Technologies Available

Thermal treatment systems are not considered new technology, but new developments in configuration and types of packing have improved reliability and reduced operating costs. Direct thermal oxidizers provide effective control for small volumes of air, but are typically too expensive for large sources. When large volumes of odorous air must be treated, heat recovery is more cost-effective. The most common heat recovery systems are the recuperative and the regenerative oxidizers.

3.2.1 *Direct-Fired Thermal Oxidizer*

Direct thermal oxidizers are recommended for highly concentrated exhaust streams and are mainly used in petrochemical and pharmaceutical industries. Where feasible, odorous exhaust streams can be used as combustion air at cogeneration engines or individual HVAC fuel-burning equipment.

3.2.2 *Recuperative Thermal Oxidizer*

Recuperative thermal oxidizers with shell-and-tube heat exchangers are capable of recovering up to 70% of the energy of oxidation. However, these units still require a large amount of supplemental fuel to fire the oxidizer. Recuperative thermal oxidizers provide up to 70% thermal efficiency and 99% destruction efficiency, and have been applied primarily in the automobile and printing industries.

3.2.3 *Regenerative Thermal Oxidizer*

Regenerative thermal oxidizers (RTOs) have multiple beds of ceramic heat transfer media capable of delivering up to 95% thermal efficiency and are considered the most cost-effective thermal treatment systems. Three-bed RTOs have higher capital costs, but probably the greatest reliability. RTOs provide 99% or higher destruction efficiency and are recommended for large-volume exhaust streams, a wide variety of pollutants, fluctuating concentration levels, and high heat recovery. Because of their high thermal and destruction efficiency, RTOs are preferred over recuperative units for WRRF applications.

Another type of thermal oxidizer to be considered is the rotary thermal oxidizer. It has a refractory wheel that rotates through two chambers, storing heat from the exhaust-side chamber and delivering heat to the supply-side chamber. This is a relatively new approach intended to provide the type of redundancy of a three-RTO system, but at lower costs.

3.2.3.1 Catalytic Operation of Regenerative Thermal Oxidizers

A catalytic oxidizer can be used to enhance the operation of an RTO. Noble metals are used as combustion catalysts. The use of catalytic oxidizers might be limited because contaminants present in a foul airstream can "poison" the catalyst and compromise the treatment process.

3.2.4 Other Technologies

3.2.4.1 Flares

A simple open flare is generally not considered by regulators to be an oxidizer, although, by strict interpretation of the word, it is. Regulators typically recognize enclosed flares with automated combustion controls to be a controlled oxidation device, also called "waste-gas burner". In those devices, the combustion, and thereby emissions are controlled.

Flares represent a low-cost method of treating airstreams with high concentrations of combustible gases. Flares combust gases by using atmospheric air in an open flame. As such, energy cannot be recovered from them. Most airstreams at WRRFs contain relatively low concentrations of combustible gases, except for digested gas. The VOC destruction efficiencies of flares are typically in the 90 to 98% range.

Hydrogen sulfide is converted to sulfur dioxide (SO_2). Sulfur dioxide is an air pollutant with established air quality standards. While emitting sulfur dioxide is bad, emitting hydrogen sulfide is worse from an odor standpoint. For sulfur dioxide, the National Ambient Air Quality Standard under the Clean Air Act 3-hour standard is 0.5 ppm. The detection threshold for hydrogen sulfide is 0.0005 ppm. Thus, oxidizing hydrogen sulfide to sulfur dioxide reduces a significant odor problem even if it creates an increase in sulfur dioxide emissions. Sulfur dioxide emissions must be permitted through the preconstruction permit process as would any other air pollutant. Depending on the magnitude of the emissions, dispersion modeling or alternative treatment systems may need to be considered.

Because typical foul airstreams from most WRRF processes lack sufficient heat content (i.e., they exhibit low British Thermal Unit [Btu] value), flares are typically a poor choice for combusting odorous gases because of their high fuel cost. They are used mainly to burn methane or digester gas produced under anaerobic conditions in biological processes in WRRFs.

3.2.4.2 Incinerators

Incinerators may be used for odor control under some circumstances. Fluidized bed incinerators typically operate at sufficiently high freeboard temperatures (greater than 760 °C [1400 °F]) and have several seconds of detention time. Existing incinerators can provide cost-effective odor control with low odorous airflows because they require fluidizing and combustion air anyway. Thus, the cost of treatment is only that associated with the drop in static pressure caused by the foul air collection system.

3.2.4.3 Catalytic Incinerators

Catalytic incinerators use a catalyst bed of noble metals (e.g., platinum, palladium, or rhodium) to promote oxidation at reduced temperatures, typically in the range of 371 to 482 °C (700 to 900 °F). Oxidation occurs on the surface of the catalyst. The actual steps of reaction are diffusion into catalyst pores, adsorption, oxidation, desorption, and diffusion back into the airstream.

VOC destruction efficiencies can be greater than 90%, but they decrease as the catalytic material "ages" or becomes poisoned. Thermal efficiencies depend primarily on the composition of the foul airstream, destruction efficiency, and the type and condition of the catalyst.

3.2.4.4 Microturbines

Microturbines are small-scale gas turbines used to convert energy (gas fuel) to electricity with opportunities for heat recovery. Microturbines represent a form of thermal oxidation and have been applied in large industries for VOC destruction. Recently, microturbines have been tested for odor control in WRRFs.

3.3 Design Considerations

3.3.1 Minimum Destruction Efficiency

In WRRF applications, manufacturers will guarantee achieving "not-to-exceed" concentrations rather than percentage removal. Destruction efficiencies of 97 to 99% or higher are achieved in many WRRF applications. In addition, thermal efficiencies of 70 to 95% are achievable.

3.3.2 Combustion Chamber Temperature

Destruction temperature, together with the appropriate detention time, is a function of the components in the airstreams. The technologies indicated herein (except for catalytic technologies) will operate at the same temperatures for the same airstream components. A minimum of 815 °C (1500 °F) is recommended for WRRF applications.

3.3.3 Detention Time

Detention time, like the combustion temperature, is a function of the components in the airstreams. A mean combustion chamber detention time in the range of 1 to 2 seconds (2 being very conservative) is recommended for WRRF applications. For any design situation, the designer should work closely with a vendor in defining temperature and detention time parameters when stipulating the required "not-to-exceed" emissions.

3.3.4 Composition of Inlet Air

Ideally, an odor source characterization study should be conducted to identify and quantify the compounds and characteristics of the inlet airstream and to evaluate the best technology available.

3.3.5 Pretreatment (Specific to Regenerative Thermal Oxidizers)

Particulates should be removed from the inlet airstream; otherwise, they will deposit on the ceramic media and increase system pressure drop, causing the media to fail. A particle size analysis and loading can be helpful in evaluating pretreatment requirements. Venturi scrubbers, either alone or combined with low-pressure-drop cyclones, are recommended upstream of RTOs for high particulate applications.

The selection of media less susceptible to plugging from particulates should consider monolithic heat recovery media. Regardless of the filtration efficiency, some materials in the exhaust will oxidize to form a fibrous particulate that would accumulate on the top (hot face) of the heat recovery beds. These materials will accumulate on the very top of the monolith, allowing it to be removed by vacuuming.

3.3.6 Media Type (Specific to Regenerative Thermal Oxidizers)

Media types include ceramic structured monolith block (preferred), structured packing, or random packing. Structured packing is used to reduce fouling and plugging in the heat exchange packing areas, which are considered critical where significant particulate carryover exists.

3.3.7 *Materials of Construction*

High-quality stainless steel construction material should be used for resistance to corrosion from sulfur-based acid attack. This should include the ductwork/manifolds, diversion valves, heat recovery beds, combustion chamber, fan, and stack.

3.3.8 *Air Permit Considerations*

Consult local and state regulatory agencies on air permitting requirements and procedures. Requirements will vary according to the industry, the application, and the location of the project. If an air permit is required, some general guidelines include:

- Estimate emissions from equipment.
- Develop a permitting strategy.
- Conduct preliminary discussions with regulators.
- Set up a meeting with regulators and the client to discuss the project and permitting requirements (e.g., extent of best available control technology [BACT] analysis, need for air modeling, timing, etc.).
- Draft permit form, BACT analysis, and modeling protocol (if required).
- Discuss and review the permit with the client.
- Submit permit application/modification to appropriate regulatory agency.
- Check with regulatory agency after 30 days for permit completeness review.
- Respond to any requests for information.
- Get draft permit from regulator.
- Review and negotiate draft permit terms with regulator and client.
- Obtain final permit before beginning construction.

4.0 REFERENCES

Apgar, D. (2014). Troubleshooting odor control efficiency problems with a vertical bed, horizontal airflow, cylindrical carbon bed scrubber. *Proceedings of the Water Environment Federation Odors and Air Pollutants Conference*, Miami, Florida, United States. Alexandria, VA: Water Environment Federation.

Apgar, D. (2016). A method for choosing between carbon media alternatives for wastewater odor control. *Proceedings of the Water Environment*

Federation Odors and Air Pollutants Conference, Milwaukee, Wisconsin, United States. Alexandria, VA: Water Environment Federation.

ARC Sewer Corrosion & Odour Research Project. (2011). Detailed Technical Subproject Research Report on ARC Corrosion & Odour Linkage Project LP0882016. Retrieved January 29, 2013.

Bashkova, S., Bagreev, A., & Bandoosz, T. J. (2002). Adsorption/oxidation of methyl mercaptan on activated carbons. *Environmental Science and Technology, 36*(12), 2777–2782.

Graham, J. (2006). Elemental odor control—New trends in activated carbons. *Water Environment & Technology, 18*(6), 46-48.

Le-Minh, N., Sivret, E., Shammay, A., & Stuetz, R. (2018). Factors affecting the adsorption of gaseous environmental odors by activated carbon: A critical review. *Critical Reviews in Environmental Science and Technology, 48*(4), 1–35.

Nicolai, D., & Smith, D. (2005). Hydrogen sulfide reduction of swine manure using potassium permanganate and hydrogen peroxide. ASAE Paper No. SD05-801. *ASAE Midwest Regional Meeting*, Brookings, South Dakota, United States. St. Joseph, MI: ASAE.

Pope, R. J. (2012). But did I learn anything?: Activated carbon miscues. *Proceedings of the Water Environment Federation Odors and Air Pollutants Conference,* Louisville, Kentucky, United States, April 15–18 (pp. 400-423). doi:10.2175/193864712811700543

Schnelle, K. B., & Brown, C. A. (2001). Adsorption for HAP and VOC control. In *Air pollution control technology handbook*. Boca Raton, FL: CRC Press.

SCORe (2014, February 7). SP3: Odour Measurement and Assessment, Evaluation of Odour Treatment Technologies. Review of the Application of Activated Carbon for Sewer Odour Abatement Sub-project, Sewer Corrosion and Odor Research Project. Australia.

Wolstenholme, P., Freedman, S., Maloney, R., Winn, R., Hopkins, K., Bowker, R., & Firmin, S. (2008). Controlling odors from a pump station surrounded by upscale development. *Proceedings of the Water Environment Federation Odors and Air Pollutants Conference*, Phoenix, Arizona, United States. Alexandria, VA: Water Environment Federation.

11

Odor Impact Assessments

1.0	INTRODUCTION	330
	1.1 Odor Impact Assessment Purposes	330
	1.2 Odor Control Goal Setting	330
2.0	ODOR SOURCE TYPES	331
	2.1 Point Sources	331
	2.2 Area Sources	332
	2.3 Volume Sources	332
3.0	ODOR DISPERSION MODELING	332
	3.1 Dispersion Modeling Theory	332
	3.2 Dispersion Modeling Process	334
	3.2.1 Odor Impact Assessment Applications	334
	3.2.2 AERMOD	335
	3.2.3 CALPUFF	335
	3.2.4 Screening Models	336
	3.3 Model Input Parameters	336
	3.3.1 Fence Line Delineation and Model Area	336
	3.3.2 Source Locations and Characteristics	337
	3.3.3 Emission Rate Calculations	338
	3.3.4 Meteorological Data	338
	3.3.5 Terrain Data	338
	3.3.6 Sensitive Receptors	339
	3.3.7 Physical Structures and Buildings	339
	3.4 Dispersion Model Output	340
	3.4.1 Isotherm Plots	340
	3.4.2 Individual Source Impacts	341
	3.5 Dispersion Model Limitations	342
4.0	AIR CONTAINMENT AND EXTRACTION DESIGN	343
	4.1 Odor Containment Methods	343
	4.1.1 Covers	343
	4.1.2 Enclosures	343
	4.1.3 Hoods	344
	4.2 Foul Air Capture Design Criteria	344
	4.2.1 Room Ventilation for Safety and Comfort	344
	4.2.2 Hood Airflow Rate Design	345
	4.2.3 Covered Processes	346
5.0	FACILITY ODOR EVALUATIONS	347
	5.1 Development of Candidate Technologies	347
	5.2 Life Cycle Cost Analyses	348
	5.3 Non-Economic Factors Analysis	349

5.3.1 Footprint 350
5.3.2 Operations and Maintenance Requirements 350
5.3.3 Safety 350
5.3.4 Familiarity 350
6.0 CASE STUDIES 351
6.1 Odor Impact Assessment Case Study 1 351
6.2 Odor Impact Assessment Case Study 2 354
7.0 REFERENCES 357
8.0 SUGGESTED READING 358

1.0 INTRODUCTION

1.1 Odor Impact Assessment Purposes

Water resource recovery facilities (WRRFs) and collection systems often present a concern in terms of their potential odorous emission impacts on residential communities and commercial areas. Reductions to current and potential odor emission impacts can be assessed using a combination of source sampling, technology evaluation, and dispersion modeling. Odor emission reduction is estimated based on the application of typical or desired efficiencies of either liquid- or vapor-phase treatment. The selected improvement(s) are often incorporated into a municipality's capital improvements program and/or annual maintenance budget, as is the case for ongoing chemical addition, for example.

This chapter discusses how odor impact assessments are implemented to benefit utility planning and resolution of odor-related complaints. These studies assume that a specific goal needs to be achieved, which is typically a function of regulatory requirements or good-neighbor policies, as described in Chapter 3. A sampling program produces excellent benefit to odor studies, but data may already be readily available, or assumptions can be made based on findings at similar WRRFs. However, using non-facility-specific data assumes a level of risk because of the uniqueness of odor profiles associated with different geographies, sources, and treatment unit parameters, for example.

1.2 Odor Control Goal Setting

Odor control goals may be related to establishing maximum stack emissions, identifying maximum odor levels at a facility fence line (property boundary), or identifying maximum odor levels at key sensitive receptors. These limits may be expressed as concentrations of specific compounds such as hydrogen sulfide (H_2S), which is a compound often written into air permits.

Alternatively, limits may be quantified using values of odor detection threshold, which is an estimate of the odor severity as perceived by the average human nose. Odor detection threshold is measured for air samples by an odor panel laboratory, with reported values given as a number of dilutions-to-threshold (D/T). The D/T value is unitless and indicates how many times an air sample must be diluted for an average human not to be able to differentiate the sample when compared to carbon-filtered ("odorless") air. For example, a detection threshold of 1000 D/T means the air sample must be diluted 1000 times to render it "odorless". (Note that odor panel detection threshold measurements are always above 0 D/T, which would theoretically equate to odorless.) The detection threshold odor limit are most often applied to facility fence line (property boundary) locations or at off-site locations identified in the odor impact study as a sensitive receptor.. Chapter 2 provides additional detail and applications of the odor panel detection threshold measurement.

Adherence to odor goals is often quantified using dispersion modeling, a tool that is discussed in Section 3.0 of this chapter. Dispersion model outputs can demonstrate whether a facility's current emissions result in compliance with odor goals, and if they do not, the same model can be used to identify which odor emission reductions would be needed for compliance (noting that different combinations of new or improved odor control likely will achieve this). This often forms the basis for odor impact assessment recommendations, where process modifications or odor control measures are identified that most cost-effectively and/or reliably accomplish required odor source emission reductions.

2.0 ODOR SOURCE TYPES

2.1 Point Sources

Process area odor emissions calculations are based on a characterization of odorous airflow into the ambient environment. Point sources are the easiest to conceptualize, and emissions calculations are the most straightforward. Stack emissions, where fans produce a desired flowrate and velocity—whether an odor control unit is present or not—represent the most common point source. Additionally, point sources may also include heating, ventilation, and air conditioning (HVAC) emissions from buildings that include odorous processes.

Point sources are characterized by the exhaust duct dimensions and the airstream exit velocity and odor concentration. The odor concentration may be expressed in terms of a specific compound, such as the commonly present hydrogen sulfide, where the value is typically expressed in units of parts per million by volume (ppmv). Alternatively, point source odor concentrations may be using the detection threshold value. For dispersion modeling inputs, this must be expressed in concentration units of D/T per cubic meter (m^3) (see Section 3.0).

2.2 Area Sources

Area source emissions are a function of the odor concentration and the size of the source. Primary clarifiers and aeration basins are commonly analyzed area sources, but solids treatment processes such as drying beds and treated air biofilter emissions are also typically considered area sources in odor impact assessments.

Area source odor emission rates are more difficult to confidently quantify than a point source, particularly for large sources. Flux hood odor sampling is often used, but each measurement only corresponds to a relatively small portion of the total surface area. Odor emission velocity is also more difficult to approximate for area sources than for a point source, where the air velocity is set by a fan and system pressure drops. Area source emission velocities can be determined by dividing the volumetric airflow rate extracted by a vacuum chamber during flux hood sampling by the flux hood opening area (Ferguson et al., 2004).

2.3 Volume Sources

Volume sources are limited to air emissions from building or structure spaces that feature a relatively large two-dimensional emission area, such as a bay door for a dewatering building. Emission rate calculations can be challenging because much depends on building pressurization and the HVAC system efficiency, which can vary significantly depending on several conditions, such as process operations, whether doors or windows are open, or even weather conditions. Fugitive (uncontrolled) emissions from buildings that contain odor sources are often a concern because of the plume's low velocity and ground-level discharge.

3.0 ODOR DISPERSION MODELING

3.1 Dispersion Modeling Theory

Air dispersion models predict ground-level airborne compound or odor concentrations given data describing facility odor sources, meteorological conditions, emissions profiles, surrounding terrain elevations, and other parameters. Listed U.S. Environmental Protection Agency (U.S. EPA) dispersion models assume straight-line Gaussian plumes, as shown in Figure 11.1.

Dispersion models use mathematically and statistically averaged approximations of plume rise and wind-driven dispersion behavior to estimate downwind impacts to odorous compound concentrations, or to perceived odor (quantified using the D/T method). The models incorporate point, area, and volume sources, and project the cumulative effect of air emissions

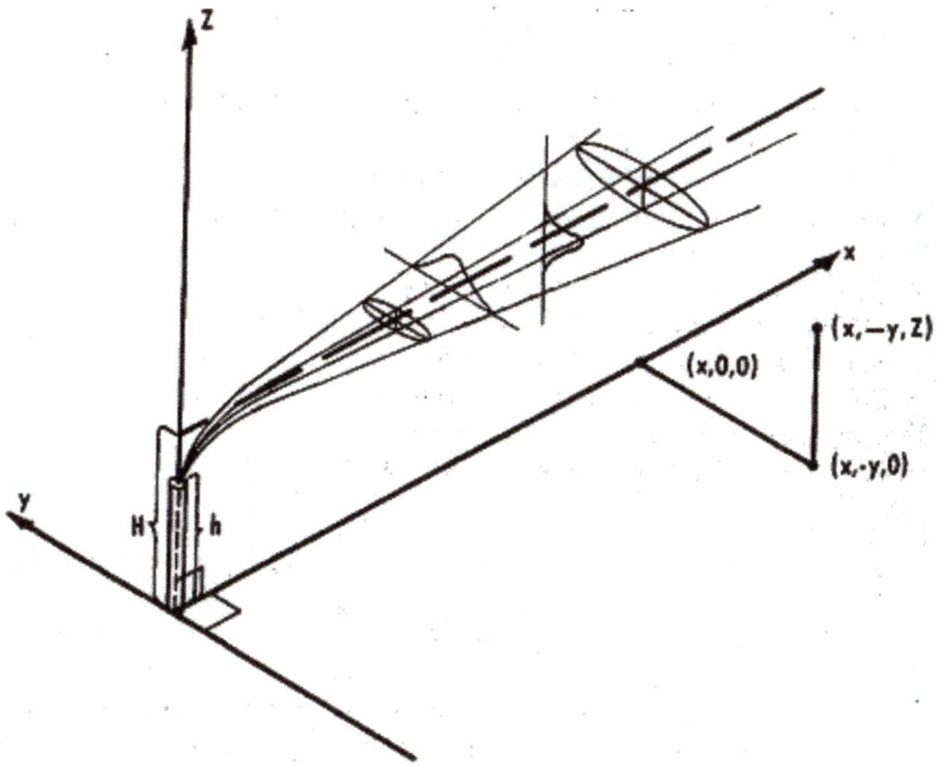

FIGURE 11.1 Gaussian plume horizontal and vertical dispersion (Turner, 1970).

from multiple sources within a WRRF by providing a two-dimensional representation of ground-level concentrations that can be plotted on a scaled map of the area. In cases where tall buildings are odor receptors, a vertical component can be added to a model to predict offsite concentrations at a specified elevation.

Buildings adjacent to the sources generate turbulence and downwash cavities; this effect is also accounted for in dispersion modeling and is discussed in Section 3.3.6. Directionally dependent building dimensions and downwash variables are required as model input.

Odor dispersion models predict off-site concentrations at receptor locations adjacent to the facility of interest, whose boundaries (the fence line) are input in the model in a defined coordinate system. Receptor locations are evenly spaced points within a grid system overlaying the facility and off-site locations, including areas of odor complaint concern. Receptor spacing is input by the modeler; smaller distances between receptor points will produce a more refined model but will increase run time. Results are presented as off-site odor concentrations calculated by the model, averaged over 1 hour. However, actual odor nuisances are often associated with exposure times on the order of minutes rather than an hour. Averaging odor levels over the course of an entire hour potentially has the effect of reducing peak odor concentrations at a receptor location in the model output.

To account for this potential lack of conservatism, each hourly odor level model output can be modified to shorter periods, often a 5-minute averaging period, which are more representative of short-term odor impacts. Average 1-hour concentrations are converted to peak 5-minute concentrations using the following power law (Wang & Skipka, 1993):

$$X_s = X_{(1\text{-hr})} \times (60 \text{ min} / t_s)^p \tag{11.1}$$

Where

X_s = the short-term (5-minute) averaging period concentration,
$X_{(1\text{-hr})}$ = the model-predicted 1-hour concentration,
t_s = the desired short-term averaging time (5 minutes in this example), and
p = the power-law exponent.

The value of the power law exponent varies according to atmospheric stability class (measurements of atmospheric turbulence). Stability classes range from A to F (with F being the most stable) and typically vary over the course of a day and night. Stability class is a function of surface wind speed, incoming solar radiation (for daytime hours), and nighttime cloud cover.

3.2 Dispersion Modeling Process

3.2.1 Odor Impact Assessment Applications

Dispersion models can be used in odor impact assessments to demonstrate compliance with air quality standards and odor nuisance criteria. Compliance with National Ambient Air Quality Standards (NAAQS) and other regulatory requirements are typically within the jurisdiction of regional or local authorities. Alternatively, utilities may complete a dispersion model absent regulatory requirements to prioritize odor sources and identify improvements that are projected to reduce off-site odors as part of being a good neighbor to the community.

When using dispersion models in regulatory applications, specific procedures must be followed. U.S. EPA-approved models to be used for regulatory applications are listed in Appendix A of the Guideline on Air Quality Models (40 Code of Federal Regulations [C.F.R.] § 51, Appendix W). Alternatively, U.S. EPA-preferred models are refined models that can be used for non-regulatory applications, typically as a planning tool for odor impact assessments. These models have undergone performance and scientific peer evaluations.

Dispersion models are consistently improved for accuracy; accordingly, the list of U.S. EPA-approved and preferred models is kept up to date as models change (U.S. EPA, n.d.). Although there are multiple U.S. EPA-approved models, some are more commonly used than others. The U.S. EPA-approved American Meteorological Society and Environment Protection Agency Regulatory Modeling System (AERMOD) model is currently most commonly used in odor impact assessments. Determining which model to use depends on the application and what type of input data are available.

New regulatory requirements and increasing reliance on air dispersion models calls for continuing improvement of accuracy; therefore, U.S. EPA guidelines are revised periodically and incorporated into model updates.

3.2.2 AERMOD

AERMOD is a commonly used U.S. EPA-approved air dispersion model. In general, AERMOD provides a good estimate of long-term odor impacts (in the range of a day to a year); therefore, it is appropriate both for regulatory applications as well as for supporting capital improvements to increase odor control. AERMOD is more limited at calculating projected short-duration off-site odors or impacts very close to the fence line (less than approximately 10 m [3 ft]).

AERMOD is a steady-state Gaussian plume model that can incorporate emissions from multiple point, area, and volume sources from surface or elevated areas in the stable boundary level. Two programs, AERMET and AERMAP, are preprocessors for meteorological and terrain data, respectively. AERMOD models use the outputs from the preprocessors and emission source data to predict downwind odor or compound concentrations. A distinct advantage of AERMOD is that it accounts for planetary boundary layer turbulence and can be used in urban applications.

3.2.3 CALPUFF

CALPUFF (currently not U.S. EPA-approved for regulatory applications) is a non-steady-state Lagrangian model. It is a U.S. EPA-preferred model and is used in odor impact assessments, particularly in applications similar to those that use AERMOD. CALPUFF is generally recognized as being more suitable than other models for long-range transport generally between 50 and 200 km from a given source. CALPUFF has the benefits of fully treating stagnation effects of air dispersion, wind reversals, and meteorological variations (Porter, 2004). Additionally, CALPUFF has an advantage of assessing impacts associated with calm or very low wind speeds (Ormerod, Galvin, & Starke, 2010).

3.2.4 Screening Models

Screening models analyze single-source release scenarios with simple terrain assumptions and non-varying meteorological conditions. The two screening model versions of AERMOD are AERSCREEN and CTSCREEN. AERSCREEN can be used in most applications, and CTSCREEN is often used for complex terrain scenarios such as hills or ridges (40 C.F.R. § 51, Appendix W, Appendix A).

Screening models are typically used to provide a rough estimate of off-site compound concentrations (often hydrogen sulfide) at a specific nearby receptor caused by the source emission. The output of the model provides compound concentrations that reduce at increasing distances from the source. A discrete distance from an odor source to a sensitive receptor (such as complainant locations) can be input to the model; the screening model will then predict the compound concentration at that location following dispersion.

Screening models are often used to predict conservative, "worst-case" dispersion scenarios, considering a combination of meteorological conditions, including stability class, wind direction, and wind speed, that result in the least amount of dispersion. A conservative scenario would be to choose a stability class F and a low wind speed. The model can be used to estimate conservative odor impacts at key receptors, the results of which may indicate that more complex multisource models (AERMOD, CALPUFF, etc.) are not needed if compound concentrations are insignificant at the receptor given these worst-case meteorological inputs.

3.3 Model Input Parameters

3.3.1 Fence Line Delineation and Model Area

A critical initial input into the dispersion model is the fence line location, defined in Universal Transverse Mercator (UTM) coordinates (X and Y). The fence line is most often the boundary of the facility (e.g., the WRRF or pump station property line). However, some model fence line coordinates will be extended beyond the actual property line to account for "green space", or uninhabited areas adjacent to the modeled facility. The reason for doing this is to provide a more reasonable output of odor impacts in critical areas (residential and commercial areas, typically). The model will predict a more realistic indication of the off-site odor impacts at critical receptors, and, therefore, the potential for complaints.

The model area is defined as a Cartesian grid including evenly spaced UTM coordinate points inside and outside of the fence line. The coordinate spacing is often closer near the fence line to produce more detailed results there. Additionally, the grid may be extended further along areas impacted by the prevailing wind direction to capture the extent of threshold odor contours away from the facility (e.g., the 10 D/T contour may need to be fully captured on an output plot because an odor goal was established to keep that contour within the facility fence line).

3.3.2 Source Locations and Characteristics

Odor sources can be point, area, or volume sources, which are defined in Section 2.0. Sources are typically selected for modeling based on general assumptions of their contributions to off-site impacts (beyond the fence line). Therefore, some sources that emit low-concentration odors or odors that are known to be minimally offensive (such as secondary clarifier or disinfection processes) may not be included in the model. However, these sources may still be included in some models where historical public complaints and/or sensitive receptors are located near the facility fence line.

The location and dimensions of point-source odor control system exhaust stacks are critical to providing sufficient dispersion. For these sources, the air velocity is recommended to be between 10.16 m/s (2000 fpm) and 15.24 m/s (3000 fpm), per American Society of Heating, Refrigerating, and Air Conditioning Engineers (ASHRAE) guidelines. The guidelines also indicate that an exhaust velocity of at least 12.7 m/s (2500 fpm) will sufficiently prevent rain from entering the stack (ASHRAE, 2015).

An exhaust stack should be constructed with an outlet height greater than the height of adjacent buildings to prevent downwash effects. Upper bound limits may be imposed by the Federal Aviation Administration (FAA) for stacks near airports. Outlet appurtenances such as rain caps and hinged covers will limit vertical dispersion and thus increase off-site odor impacts. Furthermore, outlet stacks or fans on building walls that release odorous air horizontally also will have higher off-site odor impacts than comparable vertical stacks for the same source.

High-velocity fans and stack units are manufactured to induce exhaust velocities much higher than ASHRAE recommendations in an effort to create a "virtual stack height" that will further increase dispersion. These products also will use injection of ambient air for dilution of the stack air, which also correspondingly reduces downwind ground-level odors. These fans are often used for airstreams that contain relatively dilute odors in lieu of installing more costly odor control systems, assuming this is acceptable per regulatory and/or permit limits. Some studies have found that these products are limited under meteorological conditions where an inversion layer is present, where the virtual stack height does not rise to the elevation predicted under other meteorological conditions.

3.3.3 Emission Rate Calculations

The dispersion model input requires that odor source data be inserted in concentration units. This is straightforward for modeling compound concentration emissions (such as hydrogen sulfide), where odor source emissions are defined in units of grams per cubic meter (g/m^3). Modeling odor, however, presents a challenge because the odor detection threshold results measured by an odor panel laboratory are expressed in D/T, which is unitless. To address this, a dispersion model can express these data as D/T per cubic meter to allow for model calculations. For example, an odor value of 5000 D/T would be input to the model as 5000 $D/T/m^3$. This concentration is then converted to a mass loading in units of "grams" per second by factoring in the airflow rate and using the following equation (Ferguson et al., 2004):

$$\text{Emission rate ["g"/s]} = \text{odor emission ["g"/m}^3\text{]} \times \text{airflow rate [m}^3\text{/s]} \times 10^{-6} \qquad (11.2)$$

The model output will be in concentration units, which is then converted back to D/T to predict worst-case odor impacts in off-site areas. Factors can be added to this equation to account for the degree of turbulence in the process, where higher emissions rates may not be entirely captured in only a few collected air samples. These "airflow factors" can be a function of area; for example, using 5.1 $m^3/hr{\cdot}m^2$ (0.28 cfm/ft^2) for turbulent surface emissions and 2.3 $m^3/hr{\cdot}m^2$ (0.13 cfm/ft^2) for quiescent surface emissions (Ferguson et al., 2004). The modeler should use discretion when including these factors so as not to overstate off-site impacts from either source type.

3.3.4 Meteorological Data

Nearby weather station data are acquired and input to the model. Data will include hourly wind speed, wind direction, and stability class. Often, these data are available at the nearest airport to the facility. Five consecutive years of meteorological data are recommended for the best model output resolution, though as little as one year of data can be used.

3.3.5 Terrain Data

The model input will typically include importing the appropriate United States Geological Survey (USGS) Digital Elevation Model quadrangle maps into the model. These maps are available for most areas of the United States and are directly imported into the model database. The maps include terrain elevations, which can be critical for plume dispersion projections. These maps can be eliminated in some models for simplification when the area of interest is relatively flat.

3.3.6 Sensitive Receptors

A sensitive receptor is a location that is within the vicinity of the model source(s) that is typically occupied, such as a residential home or a commercial business, in which there are people that are prone to complain about odors above a certain level. For area receptors, UTM coordinates are input to the model at each corner of the sensitive receptor polygon shape along the perimeter of a building, park, etc. In some scenarios where odor impacts may be greater at higher elevations (such as apartment buildings in an urban area), building heights may also be entered at each UTM coordinate. A sensitive receptor may also be a linear shape, such as a walking path or a road. The model output can include worst-case odor or compound concentrations at identified sensitive receptors, or the number of exceedances of an established threshold, as discussed in Section 3.4.

3.3.7 Physical Structures and Buildings

Building dimensions are similarly input to the model to account for downwash effects. These dimensions may describe polygon or circular structures, similar to odor source inputs. Examples of circular structures could include digesters, which are often not odor sources but should be included as structures in the model. The method of inputting building dimensions is identical to how the modeler inputs odor sources or three-dimensional sensitive receptors, including UTM coordinates (X and Y dimensions) and building heights (Z dimensions).

The low-pressure zone that forms when wind passes over and around obstructions also occurs at buildings. Because buildings are larger obstructions than stacks, downwash effects caused by buildings are generally more pronounced than those caused by stacks. Building downwash can direct a plume to ground level near the lee side of buildings, resulting in higher ground-level odor impacts. In addition, even if the plume is not mixed at ground level, the building downwash effects may lower the effective height of the plume in a manner similar to stack tip downwash.

Downwash effects are direction-specific; therefore, the U.S. EPA-approved Building Profile Input Program (BPIP) model is used to determine the effective building height and crosswind building width for every 10° of wind direction. AERMOD and CALPUFF use the BPIP model to calculate building downwash effects for each odor source.

3.4 Dispersion Model Output

3.4.1 Isotherm Plots

Dispersion models provide output plots that project off-site impacts given odor or compound emission rates at facility sources and the variation of meteorological conditions over the specified period. An isotherm plot of contours shows equivalent locations of off-site odor (expressed in D/T) or of the modeled compound concentration. Contour values will generally decrease with distance from the odor sources, though sometimes point sources will result in higher off-site concentrations away from the source that appear in a "bullseye" orientation.

Contours are generated by the model by calculating the highest odor or compound concentration at each receptor during the meteorological period. Therefore, the odor contour plots are, by definition, a worst-case condition, which is appropriate when evaluating odor impacts at complaint locations. Figure 11.2 shows a typical odor contour output plot.

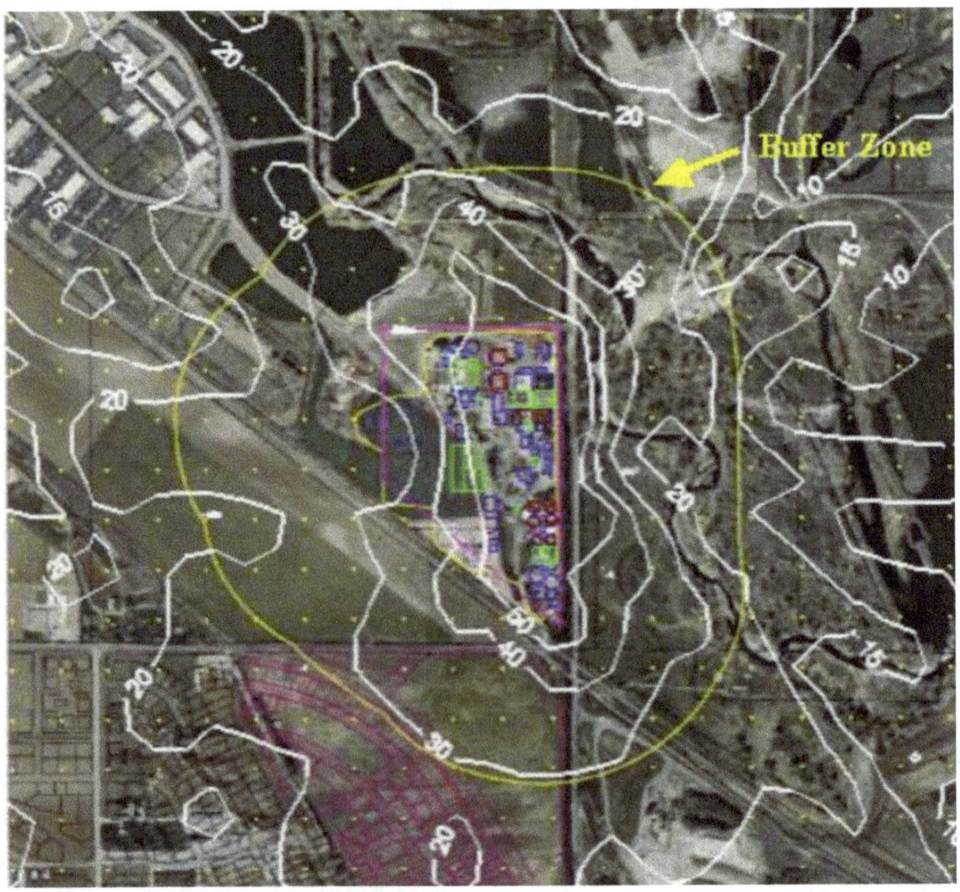

FIGURE 11.2 Odor impact contour plot (Ferguson et al., 2004).

An additional means of showing model-predicted off-site odor impacts is displaying how many times during the period that the calculated worst-case concentrations exceed a given threshold (often termed as the number of "exceedances"). This threshold is a goal used to estimate current impacts and identify what degree of odor control for priority sources is needed to meet that goal. An example goal is odor emissions from a WRRF shall not exceed 10 D/T at the fence line more than 1% (88 hours) of a given year.

Figure 11.3 shows an example threshold exceedance plot. Note that while the exceedance contours displayed are in units of hours, the actual time of exceeding the given threshold may not be a full hour, but for at least one period of consecutive minutes during that hour. This period is a function of the averaging time used for the modeling (discussed in Section 3.1).

3.4.2 Individual Source Impacts

Another valuable dispersion model output is a projection of worst-case fence line (or specific off-site receptor) odor or compound concentrations for all individual modeled sources. This provides an indication of source prioritization, which can be the basis for implementing treatment process modifications, installing new odor control, or upgrading existing odor control systems.

With respect to fence line maximum odor impacts and source prioritization, results should be analyzed by considering the nature of the off-site

FIGURE 11.3 Threshold exceedance plot (Ferguson et al., 2004).

impact. Some sources may have high fence line odor impacts only because of their proximity to the fence line. Consequently, if the fence line location with the maximum impact is in an area where the odor complaint potential is minimal (such as a green space or a body of water), prioritization of that source may be downgraded.

Table 11.1 presents an example of an individual source prioritization table. The table shows that the septage receiving and sludge dewatering sources are the greatest priority. Additionally, it shows that for an average model year, the number of exceedances of the established 10 D/T threshold for "all sources" (considering all sources rather than individual source contributions) is approximately 3300 hours (or portions of model hours, as discussed in Section 3.4.1). If the facility's goal is for the number of 10-D/T exceedances for all facility sources to be 1% of the year (88 hours) or less, this model output indicates that the goal is not currently being met and odor control improvements are needed to be in compliance.

3.5 Dispersion Model Limitations

Air dispersion modeling is the mathematical estimation of how air pollutants (including odor) disperse in the atmosphere. The results of modeling should never be interpreted or communicated to the public as a predictor

TABLE 11.1 Example off-site impacts distribution.

Source	Maximum Ground-Level Off-Site Odor (D/T)	Exceedances of 10 D/T	
		5 years[a]	1 year[b]
All sources	500	16 544	3309
Septage receiving	331	13 676	2735
Sludge dewatering	201	5156	1031
Primary clarifiers	127	391	78
Aeration basins	24	68	14
Secondary clarifiers	16	33	7
Biofilter	20	18	4
Sludge storage tanks	11	1	0

[a]5 years is the model period; the model counts the number of times 10 D/T is exceeded in this period, where the numbers shown represent the number of hours in which the 10 D/T is exceeded for at least 2 consecutive minutes.

[b]5-year number of exceedances divided by 5

of *actual* concentrations. Air dispersion models necessarily simplify complex atmospheric reactions and human-caused factors.

4.0 AIR CONTAINMENT AND EXTRACTION DESIGN

In cases where process units require new or upgraded odor control systems, a critical component of the design is identification of the air extraction rate from the source. Additionally, rooms may require ventilation for worker safety or comfort, with the exhaust air also potentially sent to odor control. The optimal design airflow rate will be a function of the odorous space characteristics and how the air is contained. Airflow rate impacts the selection and size of the odor control technology, and it will also correlate with offsite impacts. This section provides options for enclosing odorous spaces and criteria for calculating airflow rates.

4.1 Odor Containment Methods

4.1.1 Covers

Prevention of fugitive odors from wet wells, tanks, and channels can be achieved by installing covers that maintain a consistent negative pressure. Foul air from the covered headspace may then be ventilated to an odor control system. Covers need not be completely airtight. In fact, makeup air must be available through openings in the covers; otherwise, exhaust fans might place the covers under a vacuum-induced load.

These cover openings will have a small surface area compared to the cover surface area; too large of an opening surface area may result in non-attainment of negative pressure and fugitive emissions. Openings will consist of cracks between cover plates, hatch openings, and ducts that emerge from the cover to allow ambient air to passively flow into the headspace. Ducts are often goosenecks to prevent rain entry. Refer to Chapter 6, Section 5.2, for a discussion of cover design considerations.

4.1.2 Enclosures

Equipment enclosures provided for foul air containment can be unique designs or prefabricated systems provided by odor control or cover manufacturers. Enclosures greatly reduce the volume of air that is needed to be exhausted because the air is collected only from inside the enclosure rather than the entire room in which the odor source is located. A disadvantage of enclosures is that they will reduce access for operations and maintenance (O&M) staff to the equipment inside the enclosure.

Enclosures must be resistant to corrosion. They typically are constructed using fiberglass reinforced plastic, stainless steel, or aluminum. Locating appropriate foul air withdrawal points is critical to ensure that there are no dead spots where air is not sufficiently evacuated. Not obtaining this risks corrosion of sensitive equipment components. Refer to Chapter 6, Section 5.2, for a discussion of enclosure design considerations.

4.1.3 Hoods

Hoods are used to capture odorous air from a source within a room as close to the point of release as practical. Maximum foul air collection efficiency would be provided if the hood were connected to and fully enclose the piece of equipment, which would greatly reduce the potential for fugitive emissions. However, this essentially acts as an enclosure, which also presents issues for O&M access.

4.2 Foul Air Capture Design Criteria

4.2.1 Room Ventilation for Safety and Comfort

Worker safety at WRRFs is regulated at the federal level with respect to hazardous atmospheres. Odorous air that contains high hydrogen sulfide concentrations will produce a toxic environment, and sufficient ventilation will be required. In many cases, the exhaust air will be sent to odor control. Ventilation rates are governed by the National Fire Protection Association (NFPA) 820: Standard for Fire Protection in Wastewater Treatment and Collection Facilities (NFPA, 2016). The code indicates ventilation rates for various spaces to obtain reductions in electrical classification status. Refer to Chapter 6, Section 5.1, for further discussion of how NFPA 820 is used to calculate required ventilation rates for odor control and maintenance of safety.

The Occupational Safety and Health Administration (OSHA) specifies additional ventilation rate requirements in rooms as well as general procedures for employee entry into confined spaces. Regulation 29 C.F.R., § 1910.146 should be referenced for design and operational criteria (Permit-required confined spaces, 2004). Both OSHA and the National Institute for Occupational Safety and Health (NIOSH) provide limitations with respect to hydrogen sulfide exposure for workers at WRRFs. The following are the requirements posted on OSHA's website at the time of this publication:

- The OSHA standard for general industry is that hydrogen sulfide exposure must not exceed 20 ppmv (allowing for one measurement of 20 to 50 ppmv within a 10-minute period during an 8-hour shift).

- The NIOSH recommended exposure limit for ambient hydrogen sulfide concentration is 10 ppmv during a 10-minute ceiling (U.S. Department of Health and Human Services, 2007).
- The hydrogen sulfide concentration that OSHA has identified to be "immediately dangerous to life and health" is 100 ppmv.

Ventilation rates to achieve these requirements must be set for spaces, with the foul air ventilated likely being conveyed to an odor control system. Note that individual municipalities or private owners may pose stricter requirements for worker safety. Ventilation is also required for confined space entry in WRRFs. Ventilation requirements specified for confined space entry should be advised by OSHA and entities that have jurisdictional authority in that application. Ventilation requirements in confined space entry are independent of those specified for odor control.

Worker comfort is also a driver for providing an acceptable amount of room ventilation in WRRFs. Foul air removed from these rooms may also be conveyed to odor control. Chapter 6, Section 5.6.2, provides design considerations for air ventilation rates from worker-accessible and non-worker-accessible spaces.

4.2.2 Hood Airflow Rate Design

An alternative to ventilating an entire room is to construct hoods over odorous processes inside the room. The required air velocity to collect odors at the point of release should be in the range of 0.25 to 0.5 m/s (50 to 100 fpm) (ASHRAE, 2015). For example, a belt filter press hood with an inlet face dimension of 7.9 m (26 ft) × 3.05 m (10 ft) would require a minimum airflow rate of approximately 6.0 m^3/s (13 000 cfm) to sufficiently capture all odors emitted by the belt filter press. This airflow rate for a single odorous process may be undesirable because of downstream odor control cost or space limitations; therefore, an option would be to provide a lower flowrate (and thus lower velocity) with the understanding that some fugitive emissions will occur.

In situations where the bottom of the hood is located above the release point, the design airflow rate for full foul air capture would be increased according to the following equation (Wolstenholme, 2013):

$$Q_n = Q_c \times (10 \times X^2 + A) / A \tag{11.3}$$

Where Q_n = hood airflow rate, m^3/s;
Q_c = capture airflow rate, m^3/s;

A = hood surface area, m^2; and
X = vertical distance from head to point of release, m.

Even at the low end of the recommended velocity range (0.25 m/s), if the hood in the aforementioned belt filter press example were 1.0 m (3.3 ft) above the point of release, a high airflow rate of approximately 31 m^3/s (66 000 cfm) would be required for full capture of the foul air emitted by the source. In this case, a lower airflow rate can also be applied to the process unit through the hood, understanding that fugitive emissions will occur; however, diminishing returns will be realized the greater the value of X in the equation, which will call into question whether using a hood for foul air capture is reasonable.

4.2.3 Covered Processes

Exhaust airflow rates from covered headspaces have traditionally been designed by providing a required number of air changes beneath the cover. This is still the governing parameter when frequent operator access (e.g., into a dome) is required or when dilution of high-strength odors is needed. However, a typical exchange rate of 12 air changes per hour (ACH) can result in high quantities of air being exhausted from the headspace and to the downstream foul air treatment system; conversely, it may also produce too low of an airflow rate to avoid fugitive emissions.

The differential pressurization across a flow path (Δp), expressed in units of inches of water column (in. w.c.), is defined by the following equation (Wolstenholme, 2013):

$$\Delta\rho = \rho/2 \times (Q \,/\, [(C_f \times A)]^2 \qquad (11.4)$$

Where ρ = density of air entering the flow path,
Q = volumetric airflow rate,
C_f = flow coefficient, and
A = area of flow path.

For buildings, a good approximation of the flow coefficient (C_f) is 0.6 to 0.7. Using 0.65 in Equation 11.4, the equation would be expressed as follows:

$$Q = 2610 \times A \times (\Delta\rho)^{0.5} \qquad (11.5)$$

Therefore, given a desired pressure differential (often -0.1 in. w.c.), and a known area of fresh air openings in the cover, the required airflow rate can be calculated. This equation also applies to airflow rate design calculations for rooms. However, the value of A may not be easily quantified for very small cover cracks; therefore, general industry guidelines have been developed, based on pilot testing, to provide airflow rate design parameters as a function of the cover surface area. The following relationships are used (Wolstenholme, 2013):

- For loose-fitting covers, use a minimum of 18.3 m^3/hr per m^2 of cover area (1 cfm/sq ft of cover area).
- For tight-fitting covers, use 9.1 to 18.3 m^3/hr per m^2 of cover area (0.5 to 1 cfm/sq ft of cover area).
- For concrete covers, use 1.8 m^3/hr per m^2 of cover area (0.1 cfm/sq ft of cover area).

Another ventilation consideration is based on a concern for corrosion inside the covered headspace. This will be critical to address when providing ventilation of headspaces that are enclosed by uncoated concrete walls and those that are exposed to high hydrogen sulfide concentrations. To address the potential for corrosion in these areas, an industry guideline of 4 to 6 ACH for the headspace is recommended. The resulting airflow rate may be used for design unless other calculations produce higher values. Chapter 6, Section 5.6.2, provides design considerations for air ventilation rates from covered processes.

5.0 FACILITY ODOR EVALUATIONS

5.1 Development of Candidate Technologies

Several vapor-phase odor control technologies can be considered for treating source foul air emissions. Chapters 8 to 10 of this manual of practice provide descriptions of vapor-phase technologies. A general guideline (Table 11.2) identifies candidate vapor-phase odor control technologies given a general knowledge of foul air hydrogen sulfide concentration and airflow rate (WEF, 2018). The recommendations shown in the table should be considered for initial reviews of technologies in a conceptual design. The table does not provide a complete list of available vapor-phase technologies, but instead lists those identified as most commonly found in the industry (WEF, 2018).

An important consideration that may impact development of candidate vapor-phase technologies is whether upstream liquid-phase treatment exists or is being considered. Technology evaluations often include considerations of adding chemicals or oxygen upstream of a WRRF to lower liquid-phase sulfide concentrations before wastewater entry into the facility.

TABLE 11.2 Odor control technology selection guide (WEF, 2018).

Foul Airflow Rate	Foul Air Average Hydrogen Sulfide Concentration: Low Concentration (<10 ppmv)	Medium Concentration (10-50 ppmv)	High Concentration (>50 ppmv)
Low	Dry Media Adsorber Ionization	Biofilter Dry Media Adsorber	Biofilter
Medium	Dry Media Adsorber Ionization	Chemical Scrubber Biofilter Biotrickling Filter	Chemical Scrubber Biofilter Biotrickling Filter
High	Chemical Scrubber Biofilter Dry Media Adsorber Ionization	Chemical Scrubber Biofilter Biotrickling Filter	Biotrickling Filter

The desired result would be a reduction in hydrogen sulfide emissions from liquid-phase WWRF facility processes, such as a headworks. Incorporation of upstream liquid-phase treatment may modify the list of candidate vapor-phase odor control technologies because certain choices, such as activated carbon adsorption, are sensitive to high hydrogen sulfide loads. Alternatives may be developed that either do or do not include liquid-phase treatment, with life cycle cost evaluations completed for each to evaluate the benefit of the additional O&M costs versus providing greater capital investments for reduction of higher odor concentrations at the WRRF.

5.2 Life Cycle Cost Analyses

A ranking of odor control technologies based on cost-effectiveness can be produced by developing life cycle cost estimates of various odor control alternatives that are projected to sufficiently meet goals. Alternatives could include two-stage odor treatment systems if a greater amount of odor reduction and/or backup odor control units is required. Alternatives may also include upstream liquid-phase treatment combined with vapor-phase technologies. Odor control system capital costs are developed using budgetary cost estimates from system manufacturers. Costs of ancillary equipment such as ductwork, dampers, and covers enhance the accuracy of the capital cost estimate.

The costs of regular O&M requirements for the installed odor control systems are summed to produce the total annual costs for various odor control alternatives. The annual costs over a specified period (often 10 or

20 years and associated with the expected life of the odor control system) are converted to a net present value (NPV) using a discount rate. The NPV is then added to the capital cost projection to calculate the total life cycle cost, which allows for a more appropriate cost comparison between alternatives versus using the capital costs alone.

Annual O&M costs are generally calculated for the following:

- Power (electricity cost, the majority of which will be from foul air fans);
- Chemicals (for upstream liquid-phase treatment or chemical scrubbers);
- Media replacement (for activated carbon adsorbers, biotrickling filters, biofilters, and chemical scrubbers);
- Water (if potable water must be purchased for use in chemical scrubbers, biotrickling filters, or biofilters); and
- Labor (for O&M activities).

Labor costs for regular maintenance of equipment such as pumps, fans, and monitoring systems can be significant, and should be adjusted for higher cost-of-living areas. Additionally, less frequent but maintenance-intensive activities, such as media replacement, should be converted into annual labor costs.

The low-cost alternative (using the life cycle calculation) is typically recommended for conceptual development and ultimately for detailed design and construction. In cases where odor control alternatives present varying degrees of treatment, dispersion modeling can be used to evaluate a "cost per D/T reduction at the fence line" for each alternative. This method allows for a better approximation of the true cost-benefit of odor control technologies.

5.3 Non-Economic Factors Analysis

In addition to selecting the most cost-efficient odor control technology that will meet the design criteria, non-economic factors can also be evaluated. This separate analysis will allow a utility to compare alternatives with respect to parameters that may be important to O&M staff, public relations experts, nearby residents, and others. A non-economic factors analysis can be completed simultaneously with the life cycle cost analysis, or the facility may use a non-economic factors analysis to screen out undesirable alternatives before developing cost comparisons.

When scoring alternatives, non-economic factors are typically tailored to the owner or operator of the odor control system. It is generally appropriate that 5 to 10 factors be evaluated. Limiting the evaluation simplifies the process and focuses on the most important issues. The selected factors are then evaluated for each odor control alternative, potentially including a score and weighting of factors to align with expressed importance by the project stakeholders, such as managers, engineers, O&M staff, and the public. This provides an indication of which alternative is most desirable absent cost considerations. The following subsections describe some common non-economic factors.

5.3.1 Footprint

Locating the odor control system may have economic repercussions if not considered carefully. It is ideal to locate a system near the odor source to reduce the amount of required foul air ducting. However, such a location may not be feasible due to insufficient available space to site the system, equipment access concerns, or underground utility conflicts. For example, if land availability near a large odor source is limited, a bulk media biofilter may not be a reasonable technology choice because of the relatively large footprint required.

5.3.2 Operations and Maintenance Requirements

This factor reflects the ease of operating and maintaining the technology associated with an alternative. This is often associated with technologies that include many "moving parts". Chemical scrubbers are a good example because they require metering and recirculation pumps, spray nozzles, mist eliminators, and instrumentation that must be maintained.

5.3.3 Safety

Staff safety may be a significant factor when evaluating alternatives. Although there always will be mitigating additions to an odor control system to provide a safe environment, the owner may have concerns with technologies that greater safety requirements, particularly if O&M staff have experienced issues or have complained in the past. For example, operator safety may be considered for alternatives requiring chemical handling (such as chemical scrubbers), confined space entry, and tall structures (e.g., vessels in a tower orientation) that would require safety measures for falling hazards during maintenance activities.

5.3.4 Familiarity

Often, facilities will prefer odor control technologies that they have successfully implemented in the past. In some cases, the preferred technology may be less appropriate or higher cost than the alternatives. However, the owner may desire to provide the same technology for treatment of several odor sources; this may be of interest so that operators will be best equipped to work with the system. Technology uniformity may also be desired to reduce the complexity of addressing issues that may arise.

Other non-economic factors that can be included in analyses of alternatives are the following: public perception and opinion, off-site visibility of new odor control units (aesthetics), sustainability, avoidance of new and less-proven technologies, ability to treat diurnal variation or intermittency of odors, flexibility for future expansion, and carbon footprint. Table 11.3 provides an example of a non-economic factor comparison of odor control technologies.

6.0 CASE STUDIES

6.1 Odor Impact Assessment Case Study 1

A 379 ML/d (100-mgd) average dry weather capacity WRRF operated by a utility in the Midwest received numerous odor complaints from a nearby hillside residential community and surrounding area. A goal was established to limit emissions greater than 10 D/T at the treatment facility fence line to less than 1% of the year (approximately 88 hours). A facility-wide study was completed to develop an odor control program that would be designed to lower off-site impacts and meet the goal.

TABLE 11.3 Non-economic comparison of odor control technologies.

Technology	Footprint	Complexity	Capital Cost	Operating Cost
Chemical scrubber	Small to Medium	High	Medium to High	Medium
Biofilter	Large	Medium	High	Low
Biotrickling filter	Small to Medium	Medium	High	Low
Carbon adsorption	Medium	Simple	Medium	Medium to High
Ionization	Small	Medium	Low	Medium
High-velocity vertical dispersion	Small	Simple	Low	Low

In this odor impact assessment, an odor dispersion model was developed to convert measured emissions from sources into projected off-site odor contours (in units of D/T). Local meteorological conditions (5 consecutive years of hourly data) and terrain of the surrounding area created a clear prevailing wind direction. The model was used as a tool to evaluate and consider the impact of new or modified odor control systems on meeting the goal. Additionally, the model will be used for future capital improvement planning for the facility as new process units are constructed or processes are expanded.

An odor sampling program included field monitoring, sampling and laboratory analyses, and field air velocity testing. The collected data were input to the dispersion model, providing an up-to-date picture of existing processes and emissions. Field hydrogen sulfide measurements were collected by two methods: (a) instantaneous single readings with a Jerome analyzer and (b) continuous monitoring with OdaLog data loggers.

The instantaneous single measurements were used to provide an indication of odors in various facility locations as well as removal efficiencies through existing odor control units. The Jerome analyzer was also used to identify appropriate sampling locations in area and volume sources so that emissions from these sources were reflective of average conditions. However, for some critical processes, the maximum (worst-case) odor emissions scenarios were also input to the dispersion model to project whether those sources could be linked with complaints under maximum emission conditions (process unit upsets).

Once appropriate sample locations were identified, OdaLogs were deployed for continuous monitoring, and air samples were collected in Tedlar bags. The air samples were sent to laboratories for reduced sulfur compound analysis and odor panel testing. Air velocity testing was performed using a handheld instrument to estimate current stack airflow rates, which were also input to the dispersion model to reflect current conditions.

The field data and calculations were used to develop a baseline AERMOD dispersion model. The baseline model identified areas of the facility where odor control improvements could be made to bring odor contours closer to the facility and contribute to meeting goals. The odor emission rates for sources input to the model included the following two types:

- Point sources, described as emissions from existing odor control units with stack treated air releases—The odor emission rates used measured or historical odor data (in D/T), air velocities, air temperatures, and exhaust stack dimensions, including the height above grade.
- Area sources, described as emissions from large sources at or near the ground surface, such as the primary settling tanks or aeration

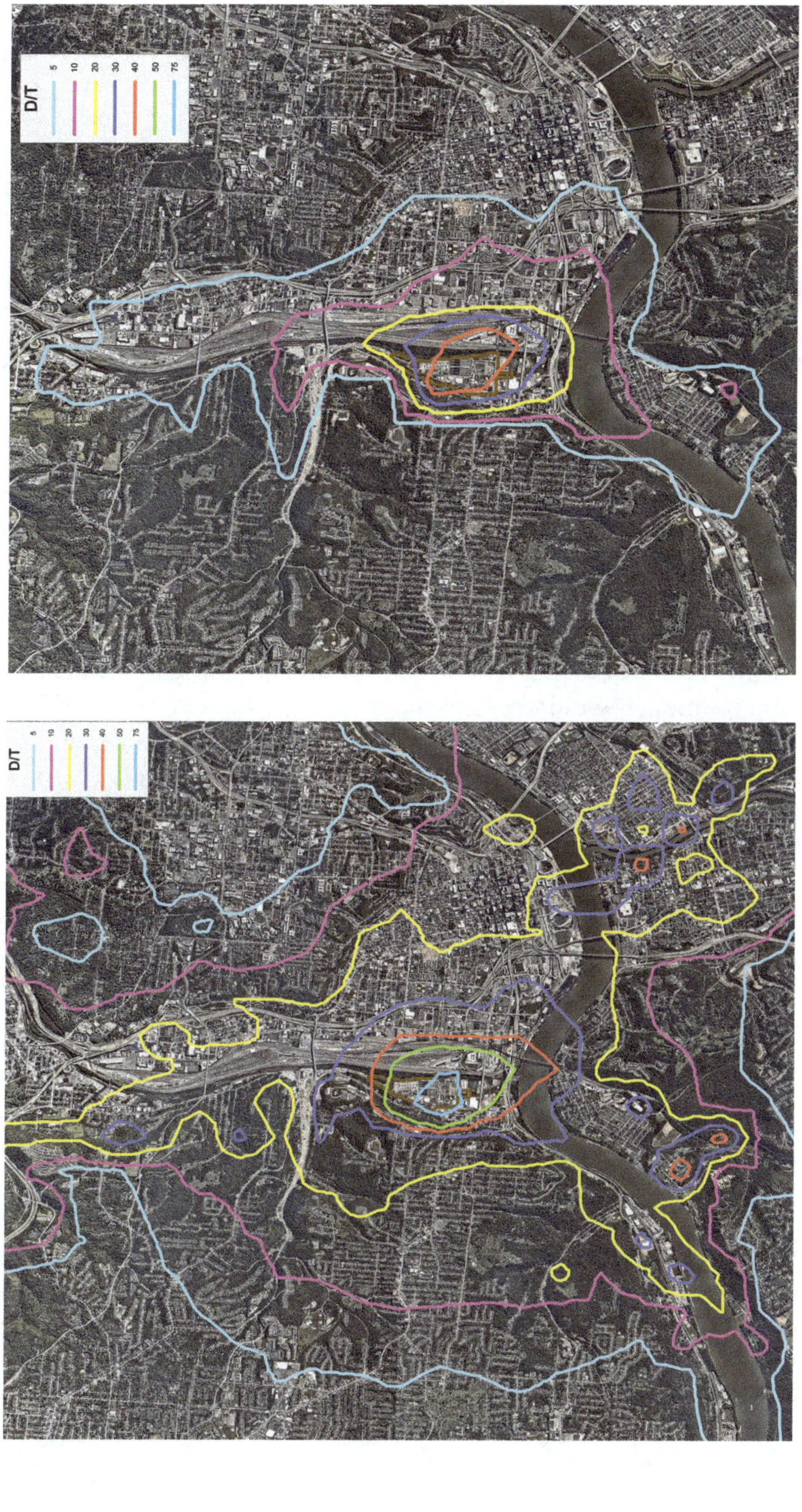

FIGURE 11.4 Dispersion model output: Case Study 1 (adapted from a Brown and Caldwell case study for unnamed utility).

basins—The odor emission rates included measured or historical odor data (in D/T), air velocities into the flux chamber used for collecting air samples, and source area dimensions.

Meteorological data and building and structure dimensional data were included in the dispersion model. Hourly meteorological data included wind speed, wind direction, and stability class. Building heights near odor sources, also input to the model, ranged from approximately 2.44 m (8 ft) to 18.29 m (60 ft).

Figure 11.4 depicts the baseline dispersion model output plot considering all continuous odor sources and their respective emission rates (plot on left) compared to resulting contours upon incorporation of odor control improvements identified in the odor impact assessment (plot on right).

Phase 1 of the recommended odor program targeted the odor sources that contribute most to off-site impacts from facility emissions. Figure 11.4 provides the output dispersion model plot associated with incorporating Phase 1 recommendations. The 10-D/T contour is the critical line to consider when evaluating the dispersion model output plots because this is the odor value above which fence line odors are acceptable for 1% of a given year.

Implementation of Phase 1 odor control improvements, projected by the dispersion model, would reduce the total exceedances of 10 D/T by the facility to 212 exceedances (hours or portions of hours) per year; this is still above the goal of 88 exceedances (1%) per year. To meet the goal, additional improvements were determined to be needed and provided upon incorporation of Phase 2 projects. The impact of the Phase 2 projects was modeled; results showed a projection of meeting the odor control goals, with 76 hours per year exceeding the 10-D/T threshold at the fence line.

6.2 Odor Impact Assessment Case Study 2

A WRRF in California had initially received minimal odor complaints after the mid-1980s expansion of its facility and installation of atomized mist chemical scrubbers (using caustic and hypochlorite solutions) that treated most odorous processes. Since 1990, however, residential development has encroached on the facility. Odor complaints began to be received intermittently starting in 2005. A facility-wide odor impact assessment was performed to understand the sources of odorous emissions and the current odor removal efficiencies through the scrubbers. Recommendations were made for odor control improvements to reduce impacts on the adjacent residential community.

Field odor sampling, laboratory analysis for odor and odorous compounds, and a ventilation assessment were conducted. The results were critical for making recommendations for improvements to the current system.

Field measurements of hydrogen sulfide concentrations were collected at the inlet and outlet of the existing scrubbers. Odor panel testing was conducted for most sources, including the inlets and outlets of several sources. Scrubber outlet air odor measurements were critical because those values were input to an odor dispersion model to project current and future (post-improvements) off-site impacts.

Odor and hydrogen sulfide removal efficiencies were calculated for the existing scrubbers by comparing sampling data for inlets and outlets. Results showed that sources at which foul air emissions contained higher hydrogen sulfide concentrations had higher scrubber inlet odor levels and that the existing mist scrubbers performed best on those foul airstreams. Processes where this was notable included the influent pumping station and headworks building. However, the hydrogen sulfide and odor removal efficiencies were still below design values.

Other downstream and solids handling processes airstreams contained lower hydrogen sulfide concentrations and odor emissions. Processes where this was the case included the sludge thickening building and the dewatering building. Because hydrogen sulfide was not a significant component of the total odor from these sources, it was determined that the current chemical scrubbing system was not effective.

A ventilation assessment and air balancing were performed for two of the buildings at the facility. Additional ventilation analyses for all other odor-scrubbed source air were conducted. Air balancing was completed to align flowrates within the foul air exhaust and supply air ducting with design values. The balancing resulted in increased supply and exhaust airflows, the latter of which subsequently increased the airflow rate to the corresponding scrubbers. The ventilation analysis results were used to recommend airflow rate increases that would result in optimal air change rates for odor removal and corrosion minimization. Stack velocities for all scrubbers were measured in the field testing. Results indicated that most of the measured velocities (generally less than 2000 fpm) were lower than ASHRAE recommendations for effective vertical dispersion away from the stack (ASHRAE, 2015).

An odor dispersion model was produced and used to predict maximum off-site impacts given the odors measured in the field testing. The model used 5 years of meteorological data from the nearest weather station to identify worst-case off-site impacts. These impact areas correlated with odor complaint locations.

Output plot odor contours showed projected areas of equal odor D/T. Off-site impacts were calculated using the model by providing the maximum odor at the fence line for all sources and for individual sources. These results indicate the overall odor impacts of the facility and the contribution of each source to that overall impact.

Additionally, the off-site impact was quantified by counting the number of houses along or inside the 30-D/T contour on the dispersion model output plots. Following review of the baseline contour locations, it was determined that the 30-D/T contour was an appropriate boundary, inside which odor complaints associated with facility odor emissions and worst-case meteorological conditions were most likely. This conclusion was based on review of the contour location with respect to historical complaint locations. The area inside the 30-D/T contour was used to compare impacts of various odor control model runs versus the baseline, with impacts for each run also quantified by counting the number of houses between the 30-D/T contour and the facility fence line.

A baseline and two odor control improvement plots were developed. Recommended odor control improvements were grouped into phases of capital improvements. Table 11.4 lists the elements of these phases (Phase 1 and Phase 2). The model results include the following: (a) the corresponding maximum odor at the facility fence line for the baseline and for each phase and (b) the number of houses that were projected by the model to experience at least a 30-D/T odor impact under worst-case meteorological conditions for each case.

The results of the dispersion model provided a projection of off-site odor reduction given implementation of various improvements, considering which scrubbers were emitting the highest odors and airflow rates, which correlated off-site impacts. Considerations were also made for testing results that showed which scrubbers were not efficiently removing hydrogen sulfide and odor. As Table 11.4 shows, 90% odor reduction applied to targeted

TABLE 11.4 Comparison of odor control improvement phases with baseline (adapted from a Brown and Caldwell case study for unnamed utility).

	Maximum Fence Line Odor		Number of Houses with at Least 30 D/T Maximum Odor Impact	
Odor Control Phase	Odor Level (D/T)	% Reduction from Baseline	Number of Houses	% Reduction from Baseline
Baseline conditions	250	—	288	—
Phase 1: Stack improvements	94	62%	166	42%
Phase 2: 90% odor reduction	52	79%	47	83%

scrubbers would result in an 83% reduction in off-site odors (based on the number of houses at or within the 30-D/T contour).

A life cycle cost analysis and assessment of non-economic factors resulted in a recommendation that the 90% odor removal efficiency for the targeted scrubbers was to be achieved by replacing the existing mist scrubbers with packed tower chemical scrubbers. The life cycle cost of the packed tower chemical scrubber alternative was lowest (compared to biological and carbon adsorption alternatives), and the non-economic factor of uniformity of odor control technologies throughout the facility was weighted heavily by the facility. Accordingly, the recommended capital improvements were construction of new packed tower chemical scrubbers at influent pump station and headworks facilities. These improvements were recommended to follow stack modifications, where cones were attached to existing stack outlets to increase velocities and improve vertical dispersion (thus reducing off-site odor impacts).

7.0 REFERENCES

American Society of Heating, Refrigerating, and Air Conditioning Engineers. (2015). *Applications handbook*. Atlanta, GA: American Society of Heating, Refrigerating, and Air Conditioning Engineers.

Ferguson, N., Bonnici, P., Holescher, C., Holland, P., Tomerlin, B., & Zebell, M. (2004). Wastewater odor modeling to prioritize process odor control and support development buffer zone. *Proceedings of the Water Environment Federation Odors and Air Emissions Specialty Conference*, Bellevue, Washington, United States, April 18–21. Alexandria, VA: Water Environment Federation.

Guideline on Air Quality Models, 40 C.F.R. § 51, Appendix W.

National Fire Protection Association. (2016). National Fire Protection Association 820: Standard for Fire Protection in Wastewater Treatment and Collection Facilities. Quincy, MA: National Fire Protection Association.

Ormerod, R., Galvin, G., & Starke, G. (2010). Improvements in odor modeling using non-steady-state models and more realistic source physics. *Proceedings of the Water Environment Federation Odors and Air Pollutants Specialty Conference,* Bellevue, Washington, United States, April 18–21. Alexandria, VA: Water Environment Federation.

Permit-required confined spaces, 29 C.F.R. § 1910.146 (2004).

Porter, R. (2004). Predicting short term impacts using AERMOD and CALPUFF. *Proceedings of the Water Environment Federation Odors and*

Air Emissions Specialty Conference, Bellevue, Washington, United States, April 18–21. Alexandria, VA: Water Environment Federation.

Turner, D. B. (1970). Workbook of Atmospheric Dispersion Estimates, (AP-26) tr. PB-191 482; U.S. Department of Health, Education, and Welfare; Environmental Science Services Administration.

U.S. Department of Health & Human Services. (2007). NIOSH Pocket Guide to Chemical Hazards (Publication No. 2005-149). Washington, D.C.: U.S. Department of Health & Human Services.

U.S. Environmental Protection Agency. (n.d.). Air quality dispersion modeling—Preferred and recommended models. Retrieved from https://www.epa.gov/scram/air-quality-dispersion-modeling-preferred-and-recommended-models.

Wang, J., & Skipka, K. J. (1993). *Dispersion modeling of odorous emissions.* Air and Waste Management Publication 93-RA-114A.05. Presentation at the 86th Annual Meeting & Exhibition, Air & Waste Management Association.

Water Environment Federation. (2018). Workshop W18: Design, startup, and optimization of vapor phase odor control technologies. *Proceedings of the 93rd Annual Water Environment Federation Technical Exhibition and Conference,* New Orleans, Louisiana, United States, October 3–7. Alexandria, VA: Water Environment Federation.

Wolstenholme, P. (2013). Even with a biological system, don't treat before collecting: Concepts and design guidelines for foul air ventilation and collection. *International Water Association World Water Congress and Exhibition.* London, UK: International Water Association.

8.0 SUGGESTED READING

American Society of Heating, Refrigerating, and Air Conditioning Engineers. (2013). Chapter 24: Airflow around buildings. *Fundamentals handbook.* Atlanta, GA: American Society of Heating, Refrigerating, and Air Conditioning Engineers.

Brown and Caldwell, Metropolitan Sewer District of Greater Cincinnati (2017). *Mill Creek Wastewater Treatment Plant Odor Dispersion Model.*

Diorsey, P. (2008). Modeling odors in the age of AERMOD. *Proceedings of the Water Environment Federation Odors and Air Emissions Specialty Conference*, Phoenix, Arizona, United States, April 6–9. Alexandria, VA: Water Environment Federation.

Porter, R., & Elenter, D. (2008). Dispersion model parameters that affect the impacts from area sources using AERMOD. *Proceedings of the Water Environment Federation Odors and Air Emissions Specialty Conference*, Phoenix, Arizona, United States, April 6–9. Alexandria, VA: Water Environment Federation.

12

Public Outreach and Involvement

1.0 INTRODUCTION 361
2.0 STAKEHOLDER GROUPS 363
2.1 Personnel (Operations, Maintenance, Pretreatment/Sampling Staff) 364
2.2 Elected Officials and Regulators 365
2.3 Local Residents/Homeowners Associations 365
3.0 ODOR COMPLAINT MANAGEMENT AND RESPONSE 366
3.1 Documentation and Response Procedures 366
3.2 Data Analysis and Use 366
3.3 Personnel Knowledge and Communication 368
3.4 Standard Operating Procedures 369
3.4.1 Improve Odor Complaint Data Level of Detail 370
3.4.2 Complete the Circle 370
3.4.3 Implement Standard Operating Procedures for Conducting and Documenting Site Investigations 370
3.4.4 Standardized Site Investigations 375
3.4.5 Collect Both Gas and Liquid Samples During Site Investigations 375
3.5 Coordination Activities 376
4.0 ENGAGING THE COMMUNITY 376
4.1 Public Outreach Meetings 376
4.2 Managing Expectations 377
4.3 Social Media 377
4.4 Coordination of Outdoor Activities 378
5.0 CASE STUDY—TRINITY RIVER AUTHORITY OF TEXAS, DENTON CREEK REGIONAL WASTEWATER SYSTEM 378
6.0 REFERENCES 381

1.0 INTRODUCTION

Solving the problem of odors at a water resource recovery facility (WRRF) involves more than just minimizing the sources of the odors. A key to any odor mitigation effort's success is communicating with neighbors so they do not get frustrated with the problem before it is resolved. The goal of the wastewater treatment industry is to have a treatment facility that puts out a quality effluent and is perceived to be a good neighbor by the community. While working to minimize odor concerns, straightforward, honest communication and proactive community relations with the treatment facility's neighbors will gain their confidence and trust until this goal is achieved.

It is a common scenario for a WRRF that was originally built alongside farmland, open space, or oil wells to become neighbors with businesses, industry, or homes as the surrounding area develops. In retrospect, all of the properties surrounding a treatment facility built years ago should have been purchased as buffer land before this development occurs. However, this is rarely done and may not be economically feasible. Now, WRRF personnel are faced with rising occurrences of neighbor complaints about odors from a facility.

This chapter lays the groundwork for establishing and implementing a communications and community relations plan that will help WRRF personnel meet with neighbors and form a common bond of trust and understanding. The methods described in this chapter will help establish this trust. Once trust and understanding are in place, the neighbors will likely be more patient and understanding as WRRF personnel work to control odors. It is important to note that this chapter is not intended to be an all-inclusive list of possible methods for communicating with the public.

This chapter includes experiences of the Trinity River Authority of Texas. Additional information may be available from large wastewater agencies in a specific area or by contacting the Water Environment Federation.

During an odor mitigation program, it is important to understand the following:

- The goal must be to solve odor problems. It may take a while, but one must show improvements and a continual commitment toward that goal and communication. If neighbors think that WRRF personnel are just talking, credibility will be lost and not easily regained. That is when lawyers get involved, and everyone loses. Expectations must be understood on both sides of the issue, and unrealistic expectations should also be addressed. It is important to remember that odor is an emotion-provoking issue that must be approached with respect and understanding.
- Management must be fully committed to supporting the effort. Promises should not be made to neighbors that management will not support. It typically takes a lot of time and money to resolve the type of odor problems that can occur in a treatment facility, and it is important to be confident that management will commit to both before embarking on a community relations program. If management is unaware or

unconvinced that an odor problem exists, one may consider a random survey of neighbors to gather data to support concerns.

Before WRRF personnel actually go into the community, they should make sure their facility is clean and representative of a good neighbor. A little landscaping and aesthetically pleasing fencing goes a long way toward improving impressions from the community. The not-in-my-backyard (NIMBY) attitude is very much alive in today's society. The best way to counter NIMBY is to avoid giving the neighbors a reason to feel that way in the first place (Sandman, 1998).

2.0 STAKEHOLDER GROUPS

One of the first steps in an effective public communications program is to identify the stakeholders. These may be WRRF personnel, members of the public perceiving and complaining about the odors, people who contribute wastewater but are far enough from the facility that they are not affected by the odors, elected officials, and regulators. Responding to an odor complaint is a complicated process, but can be very organized and professional.

Figure 12.1 presents an example flow chart of procedures for handling odor complaints, including coordination with the various stakeholders.

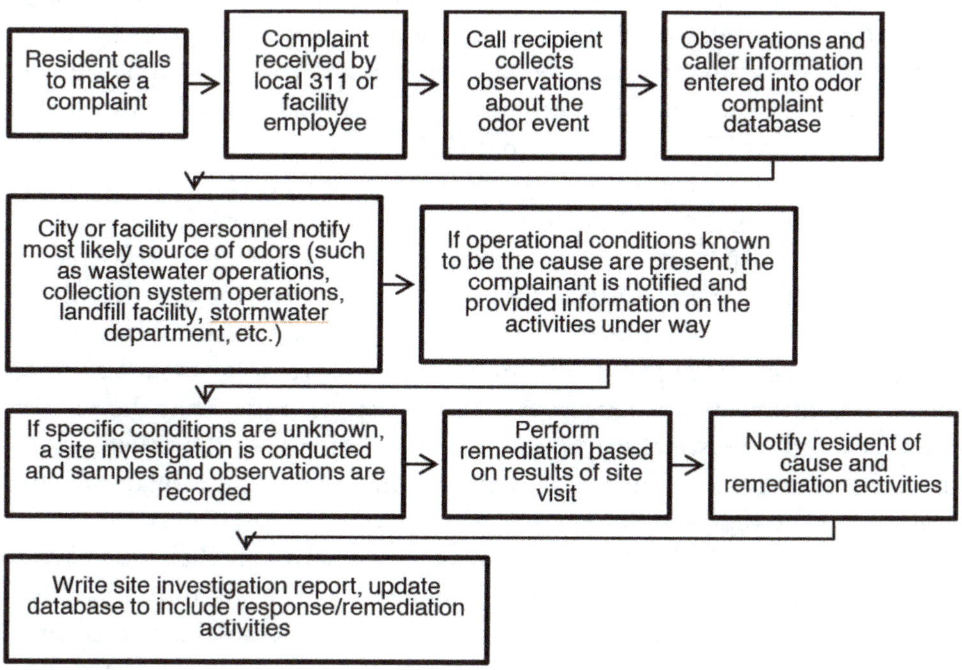

FIGURE 12.1 Example odor complaint response process.

2.1 Personnel (Operations, Maintenance, Pretreatment/Sampling Staff)

All staff who have the potential to be in contact with the facility's neighbors should be educated about sources of odors and distribution of odors in the vicinity of the facility. Staff must understand that neighbors are not the problem; rather, odors are the problem. In fact, neighbors are part of the solution. If neighbors become a WRRF's ally, they can be the best source of information concerning odor frequency, time of occurrence, and type of odor. The only way to build credibility with neighbors is to acknowledge that there is a problem and that the WRRF has a desire to address it as much as possible.

It is a good idea to hold meetings or workshops with staff to train them on how to handle complaints from neighbors. When a neighbor calls to complain about an odor, it is best to avoid denying that odors exist and, instead, be sympathetic, nonargumentative, courteous, concerned, and caring. The blame or responsibility should not be placed on the neighbor for moving next to the facility; it should be accepted by WRRF personnel. The neighbor should be shown that WRRF personnel are willing to listen. Sometimes, people just need to talk to someone who will listen with a caring ear. In many cases, this will substantially reduce the anger that the neighbor is feeling.

Role playing at WRRF staff meetings is helpful. It is good to have one staff member play the role of the irate neighbor and another field the imaginary call. Afterward, the group can critique the exercise with constructive suggestions. A person or several persons from the WRRF should be trained to meet with the public. This is often the facility manager, their assistant or the chief operator. The person designated should attend training courses on communication and dealing with people on a face-to-face basis.

The selected person should be comfortable with face-to-face situations because these are important to building credibility with neighbors. People do not trust faceless experts. Often, when someone is angry and tries to solve a problem over the phone, they can imagine all types of negative things about the person at the other end of the line that they have never met. When the caller actually meets the representative in person, seeing that the person has a face and a personality, the relationship changes. As the caller gets to know the person, he or she becomes more real; this facilitates relationships, especially if the representative is sympathetic and caring.

It is good for WRRF personnel to put themselves in their neighbors' shoes and imagine how they would feel under the same circumstances. No one wants to live next to a foul-smelling treatment facility. This consideration will go a long way toward guiding the actions of the WRRF personnel. If WRRF personnel do not like what they are about to do, there is a good chance that the neighbors will not either. This consideration will help to generate a feeling or sympathy and caring toward the neighbors, which will show in dealing with them.

2.2 Elected Officials and Regulators

Once WRRF staff have been educated in the appropriate handling of odor complaints, it may be time to involve elected officials and regulators. This is an important but often forgotten step in developing an odor control plan. Local elected officials should be contacted and educated because if wastewater treatment and odor control are not addressed properly, they can become a political problem. If the neighbors are calling the WRRF, they may be calling the elected officials and/or the regulatory agencies. By contacting the elected officials as soon as there is a concern, future problems may be avoided, such as neighbors thinking that the WRRF personnel do not care. Elected officials and regulators should be told about the concerns and involved in developing the community outreach program. By being involved in the program at the start, they become stakeholders in the solution and will be more apt to support and defend the program in the future. Selected officials should be put on the mailing list and kept informed.

2.3 Local Residents/Homeowners Associations

Local residents are the ones affected by odorous conditions at WRRFs. They are the ones who cannot cook in their backyards because of odors. They have to sleep with their windows shut because of odors. And they worry about how the odors they are smelling affect their health or their children's health. They have a problem, and WRRF personnel are the ones who get the angry phone calls describing how the odor is affecting their quality of life. Taking these complaints seriously is a good start at resolving the issues. A public meeting gives the neighbors a place to vent their frustrations, meet WRRF staff and management, and hear about the facility's odor control plan. Elements of a presentation at a homeowners association (HOA) or neighborhood meeting could involve the following:

- introduction of WRRF staff and management,
- introduction of the audience and perhaps where each member is located relative to the WRRF,
- description of facility and wastewater treatment processes,
- types of odors generated by the various facility processes,
- levels at which odorous compounds can be perceived (people are often surprised by the low concentrations),

- levels at which odorous gases become threatening to health,
- odor control implemented or soon to be implemented at the facility,
- data collected before and after odor control (if available), and
- questions and answers.

The meeting can end with an invitation to tour the facility to see the wastewater processes firsthand. Showing the neighbors the influent and the effluent can completely change their attitudes.

3.0 ODOR COMPLAINT MANAGEMENT AND RESPONSE

3.1 Documentation and Response Procedures

An odor control program begins with the collection of data relative to the odor complaints. An odor hotline number or website should be established and distributed to stakeholders. Important data include a description of the odor, climatological data when the odor is perceived, and other odor sources in the area. A WRRF may be located in an area that has other strong sources of odors (e.g., landfill, confined animal feeding operation, slaughter house, waste transfer station, or meat processing plant). A WRRF may find that up to half of the odor complaints are coming from other sources. Thus, it is important to develop information on other potential odor sources in an area as well as wind direction and speed during an odor event.

Figure 12.2 presents a typical odor report that can be used as either a call-in sheet or a web-based form. Note that it is important to educate the public on descriptions of odors to minimize the variety of descriptors you receive. Climatic data can be keyed to the time of the odor complaint. Climatic data are available from local National Oceanic and Atmospheric Administration (NOAA) offices, airports, or the facility. Recording and responding to odor complaints is one of the first steps in convincing neighbors that you are listening to them.

3.2 Data Analysis and Use

As odor complaint data are collected, a map of the complaints can be developed. Figure 12.3 shows an example of an odor complaint map. Dots of different sizes may be used to identify the frequency of the odor complaints. In Figure 12.3, the dot densities are used to develop a map to compare the complaint densities and identify key areas. The odor complaints can be compared to wind direction at the time of the complaint to determine

Odor Log

Date	Time	Location	Weather				Odor Observations		Comments
			Temp °F	Wind		General	Odor strength (Note 1)	What does the odor smell like? (Note 2)	
				Direction	Speed	Rain, sun, clouds?			
Additional Comments:									

Note 1: Scale ranges from 0 to 5: 0 – no odor, 1 – very faint, 2 – faint, 3 – distinct, noticeable, 4 – strong, 5 – very strong

Note 2: Odor descriptors included, but are not limited to: Sewer, rotten eggs, putrid or foul, chemical, fecal (like manure), garbage, burnt or smoky, musty, earthy

Name:______________________________ Signature:______________________________

FIGURE 12.2 Odor complaint data log.

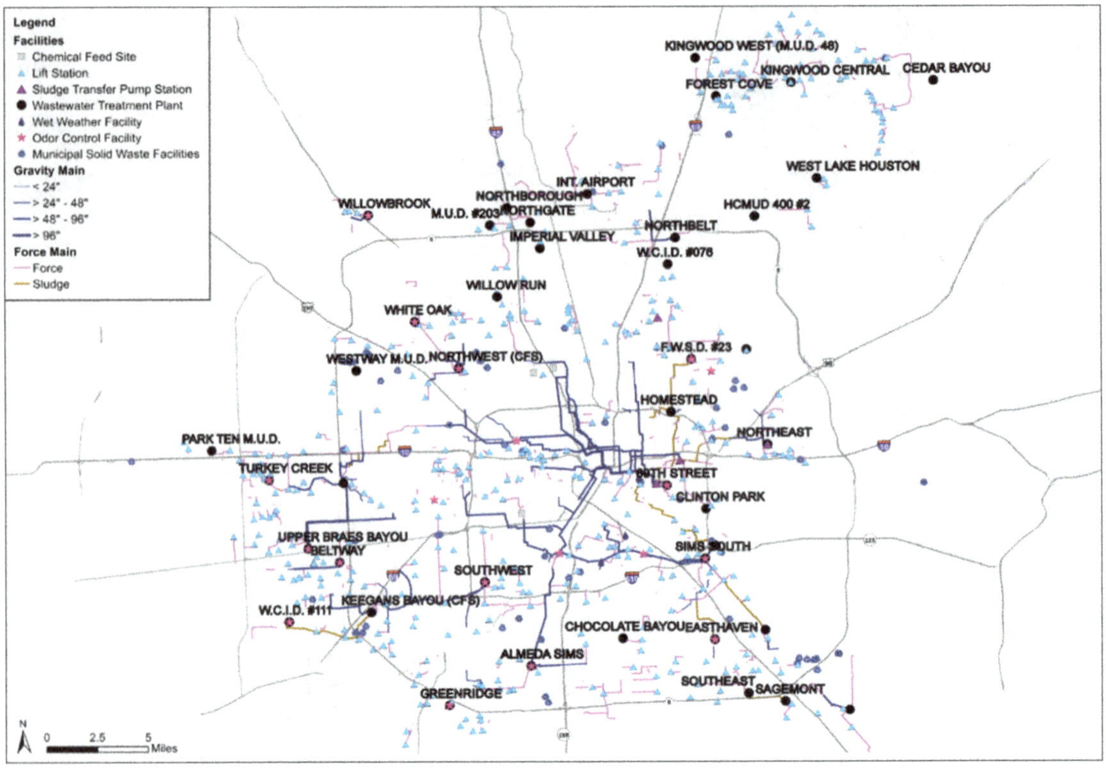

FIGURE 12.3 Typical odor complaint map (Nagle, 2017).

whether the complaint is potentially directly associated with the WRRF or whether it could have originated from another odor source in the vicinity of the perceived odor. Figure 12.4 shows a typical wind rose for a given area.

Odor complaint logs can be used at public meetings to show that the neighbors' complaints are being heard and evaluated. They can also be used to identify other potential odor sources in the vicinity of the facility and its associated facilities. A comparison with the wind direction and the description of the odor can be used to confirm whether the odor was originating from the facility or not.

It can also be useful to map locations of odor complaints against the collection system and its related components (e.g., pipelines, siphons, lift stations, air release valves, drop manholes, force main discharge locations, etc.).

3.3 Personnel Knowledge and Communication

To better characterize the current state of odor and corrosion conditions in the project area, it is helpful to have key personnel, such as operations, maintenance, management, and engineering, to complete surveys about the system. The results may reveal important details or provoke investigation of areas presumed to be irrelevant. Showing key personnel the odor complaint

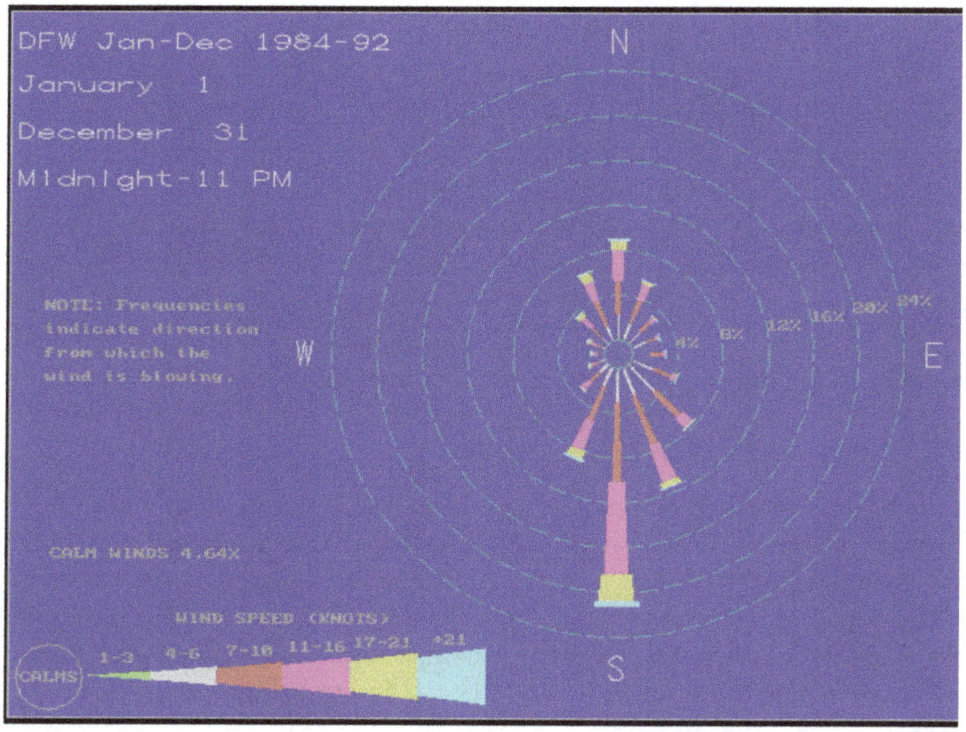

FIGURE 12.4 Example of prevailing winds wind rose (Texas Commission on Environmental Quality, n.d.).

figures and giving them a visual tool can be thought-provoking and trigger key details on the system, which may have otherwise been considered unrelated. Operational or physical changes to the system may also be identified that have not been recorded and carried through the historical maps. Oftentimes, these personnel may have even been the ones to respond to the odor complaints, and they may have details that do not show up on the odor complaint reports.

Operations personnel are invaluable for capturing "legacy" information. It is important to maintain records of the system in an accessible central location that is regularly updated. Reviewing and confirming the historical changes in the system with key personnel can help to identify changes that may not have been recorded previously.

3.4 Standard Operating Procedures

When personnel capable of conducting odor investigations are limited, timely site investigations can be difficult to perform. Implementing standard operating procedures (SOPs) and templates can ease delegation of these activities to other personnel who may be equally familiar with the system. The activities or standards outlined in the following subsections may be useful in capturing on-site information as quickly as possible.

3.4.1 Improve Odor Complaint Data Level of Detail

Include thorough data collection in initial odor complaint and response data sets. Have clear questions for callers to specifically identify the location where odors were observed (not just their address or somewhere nearby), the time/date, the weather-related observations at the time of the observation, descriptors of the odor(s), and so on. Providing examples of the types of observations and odor descriptors will improve the results. The use of an online and/or mobile app-based format will streamline data collection. Requiring users to register with an email address and their name is strongly recommended to limit abuse that potentially comes with anonymous observers.

3.4.2 Complete the Circle

Update the odor complaint database after the investigation is complete so that patterns can be derived. Over time, common odor sources may be more readily identified, which may reduce the time required to address complaints or allow more time to schedule preemptive activities. The database should include categories for system component(s) directly related to the odor source and data points that can be used for tracking progress.

The following are examples of metrics that can be developed from the dataset to evaluate progress:

- odor complaints per month, per year, per drainage basin, per 1000 LF (leaching fraction) of pipe;
- sewer backups per 1000 LF of pipe;
- odor complaints in areas with or without odor control, during average and peak wet weather events;
- odor complaint or work order response times and follow-up;
- odor complaint or work order repetition intervals;
- ratio of capital to operations and maintenance expenditures; and
- asset inventory and preventative maintenance activities.

3.4.3 Implement Standard Operating Procedures for Conducting and Documenting Site Investigations

Having standard forms and procedures for site investigations allows delegation of these activities to other personnel, if necessary, for a timelier response. Having additional staff or the ability to cross-train other personnel helps to streamline responses and improve the ability to address the sources of the complaints. Figures 12.5 through 12.8 present example templates and forms.

311 Help Desk Odor Complaint Record

Date, Time of Complaint: ______

Representative Name: ______

1. When the odor was first detected?

___ / ___ / ___ ___ : ___ AM/PM

2. How long did the odor persist?

______

3. Where was the odor perceived? Please provide address, if available, and if indoors/outdoors.

______

______

______

4. Weather conditions at the time when the odor was detected (approx.):

Temperature: ______ Deg F

Wind Direction: ______

Wind Speed: ______ mph

Humidity: ______ %

5. Have you experienced this same type of odor in this location before? How often?

______

______

7. Was the odor continuous or intermittent? (Please give as much detail as possible)

______

______

______

8. How would you describe the odor? (Please circle one or multiple options, or write in a description)

Sewer	Fecal (like manure)	Earthy
Rotten Eggs	Garbage	Chemical
Putrid or Foul	Burnt or Smoky	Musty

Other: ______

9. How strong was the odor? (Please use the scale range 0-5 as follows: 0-no odor, 1-very faint, 2-faint, 3-distinct, 4-strong, and 5-very strong.)

______

10. Contact information

Name: ______

Phone Number: ______

Email: ______

FIGURE 12.5 Odor complaint record (Nagle, 2017).

Vapor-Phase Grab Sample Results

Investigator Name:______________________

Sample ID	Date	Time	Sample Description (i.e. manhole, ambient, vent, etc.)	Ambient Odors		H_2S Conc[1] (ppm)			Additional Notes/Comments (i.e. color observations
				#[2]	Descriptors[3]				

Weather Conditions[4], Additional Site Notes: ____

____

____

FIGURE 12.6 Vapor-phase grab sample results (Nagle, 2017).

Odalog SN	Start Date	Start Time	Stop Date	Stop Time	Installation Location (i.e. manhole ID, depth below manhole rim, vent, etc.)	Additional Notes/Comments

[1] If using Jerome Meter, discard first reading, then record the next three (3) readings.
[2] Odor strength (0-no odor, 1-very faint, 2-faint, 3-distinct/noticeable, 4-strong, 5-very strong)
[3] Odor description (sewer, rotten eggs, putrid, foul, chemical, fecal/manure, garbage, burnt, smokey, etc.)
[4] Note ambient temperature, wind direction, wind speed, precipitation, etc.

FIGURE 12.7 Odalog monitoring record (Nagle, 2017).

Liquid-Phase Grab Sample Results

Investigator Name: ______________________

Sample ID	Time	Sample Location	Water Temp (°C)	Cond. (µS/cm)	DO (mg/L)	pH (SU)	ORP (mV)	Total Sulfide (mg/L)	Diss Sulfide (mg/L)	Notes / Comments

Weather Conditions[§], Additional Site Notes: ______________________

[§] Note ambient temperature, wind direction, wind speed, precipitation, etc.

FIGURE 12.8 Liquid-phase grab sample results (Nagle, 2017).

3.4.4 Standardized Site Investigations

Site investigation SOPs should also include monitoring equipment placement procedures, expected levels of documentation detail, and photographs, and they should explain the purpose of data collection. Data collection requirements should include weather conditions, wind speed and direction, temperature, barometric pressure, and precipitation, at a minimum. Correlation of monitoring results with real-time weather and lift station operation data is recommended to evaluate if odor event patterns can be attributed to environmental or operational conditions, when available.

3.4.5 Collect Both Gas and Liquid Samples During Site Investigations

The wastewater quality at the time of the odor event (or as close to that time as possible) in the immediate vicinity of the complaint provides clues to the source or cause of the odor event. In particular, the total and dissolved sulfide concentrations, pH, oxidation–reduction potential, dissolved oxygen concentration, and temperature are valuable parameters that can also be used to verify and optimize chemical feed rates of chemicals being used for odor control.

3.5 Coordination Activities

During evaluation and design phases of a project, it is important to include operational personnel in review meetings and project/construction sequencing discussions. Personnel can communicate the impacts of taking odor control units offline during construction and resulting safety or corrosion risks. They can provide observations on corrosion and changes over time. Operations personnel can provide a reality check of actual conditions compared to how the system is assumed to be operated or how the system was designed compared to how it is actually operated on a daily basis. Personnel can provide feedback on what has been tried in the past and what has not been effective as well as identify shortcomings in how previous systems were designed or applied. They can also provide a history of which actions personnel have observed as having the greatest impact on odors in the collection system or treatment facility.

4.0 ENGAGING THE COMMUNITY

4.1 Public Outreach Meetings

Public concern about odors frequently shows up in permit hearings. People are concerned that expanding a facility will increase the odors in the neighborhood. By holding public meetings at key intervals around a project's design and construction activities, WRRF personnel can mediate opposition to the permit and the facility. A public meeting is also a good place to introduce an odor complaint hotline or other mechanism of reporting and educate the public on odor descriptors. The following are typical slides shown at a public meeting:

- introduction of the owner and what they do,
- figures showing the facility and where certain odors originate,
- table of typical odors perceived at a WRRF and the levels of perception,
- graphic illustrating parts per billion and/or parts per million,
- table showing levels at which hydrogen sulfide is dangerous to health,
- figures showing the odor control practices and equipment already in place,
- questions and answers, and
- invitation to tour the facility.

Typically, a number of people will sign up to tour the facility. It is good to have a process brochure with a site plan to hand out to those touring. One facility manager asked tour-goers to let him know when they smelled the odor at the facility that they were smelling in their neighborhood. Their responses helped him identify areas of odor he needed to address, and his approach involved the neighborhood into the solution.

These types of activities and improved communication can enhance neighbors' system familiarity, and proactive "good neighbor" behavior and outreach can improve neighbors' perceptions.

4.2 Managing Expectations

No one likes surprises. One of the ways to manage expectations of neighborhoods is to send pre-shutdown notifications to neighbors and help them understand the nature of the activities and the reason/benefits of the improvements taking place. Manage expectations by letting them know whether they will experience odors. Communicate the duration of the activities and, with specificity, how "temporary" the activities will potentially be observed. Reassure residents and/or local businesses that there are no safety or health impacts and that precautions are being taken to limit the odor-related impact.

4.3 Social Media

Because of the nature of odor complaints and the emotional response they often evoke, an odor issue can balloon quickly into a political movement. Social media outlets, such as Facebook, Nextdoor, and Twitter, are useful communication methods that can be used to notify residents of upcoming meetings and remind them of proper grease disposal, stormwater protection, and so on. However, social media types differ in reach and accessibility, which should be taken into account when used for public outreach.

Municipalities have had success with online forums, such as Facebook, as an outlet for being notified of odor complaints and for communicating response activities. Facebook is a public network and, as such, there are limits to whether a municipality can limit access to users who are actually affected by the odor-related conditions being raised.

Similarly, Twitter notifications can be received by anyone who is connected to specific users or topics, allowing involvement from users who may not have direct access to the area in question. Conversely, Nextdoor uses address verification to connect users within specific neighborhoods and location-based communities within a city or region. Notifications may be more effective when directed at the specific areas related to an odor concern.

However, when issues are initiated by complainants on social media sites where geography is not a factor, it is important to respond via that same media type so that is may be received more readily. Because strong emotions are often involved in social media, it is important to be proactive and consistent. When activities are underway on a particular issue or area, it is critical to maintain regular contact and update those affected. Progress should be communicated clearly and accurately to maintain realistic and manageable expectations.

4.4 Coordination of Outdoor Activities

It can be difficult to predict how maintenance activities can affect outdoor activities by neighbors. For systems with residents and businesses nearby, maintenance activities such as basin washdown and operational changes can result in complaints. Coordinating maintenance activities to avoid weekends and holidays should be standard practice, but not all events may be as predictable. Having a regular, open dialogue with neighbors is a valuable part of any proactive outreach program.

Organizations such as HOAs can be helpful outlets to coordinate communication or provide a means to notify residents of activities or responses to odor events. Neighbors may ultimately have a spokesperson who can coordinate with the wastewater facility to notify them of scheduled events. Facility managers can improve relationships with neighbors by notifying them about potentially odor-causing activities in advance. Notifications can help to manage neighbors' expectations and let them know that the facility is aware and sympathetic to the impacts it may have on others.

As coordination with neighbors is maintained over time, the impacts to neighbors may be more readily communicated directly to the treatment facility rather than to regulatory agencies. If preemptive measures such as short-term odor control solutions are used, following up to receive feedback from neighbors can be useful to evaluate whether those methods are having the desired effect. Again, these types of communication help to build better relationships with neighbors.

5.0 CASE STUDY—TRINITY RIVER AUTHORITY OF TEXAS, DENTON CREEK REGIONAL WASTEWATER SYSTEM

John Bennett, Project Manager at the Trinity River Authority of Texas (the Authority), Denton Creek Regional Wastewater System (DCRWS) WRRF, showed how to handle odor concerns and turn complainants into friends and advocates of the facility. DCRWS was originally constructed far from housing and commercial areas. However, major housing developments were constructed around the facility, one sharing the fence line with the facility. With the residential developments came businesses and a new elementary school, which was also built directly adjacent to the facility. With the developments came odor complaints.

The Authority's first line of defense was to hold a public meeting along the line of the outline presented in Section 4.1. John explained what the Authority does and which treatment units could potentially generate odors, the levels at which odor might be perceived, the levels at which the odorous compounds became health concerns (a concern of the residents), and what the facility had done for odor control. In their most recent expansion, the facility had built the solids treatment units at the back of the facility, away from the neighbors. They also planted about 150 cedar trees on the fence line between the residential area and the elementary school to improve odor dispersion. The information was well received and there were a lot of questions and discussion after the presentation. The neighbors were treated with respect and sympathy.

John invited them to tour the facility, and several wanted to take the tour. John provided the brochure presented in Figure 12.9 to those who attended the meeting. The first tour was a success and was followed by several other tours. One of the most productive elements of the tour was that John asked people to let him know when they smelled the odor that

FIGURE 12.9 Typical brochure for odor control (Trinity River Authority, 2013).

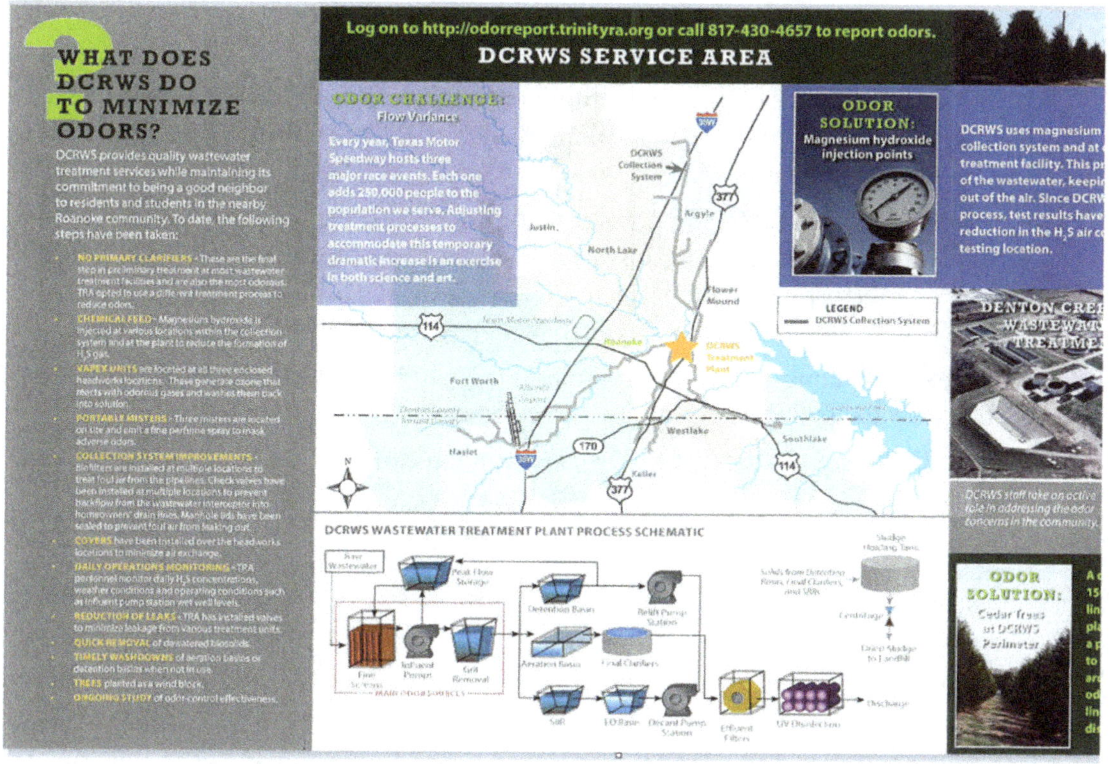

FIGURE 12.9 (*Continued*)

they smelled in the neighborhood. In this way, he confirmed the next areas in which he needed to address odor.

John and the Authority also developed an online odor complaint form. The form was a success and initiated an online database for the Authority's use in tracking complaints. John's quick and courteous response to odor complaints won the trust of residents, who ultimately became advocates of the facility and its personnel.

Through the online odor complaint form, John was able to identify the area in which foul air from the collection system was venting into home intake vents from the collection system. The houses in a particular area did not have P-traps in their vents, so the Authority installed duckbill check valves at the collection system laterals to prevent the foul air from being purged into those vents. John was also able to identify odors from the headworks and the anaerobic zone of the biological phosphorus basins as sources of odors based on the complaint descriptions.

The residents recognize that he is addressing these sources of odors and rarely complain now. John has held several public meetings explaining what he and the Authority are doing and getting input from the neighbors. He is no longer the villain, but the hero.

6.0 REFERENCES

Nagle, L. (2017). Phase I preliminary engineering services: Odor & corrosion control management plan (WBS No. R-000020-0010-3). Alan Plummer Associates, Inc.

Sandman, P. M. (1998). *Six principal strategies.* Presented at AMSA Public Affairs Seminar Risky Business Effectively Communicating Risk, Reality, and Reform, Providence, Rhode Island, United States, Aug 31–Sept 1. The National Association of Clean Water Agencies.

Texas Commission on Environmental Quality (n.d.). https://www.tceq.texas.gov/assets/public/compliance/monops/air/windroses/dfwall.gif

Trinity River Authority. (2013). Denton Creek Regional Wastewater System.

Index

A

Accuro® hand pumps, 68
Acid-impregnated carbons, 311
AcruLog™ units, 69, 82–83
Activated carbon media, 314
  advantages, 313
  disadvantages, 313
Activated carbon units, 134
Activated sludge, 270
Activated sludge diffusion (ASD), 232, 271–272
Activated sludge processes, 153
Activated sludge recycling (ASR), 232, 268
Actual gas contact time (AGCT), 241
Advanced vapor monitoring, 85
AERMOD models, 335
Air containment and extraction design, 343
Air dispersion modeling, 342
Air dispersion models, 332, 343
Air-distribution system, 234, 236
Air extraction system, 235
Airflow distribution, 241
Airflow measurement, 99
Airflow modeling, 100–101
Air inlet concentrations, 260
Air jumpers, 91
Air movement, 98
  air drag, 95
  artificial ventilation, 96
  corrosion management and mitigation, 96
  modeling hydrogen sulfide generation, 101
  occupational health and safety, 96
  physical assessment, 98
  wet well filling, 95
Air pressure measurement, 100
Air-supported canvas, 181
Air-vacuum valves, 90
Aluminum covers, 176
Aluminum flat cover, 178
Aluminum geodesic dome cover, 177
Ambient air, 2
Ambient odor intensity, 26–27
  ASTM E544, 26
  OIRS, use of, 27
Ambient odors, 26–31
  grid method, 27
  concentration, field olfactometer, 28–29
Ambient odor limits, examples of, 48–49
American Society of Civil Engineers (ASCE), 87
American Society of Heating, Refrigerating and Air-Conditioning Engineers (ASHRAE), 195, 337
American Society for Testing and Materials (ASTM)
  ASTM D1391, 13, 38
  ASTM D5466, 35
  ASTM D5466-15, 39
  ASTM D5504, 33
  ASTM D5504-12, 39
  ASTM E544-18, 19, 23, 26–27, 39
  ASTM E679-19, 13, 14, 39
  ASTM procedure B, 20
Amines, nitrogen compounds, and ammonia, 149
Ammonia, 63, 68

Anaerobic (low ORP) conditions, 208
Anaerobic digestion processes, 63, 211
Anemometer's probe, 69
Anoxic basins, 154
Area and point source sensory sampling, 67
Area source emissions, 332
ASD operation, 273
Authority having jurisdiction (AHJ), 185
Autothermal thermophilic aerobic digestion (ATAD), 63, 212

B

Barrel arches, 166
Bed depth, 317
Bed smoldering, 319
Bed velocity, 318
Belt filter press with integral enclosure system, 220
Best available control technology (BACT), 40, 327
Biochemical oxygen demand (BOD), 85
Biofilter design guidelines
- airflow distribution, 241
- gas contact time, 241
- pressure drop, 240

Biofilter media, 239
Biofilters and biotrickling filters, 132
Biofilters system, 235–240, 243
Biological activated sludge tanks, 153
Biological gas treatment techniques, 244
Biological nutrient removal processes, 73
Biological odor control system, 229, 263
Biological odor treatment
- biochemical processes, 230–232
- operating parameters, 232
- treatment process, 229–230

Biological ORP scale, 208
Biological processes, 229
Biological systems, 257
Biological treatment methods, 121–122
Biological vapor-phase treatment technologies, classification, 246
Biological wastewater treatment processes, 152
Bioscrubber, 250
Bioscrubber configuration, 247
Bioscrubbers, 245, 249
Biosolids handling processes, 73–74
Biotrickling filter, 133, 247, 249, 260, 262
- liquid phase, 253
- pressure drop, 250
- reactor configurations, 245
- reactor, 247

Biotrickling odor treatment system, 259
BPIP model, 339
British thermal unit, 325
Bulk gas phase, 281

C

CALPUFF, 335
Canvas style covers, 180
Carboxylic acids, 146, 150
Cartesian grid, 336
Case studies
- Coney island water resource recovery facility, 199
- Denton creek regional wastewater system, 377
- North river water resource recovery facility, 199
- odor impact assessment, 351, 354–357
- Trinity River Authority of Texas, 377

Catalytic incinerators, 325
Catalytic oxidation, 305
Catalytic oxidizer, 324
Chemical adsorption, 309
Chemical oxidants, 119
Chemical oxygen demand (COD), 256
Chemical probes, 283
Chevron blade mist eliminators, 301

Chlorine contact tanks, 154
Closed circuit television (CCTV), 111
Coal-based carbon, 311
Coconut fibers, 238
Coconut shell carbon, 311
Code of Federal Regulations, 105
Collection systems, 79–136
  ancillary equipment, 95
  evaluation and design, 85
  flow monitoring, 85–86
  force main design, 91
  hydraulic effects, 86–87
  inverted siphon design, 91
  lining and coating systems, 92
  material selection, 94
  odor capture, 91–92
  pumping station, 95
  site selection, 94
Colorimetric sulfide test kit, 65
Compost, 238
Computational fluid dynamics (CFD) models, 102
Concurrent flow scrubbers, 285
Coney Island Wastewater Treatment Plant, 199
Conservation Environmental Response, Compensation, and Liability Act (CERCLA) lists, 123
Contour values, 340
Conventional scrubber theory, 288
Corrosion-causing compounds, 85
Countercurrent operation, 248
Crossflow scrubbers, 285
Custom-made in-ground biofilter system, 236
Custom-made in-vessel biofilter system, 236

D
Data logger and housing, 101
Data loggers, 66
Debris accumulations, 112
Deep sludge blankets, 141
Dense-phase carbon conveyance system, 322
Denton Creek Regional Wastewater System (DCRWS), 377
Desulfovibrio bacteria, 62
Detection threshold, 16, 40
Detention time, 326
Dewatering processes, 74
Digested sludge, 212
Digestion methods, 209
Direct thermal oxidizers, 323
Dispersion model, 331
Dispersion modeling theory, 332–334
Dispersion models, 334, 335, 340
Dome Enclosure, 170–171
Downwash effects, 339
Dräger-Tubes®, 68
Drag shear, 87
Drain water, 256
Dry media adsorption, 308, 321
  chemical, 309
  physical, 309
  reactive, 309
Dry media mass transfer, 319
Duct materials, 194
Duct-sizing criteria, 195
Ductwork, 196
Durability, 161

E
Electronic noses, 69
Electronic odor monitoring system, 70
Emission rate calculations, 338
Empty bed gas contact time (EBCT), 241, 251
Engineered biological systems, 262
Enhanced buffer capacities, 240
E-Nose, 69
Equipment enclosures, 343
Excess biomass, 261
Exhaust airflow rates, 346
Extension wand, 316

F
Facebook, 376
Feast–famine conditions, 261
Federal air emissions, 92

Federal Aviation Administration (FAA), 337
Fiberglass covers, 177
Fiberglass reinforced plastic (FRP), 176, 177, 178, 288, 320
Field olfactometry, 28–31
- citizen monitoring, 30–31
- complaint response, 31
- intensive odor survey, 30
- on-site monitoring, 29–30
- plume profiling, 31
- random off-site monitoring, 30
- scheduled monitoring, 30

Field parameters
- conductivity, 81
- dissolved oxygen, 81
- oxidation–reduction potential, 81
- pH, 81
- sulfate, 81
- sulfide, 81
- temperature, 81

Field sampling program, 82
Fine-mist droplets, 243
Flat aeration tank cover, 163
Flat circular tank cover, 164
Flat covers, 162
Flat headworks channel cover, 162
Flat influent channel cover, 163
Flat rectangular tank cover, 164
Flushing systems, 97, 113
Foul air capture design, 344
Foul air contains, 315
Frequency, Intensity, Duration, Offensiveness, and Location (FIDOL), 51

G

Gas contact time, 241
Gas distribution, 252
Gas-gas oxidation, 279
Gas-phase treatment, 219, 221
Gas velocity, 251
Geodesic dome enclosure, 170, 172
Geographic information system (GIS) database, 81
Gravity Sanitary Sewer Design and Construction, 94
Grit tanks, 157
Guides for the Design of Wastewater Treatment Works, 220

H

Hazardous air pollutants (HAPs), 209
HDPE slip lining, 93
Heating, ventilation, and air conditioning (HVAC), 331
Henry's law, 142
High-capacity carbon, 312
High-density polyethylene (HDPE), 93, 194
High-strength odors, 285
High-velocity fans, 337
Homeowners association (HOA), 365
Homeowners Associations, 365–366
Horizontal airflow, 316
Hot wire anemometer, 69
HVAC fuel-burning equipment, 323
HVAC systems, 69
Hydraulic jumps, 158
Hydrogen peroxide, 282
Hydrogen sulfide, 62, 218, 324
- gas, 62
- low-profile three-stage scrubber, 286
- sulfide generation parameters, 62

Hydrogen sulfide generation, 101–102
Hyperion Water Reclamation Plant, 258
Hypochlorite-containing cleaning agents, 260

I

Incinerators, 325
Inert biofilter media, 239
Influent diversion structures, 157
In-ground biofilters, 235
Inlet air, 243
Instantaneous hydrogen sulfide sampling, 66

Institute of Air Quality Management (IAQM), 51, 53
Instrumentation and controls, 256
In-vessel biofilter system, 265
Ionization systems, 136, 304–305
Ionization technology, 136
Iron salts, 127
Iron sponge, 312
Isotherm plots, 340

L

Lance injectors, 128
Larger-sized biofilter media materials, 243
Laser scan inspection techniques, 111–112
Life cycle cost analyses, 348
Lime stabilization, 213
Lining and coating systems, 92
Liquid-phase dosing, 219
Liquid-phase grab samples, 64
Liquid-phase odor control methods, 218
Liquid-phase sludge treatment, 218
Liquid-phase treatment, 131
  adjustment of pH, 129–131
  biological treatment methods, 121–124
  chemical oxidants, 119
  iron salts, 124–127
  oxygenation, 128–129
  peroxide regenerated iron, 127
Liquid sampling, 81
Liquid scrubbant, 303
Liquid sidestreams, 234
Low-efficiency humidification system, 243
Lower explosive limit (LEL), 104, 185
Lower rates, 186
Low-profile multi-stage scrubbers, 285

M

Magnesium hydroxide dosage, 130
Manhole hydrogen sulfide monitor, 84
Mass transfer front, 317
Mechanical aerators, 73
Mechanical surface aerators, 153
Media irrigation, 243
Media-supported air plenums, 242
Media types, 326
  Intalox Snowflakes, 249
  Jaeger Tri-packs, 249
  Tellerettes, 249
  virgin carbon, 310
Mesh pad mist eliminators, 303
Metal impurities, 315
Metal sulfide, 311
Methyl mercaptan (MM), 315
Microbial degradation activity, 234
Microturbines, 325
Minimization of Odors and Corrosion in Collection Systems, 98
Minimum destruction efficiency, 325
Mist eliminators, 301–302
Mist scrubber systems, 303
Modeling hydrogen sulfide generation, 101
Modeling odor, 338
Model input parameters, 336–337
Modular biofilter systems, 237
Moisture control, 243
Multi-bed units, 317
Multilayered biotrickling systems, 251
Multiple configurations, 316
Multi-stage biotrickling filter system, 265
Municipalities, 376
Municipal systems, 91

N

National Ambient Air Quality Standards (NAAQS), 324
National Fire Protection Association (NFPA), 96, 184, 220, 280
National Institute of Occupational Safety and Health, 106
National Oceanic and Atmospheric Administration (NOAA), 366

Natural materials, 234
Net present value (NPV), 348
New York City Department of Environmental Protection, 199
Nextdoor, 376
Nitrogen compounds, 63
Non-economic factors analysis, 349–350
Nonstructural lining systems, 93
Nonworker-accessible spaces, 186
Northwest Wastewater Treatment Plant, 129
Not-in-my-backyard (NIMBY), 363
Nuisance levels, 40
Nutrient dosing pump, 259

O

Occupational Safety and Health Administration (OSHA), 98, 344
Odor, 2, 5
Odorant concentration limits, 44–46
Odorant production pathways, 210
Odorants
  compounds, 2
  concentrations, 3
  matrix of, 36–38
Odor capture and containment
  cover alternatives, 159
  progressive odor control process, 160
Odor characterization, 23–24, 81, 140
  advanced testing and monitoring, 83–84
  air monitoring, 83
  ascending concentration series, 17, 18
  assessors, role of 23
  derivation from collection systems, 141
  derivation from water resource recovery facilities, 141
  instrumentation, 84–85
  liquid sampling, 81
Odor complaint data log, 367
Odor complaint logs, 367
Odor complaint record, 371
Odor concentration, 12, 15–19
  best-estimate threshold, 18
  olfactometer, 15
  triangular forced-choice, 16
Odor containment, 184, 343
Odor control
  airflow conditions, 194
  brochure, 378–379
  dry adsorption medias, 318
  goals, 330
  improvement phases, 356
  performance, 272
  systems, 255, 348
Odor control technology selection guide, 348
Odor dispersion model, 352
Odor dispersion modeling, 332
Odor emission rate (OER), 67
Odor impact assess, 330
Odor impact contour plot, 340
Odor intensity, 12, 19–20
  dynamic scale method, 19
  n-butanol concentration, 19
  odor intensity result, 19
  static scale method, 19
Odor intensity referencing scales (OIRS), 19–20
Odor limits, 40
Odor management plans, 55–56
Odor mitigation, 61
Odor nuisance, 51
  persistent odor, 51
  unpleasant odor, 28
Odor nuisance regulations, 51
Odor nuisance response protocols, 51–55
Odor parameters, 3–4, 12–26
  assessor, 15
  character descriptors, 12
  laboratory olfactometry standards, 13
  panels, 14–15
  persistency, 12, 20–23
  subjective parameters, 12–13

whole-air odor samples, odor evaluation of, 14
Odor results, confidence interval of, 24–25
Odor standards 40–43
averaging time 43
concentration, 13
concentration-based standards, 43
detection thresholds, 40
frequency, 43
impact location, 43–44
olfactory sensory system, 43
Odorous air, 219
Odorous air emissions, 2
Odorous air samples, 6–12
odor and odorant sampling, 6
sampling exhaust stacks and vents, 7–9
sampling protocols, 6–7
sampling surfaces, 10–12
whole-air samples, 6
Odorous compounds, 62, 315
alcohols, 62
aldehydes, 62
dimethyl disulfide, 63
dimethyl sulfide, 63
indole, 62
mercaptans, 63
organic acids, 62
skatole, 62
Odor reduction efficiency, 25–26
Odor regulations and approaches, comparison of, 57–58
Odor source characterization
cake storage, 213
composting, 213
dewatering, 212
digestion, 211
lime stabilization, 213
thickening, 211
Odor sources, 337
Odor testing, 12
Odor threshold value (OTV), 15
Olfaction, 4
chemical odorants, 5
normal nose breathing, 4
olfactory nerve (first cranial nerve), 4
olfactory signals, 5
sniffing action, 4
trigeminal nerve (fifth cranial nerve), 4
Olfactory system, 2, 4
Onsite olfactometry, 32
Open air plenums, 242
Operations and maintenance (O&M) requirements, 256
Organic biofilter media, 238
Organic sulfur-based compounds, 63
Oxidants, 311
Oxidation catalysts
metallic oxides, 311
noble metals, 311
Oxidation–reduction potential (ORP), 64, 208, 281
Oxidized ammonium recycling (OAR), 274–276

P

Packed tower scrubber, 287
Packing manufacturers, 291
Parshall flumes, 157
Passive *versus* active systems, 320
Pelletized media, 315
Penthouse enclosure, 168
Performance parameters, 257
Periodic slug dosing, 130
Permanganate-impregnated media, 312
Peroxide regenerated iron, 127
Physical adsorption, 308, 309
Pipe and wet well cleaning, 97
Pipe cleaning, 112–113
Pitched covers, 168
Plastic materials, 239
Plastic resin-wetted fiberglass filaments, 288
Plow–furrow–cover techniques, 75
Point-source odor control system, 337
Polarity, 309
Polluted airstreams, 248

Polymeric materials, 239
Polyvinyl chloride (PVC) strip, 84, 174
Portable personal gas monitors, 66
PortaSens II, 66
Pre-humidification systems, 243
Pressure drop, 240
PRI-SC process, 127
Process area odor, 331
Process improvement and odor mitigation
  odor prevention, 155–156
  physical system layout, 156–158
Process integrated biotechnology, 264–268
Psychophysics, 20

R
Reactive adsorption, 309
Recognition threshold, 16
Regenerative thermal oxidizers (RTOs), 323
Relative humidity (RH), 294
Reliability, 257
Reliability of biological odor treatment systems, 266
Resilience, 257
Resource Conservation and Recovery Act (RCRA), 126
Robustness, 257

S
Sample methods, 64
  liquid phase, 64
  vapor phase, 65
Sample sources, 70
  collection systems, 70
  gravity sewers, 70–71
  lift stations, 70–71
  pumping stations, 70–71
Sample taps, 320
Sampling exhaust stacks and vents, 7–9
  dynamic dilution sampling probe, 8
  tedlar sample bag, 8
  teflon sample tubing, 7
  whole-air sample, 8
Sampling surfaces, 10–12
  area sources, 10
  flux chamber, 11
  surface emission isolation chamber, 11
  odorous emission rate, 12
Screening models, 336
Scrubber pollutant gas absorption diagram, 289
Scrubber theory, 289
Scrubbing liquid evaporation, 295
Scrubbing solution, 283
Sensitive receptor, 339
Sensor technology, 69
Sensory samples, 67
Septage-receiving stations, 113
Several vapor-phase odor control technologies, 347
Shear stress, 87
Sheet Metal and Air Conditioning Contractor's National Association (SMACNA), 198
SilcoCan glass-lined canisters, 68
Single-bed units, 317
Sludge drying beds, 75
Small-scale gas turbines, 325
Smell, sense of, 4
Smoke candles, 69
Smoke testing, 112
Solids process configurations, 217
Sonic calipers, 112
Sorbent detection tubes, 68
Source control, 103
  budgeting, 117
  corrosion control, 114–115
  equipment maintenance, 113–114
  financial management, 117
  industrial waste pretreatment, 109
  inspections, monitoring, and enforcement, 108
  laser scan inspection techniques, 111–112
  odor emissions, 103

- permits and ordinances, 108
- pipe cleaning, 112
- protective coatings, 114–115
- Radar, 111–112
- recordkeeping, 116
- safety precautions, 110
- ultrasonic, 111–112
- visual and closed circuit television inspections, 110

Specific odorants, sampling and chemical analysis of, 31–36
- amine compounds, methods for, 34
- ammonia, 35
- diamines, 36
- formaldehyde, 36
- indoles, 35
- reduced sulfur compounds, 32
- target compounds, 33
- volatile fatty acids, 33–34
- volatile organic compounds, 34–35

Stabilization ponds/lagoons, 154
Standard operating procedures (SOPs), 369
Standard ventilation rates, 186
Structural integrity, 161
Structured pads, 303
Subsurface drip irrigation, 243
Sulfate-reducing bacteria (SRB), 123
Sulfide equilibrium diagram, 282
Sulfide removal scrubbers, 289
Sulfur-based compounds, 147–148
Sulfur dioxide, 324
Super oxygenation system, 129
Surface-based process, 309
Surface irrigation, 243
Surface reaeration, 87

T

Tank ventilation rates, 190
Tedlar® bags, 8, 10, 68
Testing, adjusting, and balancing (TAB), 197
Thermal hydrolysis predigestion processes, 212
Thermal oxidation, 322
Thermal treatment systems, 323
Threshold exceedance plot, 341
Total suspended solids (TSS), 85
Tractive tension, 87
Triangular forced-choice test, 67
Trickling filters, 153
Tri-methyl amine (TMA), 314
Trinity River Authority of Texas, 362, 377
Turbulence, 142–143
Turbulence reduction, 88
Turbulent processes, 72
Twitter, 376

U

Ultrasonic techniques, 111
Ultraviolet–visible (UV–VIS) spectroscopy, 85
United States Geological Survey (USGS), 338
Universal Transverse Mercator (UTM) coordinates, 336
U.S. Environmental Protection Agency, 67, 290

V

Vacuum-release valves, 128
Valuable dispersion model, 341
Vapor-phase biological odor treatment systems
- advantages, 254
- disdvantages, 254

Vapor-phase monitoring, 83
Vapor-phase sampling, 65
Vapor-phase techniques, 62
Vapor-phase technologies, 347
Vapor-phase treatment, 88, 131, 135
- adsorption, 134–135
- biological, 132–133
- chemical, 133–134

Vapor-phase treatment system, 97
Variable frequency drive (VFD) controls, 260
V-bank, 317

Vector attraction-reduction requirements, 209
Ventilation rate, 184
Ventilation requirements, 345
Ventilation systems, 191
Vertical vessel configurations, 316
Virgin activated carbon, 310
Virgin carbon, 311
Virgin carbon media, 134
Volatile organic compounds, 314
Volatile organic compounds (VOCs), 63, 106, 146, 209
Volume sources, 332

W

Waste sludge, 113
Wastewater collection and treatment, 2
Wastewater collection systems, 70
Wastewater odorants, 63
Wastewater odorous compounds, 145
Wastewater pipes, 91
Wastewater pumping station, 98
Wastewater treatment and collection facilities, 184
Wastewater treatment processes, 252
Wastewater turbulence, wet wells, 156
Wastewater velocity, 87
Water Environment Federation (WEF), 87, 362
Water-regenerable activated carbons, 312
Water Research Foundation (WRF), 98, 209
Water resource recovery facility (WRRF), 2, 71–72, 140, 207, 279, 233, 308, 330, 366
Water supply failure, 258
Water supply sources, 255
Weirs, 157
Wet chemical gas adsorption systems, 280
Wet chemical odor scrubbing systems, 280
Wet chemical scrubber, 283
Wet chemical scrubber configurations, 283
Wet chemical scrubber mass balance, 285
Wet wells, 96
Worker-occupied buildings, 189

CPSIA information can be obtained
at www.ICGtesting.com
Printed in the USA
LVHW061222250221
679918LV00002B/5

9 781572 783676